THE BODY OF THE MUSICIAN

WORLDS OF SOUTH AND INNER ASIA
WELTEN SÜD- UND ZENTRALASIENS
MONDES DE L'ASIE DU SUD ET
DE L'ASIE CENTRALE

Band / Vol. 3

Edited by / Herausgegeben von / Edité par
JOHANNES BRONKHORST
KARÉNINA KOLLMAR-PAULENZ
ANGELIKA MALINAR

PETER LANG
Bern · Berlin · Bruxelles · Frankfurt am Main · New York · Oxford · Wien

THE BODY OF THE MUSICIAN

An Annotated Translation and Study of the Piṇḍotpatti-prakaraṇa of Śārṅgadeva's Saṅgītaratnākara

MAKOTO KITADA

PETER LANG
Bern · Berlin · Bruxelles · Frankfurt am Main · New York · Oxford · Wien

Bibliographic information published by Die Deutsche Nationalbibliothek
Die Deutsche Nationalbibliothek lists this publication in the Deutsche Nationalbibliografie; detailed bibliographic data is available on the Internet at ‹http://dnb.d-nb.de›.

British Library Cataloguing-in-Publication Data: A catalogue record for this book is available from *The British Library*, Great Britain.

Library of Congress Cataloging-in-Publication Data

Kitada, Makoto.
The body of the musician: an annotated translation and study of the Pindotpatti-prakarana of Sarngadeva's Sangitaratnakara / Makoto Kitada.
p. cm. – (Worlds of South and Inner Asia, ISSN 1661-755X; vol. 3)
Includes bibliographical references and index.
ISBN 978-3-0343-0319-4
1. Sarngadeva. Sangitaratnakara. Pindotpattiprakarana. 2. Music–Physiological aspects–Early works to 1800. 3. Music theory–India–Early works to 1800. I. Sarngadeva. Sangitaratnakara. Pindotpattiprakarana. English. II. Title.
ML3820.K57 2012
781'.1–dc23

2012015470

The present work has been accepted as a dissertation by the Faculty of Philosophy I at Martin-Luther-Universität Halle-Wittenberg.

Cover illustration: The celestial musician, Nārada. Courtesy of the Patan Museum, Nepal.

ISSN 1661-755X
ISBN (pb.) 978-3-0343-0319-4 ISBN (eBook) 978-3-0351-0417-2

Hochfeldstrasse 32, CH-3012 Bern, Switzerland
info@peterlang.com, www.peterlang.com

For Sumiko Kitada

Contents

Preface

I would like to express my cordial thanks to my teacher Prof. Dr. Rahul Peter Das, who inspired and encouraged me in writing this thesis. I also wish to thank Prof. Dr. Eli Franco, Prof. Dr. Minoru Hara, Prof. Dr. Hiroshi Kumamoto, Prof. Dr. Hiroshi Marui, Prof. Dr. Walter Slaje and Prof. Dr. Michio Yano for their most valuable help and corrections.

I am grateful to the Deutscher Akademischer Austauschdienst and the Iwanami Fujukai Foundation, which granted me financial support to study in Germany, as well as to Prof. Dr. Sengaku Mayeda, the Toho Kenkyukai (Eastern Institute, Tokyo), Prof. Dr. Kei Kataoka, Prof. Dr. Shinya Moriyama, Dr. Takahiro Kato, Prof. Dr. Hans Harder, Mr. Robert Siegfried M.A. and his family (especially Charlotte) and Ms. Carmen Brandt M.A. Special thanks to Mr. John Perkins M.A., who read through the whole text with me, making many helpful suggestions. I express my appreciation to Dr. Ananda Samir Chopra for granting me the opportunity to visit the Habichtswald Klinik Ayurveda in Kassel, to Mr. Amit Roy for his instruction in the sitar and Hindustani music for many years, and to Ms. Chiharu Ebisawa M.A., Mr. Niranjan Banerjee B.A. and Mr. Yagyaswor Joshi B.A.

I thank the Patan National Museum for allowing me to use the picture of the unique statue of Nārada on the cover.

I am beholden to Prof. Dr. Johannes Bronkhorst for accepting my work for the series “Worlds of South and Inner Asia”, and to Ms. Martina Fierz, Ms. Martina Räber and the rest of the Peter Lang staff for their copy editing, advice and cooperation.

Abbreviations

Abbreviations for the texts

AgniP	Agnipurāṇa
AH	Aṣṭāṅgahṛdaya
AS	Aṣṭāṅgasaṅgraha
AV	Atharvaveda
CA	Carakasaṃhitā
cikitsā.	cikitsāsthāna/cikitsitasthāna
GaruḍaP	Garuḍapurāṇa
comm. K.	The Kalānidhi commentary on SR
MBh	Mahābhārata
nidāna.	nidānasthāna
PadmaP	Padmapurāṇa
śārīra.	śārīrasthāna
ŚB	Śatapathabrāhmaṇa
ŚG	Śivagītā
comm. S.	The Sudhākara commentary on SR
SR	Saṅgītaratnākara
SU	Suśrutasaṃhitā
siddhi.	siddhisthāna
sūtra.	sūtrasthāna
vimāna.	vimānasthāna
ViṣṇudhP	Viṣṇudharmottarapurāṇa
YS	Yājñavalkyasmṛti
YS, prā., yati.	prāyaścitta-adhikaraṇa yatidharma of the Yājñavalkyasmṛti
YY	Yogayājñavalkya

Abbreviations for the technical terms

comm.	commentary
ed.	edition
f.	feminine
lit.	literally
m.	masculine
n.	neuter
opt.	optative
pl.	plural
prā., yati.	Cf. YS, prā., yati.
rep.	reprint
sg.	singular
Skt.	Sanskrit
tr.	translation, or translated
Up	upaniṣad
v.	verse

Abbreviations for the manuscripts

ka, kha, ga, gha, ṅa	Manuscript readings given as footnote in the Ānandāśrama edition. These are also noted in the Adyar edition.
A, B, C, D, E	Readings of the manuscripts consulted by the editor of the Adyar edition himself.
C.E.	Presumably the edition by Kālīvara Vedānta-vāgīśa and Śārada Prasāda Ghoṣa in Calcutta, 1879.

Preliminary remark on citing śloka-s

Most of the SR verses dealt with in this study belong to the section entitled "Piṇḍotpattiprakaraṇa", i.e. the second section of the first chapter of the SR (i.e., SR adhyāya 1, prakaraṇa 2). So the SR verses are very often referred to only with the verse number, omitting the number of the chapter (*adhyāya*) and that of section (*prakaraṇa*). E.g., SR adhyāya 1, prakaraṇa 2, śloka 3, is usually referred to simply as SR śl.3.

In contrast, I give the full reference of the verses from other sections, i.e. SR adhyāya 2, prakaraṇa 3, śloka 4 is referred to as SR 2,3,4.

Prologue

§1. Saṅgītaratnākara (SR) and Śārṅgadeva

In Indian musicological literature, the Saṅgītaratnākara[1] (SR) of Śārṅgadeva, a Sanskrit treatise in seven chapters on music and dance, is considered to be second only to Bharata's Nāṭyaśāstra in importance. This work comprehensively covers the whole early Indian musical theory and contains an abundance of instances of musical notation.[2] With its many citations from earlier authorities, it offers much information on the early musicological history of India. This work had profound influence on many later treatises in Sanskrit, Persian and other South Asian languages.[3]

Śārṅgadeva, the author of the SR, was the Auditor-General[4] at the court of King Siṅghaṇa II (reigned AD 1210–1247) of the Yādava dynasty in the Deccan. This was the time shortly before the onset of Muslim hegemony in Northern India. Qutb ud-Din Aibak (Quṭb al-

1 For general information on the SR, cf. the preface of the Adyar edition by S. Subrahmanya Sastri (noted under "Saṅgītaratnākara" in my bibliography); SHRINGY 1999 (vol. I), pp. vi–xxxiii; NIJENHUIS 1977, pp.12–13.

2 Chapter 7 deals with dance.

3 WIDDESS 1995, p.161. However, WIDDESS' (ibid.) opinion that the SR has a profound influence on "musical practice down to the present day" is problematic. We should consider the fact that musical practice does not necessarily orientate itself on theoretical texts. I myself, as a practicing musician, have found that music practice develops almost independently from written manuals in theoretical works. RAU 1986 discusses discrepancies between theory and practice in the exact sciences in traditional India. He points to the discrepancies between works being in Sanskrit, whereas artisans, performers etc. more often than not belonged to strata not conversant with the language or its intricacies. On the other hand, he also reports cases in which illiterate artisans of the lowest castes brought about technical innovations.

4 Auditor-General (śrīkaraṇādhipati). Cf. Ānandāśrama ed. (reprint in 1993), p.i. NIJENHUIS 1977, p.12, translates the same term as "royal accountant".

Dīn Aybak) founded the Slave Dynasty (AD 1206–1290).[5] The city of Devagiri, the capital of the Yādava dynasty, fell to Alāuddīn Khān in AD 1294.

Since the city of Devagiri (near modern Aurangabad) lay in an area where the northern and southern styles of music met, Śārṅgadeva was surely well versed in both styles of music. Besides, he might also have been initiated into the music of Kashmir through his grandfather and father[6]; Śārṅgadeva's grandfather Bhāskara was originally a physician from Kashmir[7] and later under the patronage of King Bhillama V (reigned AD 1185–1193) of the Yādava dynasty.[8]

The contents of the SR are as follows[9]:

1. The Svarādhyāya ("chapter on musical tones") with seven sections (*prakaraṇa*): it contains introduction, the author's genealogy, production of the human body (embryology, anatomy, Haṭhayogic anatomy), production of sound (*nāda*), musical tones, octave, basic scales (*grāma*), melodic types (*jāti*), embellishments (*alaṅkāra*), manners of singing (*gīti*) etc.[10];
2. The Rāgavivekādhyāya "chapter on discernment of modes (*rāga*)" containing two sections;
3. The Prakīrṇakādhyāya "chapter on miscellaneous topics": definition of musician-composers (*vāggeyakāra*) and singers, quality of voice, various vocal techniques, interpretation of the

5 FUNATSU 1991, p.83.

6 FUNATSU ibid., p.83. This is FUNATSU's speculation, but it is intriguing. The contents of the SR itself represent a music theory of a pan-South-Asian character, including pieces of information from all over the Subcontinent. But I did not find any trace of folk music from Kashmir in the SR.
However, about 300 years later, some Kashmiri musicians made efforts to integrate Indian music with "foreign" music, adopting foreign instruments like *rabāb*, at the birth-ceremony of Muḥammad (AD 1478), the son of King Ḥāsan Šāh, according to Śrīvara's Rājataraṅginī (3,235ff).

7 SR 1,1,2ab: asti svastigṛhaṃ vaṃśaḥ śrīmatkāśmīrasambhavaḥ.

8 Cf. The Adyar ed. of the SR, xiii–xiv. Also see FUNATSU 1991, p.83.

9 Cf. SR 1,1,31–49ab which is a table of contents. Also cf. SHRINGY 1999 (Vol.I), xxiii–xxv, enumerating the contents.

10 The system of *khaṇḍameru* (i.e., permutation-indicator) discussed in this chapter is an importation from mathematics into musicology. It is an exercise in determining the number of possibilities in combinations of notes. Cf. KATZ 1983, p.72 (note 25).

rāga in an unmeasured, free form (*ālapti*) etc.;
4. The Prabandhādhyāya "chapter on compositions";
5. The Tālādhyāya "chapter on rhythms";
6. The Vādyādhyāya "chapter on instruments";
7. The Nartanādhyāya "chapter on dancing": it defines and illustrates dramatics and treats the theory of aesthetic sentiments (*rasa*).

Thus, the SR is an exhaustive inquiry into various aspects of music by giving examples and notes. It also deals with dancing and dramatics in the seventh chapter.

§2. Two Commentaries: Kallinātha's Kalānidhi and Siṃhabhūpāla's Sudhākara

The preface of the Adyar edition (1943) of the SR informs us that there are several commentaries on the SR. Here, I have considered only two of them, the Kalānidhi (comm. K) by Kallinātha and the Sudhākara (comm. S) by Siṃhabhūpāla, as only these are readily accessible through the Adyar edition of the SR. I give some pieces of information on these two commentators in the following.

Siṃhabhūpāla was a king of the Recerla dynasty in Andhra in the 14th century. His commentary is mostly brief, but sometimes provides valuable information.[11]

11 The Introduction of the Adyar edition of the SR contains very detailed information on Siṃhabhūpāla, his works and biographical data (cf. Adyar ed., pp.xvii–xxiii). NIJENHUIS 1977, p.15 seems to be based on it. However, KUPPUSWAMY 1984, p.37ff, makes a contradictory statement that Siṃhabhūpāla is mentioned in an inscription of Nepal as a Mithila ruler after Śaktisiṃha and Harisiṃhadeva; besides his commentary on the SR, he is said to have written a short treatise on poetry, Rasārṇavasudhākara. KUPPUSWAMY also compares the original text of the SR and its commentary by Siṃhabhūpāla, and clarifies the points of theoretical discrepancies between the original and the commentary.

A century later, Kallinātha wrote his commentary on the SR under the king Immadi Devarāya (reigned AD 1446–1465) in Vijayanagara.[12] NIJENHUIS 1977 (p.16) considers it the best commentary on the SR, stating that "it adds to its lucid explanation of the original work also some interesting observations regarding contemporary music, especially when commenting upon the *rāga*-s of ch. [= chapter] 2 of the SR". Of course, this does not automatically mean that all the interpretations are correct, and we should sometimes remain skeptical. The time gap between Śārṅgadeva and Kallinātha is about two hundred years.

§3. Piṇḍotpattiprakaraṇa of SR

The SR deals not only with purely musical topics like octave, musical scales, melody and rhythm, notes etc., but also metaphysical discussions on music. Namely, the second section (*prakaraṇa*) of the first chapter (*adhyāya*), titled Piṇḍotpattiprakaraṇa: the "Section of the Origin of the Human Body", discusses Brahman, the self (*ātman*), the individual self (*jīvātman*), reincarnation, embryology, physiology, phrenology, anatomy, Haṭhayogic anatomy of *cakra* and *nāḍī*, meditation etc.[13] In this section, the attempt is made to authorise and sanctify musical practice as religious exercise, through relating music

12 The Introduction of the Adyar edition of the SR contains detailed information on Kallinātha (cf. Adyar ed., pp.xxiii–xxiv). Also cf. NIJENHUIS 1977, p.16; KUPPUSWAMY 1984, p.44. NIJENHUIS' information that Immaḍi Devarāya was a Yādava king seems to be based on the Adyar edition of the SR. In his commentary on SR 1,1,5–14, Kallinātha gives information about himself (cf. the Adyar ed., p.xxiii), and mentions Devarāya and Immaḍi Devarāya; he considers Devarāya to be the son of Vijaya of the Yādava Dynasty (cf. Adyar ed., p.xxiv). However, this statement is problematic. Devarāya's reign AD 1446–1465 actually fits in the period of the Saṅgama dynasty, the first dynasty of Vijayanagara. Although a theory considers Harihara I and Bukka, the founders of Vijayanagara and this dynasty, as belonging to a Yādava clan, this does not mean that the Saṅgama dynasty is a successor to the former Yādava dynasty, even if it might claim to do so.

13 FUNATSU 1991, p.84.

to the whole world, and the world to sound (*nāda*). Music is said to consist of sound (*nāda*), particularly in its essential form, *anāhata nāda* ("sound not struck"). Music is thus related to the absolute of the *nāda-brahman*, which is originally an analogue of the *śabda-brahman* of grammatical theory.[14] The concept of *nāda-brahman* is that sound as fundamental cosmic vibration is identified with the creative principle of the universe: "Making music is considered to be a creative process comparable to *yoga*, the religious exercise repeating cosmic creation on a human level."[15]

My study chiefly aims at an investigation of the text part in the SR related to Indian medicine[16], i.e. the embryologico-anatomical verses mentioned above, which has never become the object of research or consideration of scholars until now.[17] In the following, however, we shall take a look at the whole section and its background, because this is indispensable in comprehending the position of the embryologico-anatomical part of the SR, which is not, after all, a medical, but a musicological work. A medical statement which is put into a new context, totally different from a medical one, could assume values or functions different from those which it originally assumed in a medical context.

§4. Studies on Piṇḍotpattiprakaraṇa: SHRINGY 1999 and FUNATSU 1991

There is an abundance of studies on the theoretical aspect of music in the SR. Those referred to in this work are the following:

The most representative of these studies is SHRINGY Vol.I (1999) and Vol.II (1989), which consists of an English translation of the original Sanskrit text (Chapter I–IV) with Shringy's own comments

14 KATZ 1983, p.66.

15 NIJENHUIS 1992, p.4.

16 I avoid using the term *āyurveda*, because the medical works which I am here speaking of do not restrict themselves to the works usually considered as "standard".

17 Except for SHRINGY 1999.

and notes. TĀRAḶEKARA 1975 is a Marathi translation of the SR text and the comm. K, and contains the translator's own commentary which is sometimes informative. WIDDESS 1995 is an interesting attempt to reconstruct the musical practice in ancient times, based on the musical notes contained in the musicological works in Sanskrit, including the SR. Although NIJENHUIS 1992 and NIJENHUIS 1970 are research-works on musicological texts, Saṅgītaśiromaṇi and Dattila, respectively, NIJENHUIS very often refers to and explains the statements of the SR, because these two works contain many parallels to the SR. For the seventh chapter of the SR on dancing, there is an English translation by RAJA & BURNIER 1976.

In contrast to the abundance of such research-works on the technical aspects of music in the SR, little effort has been made to elucidate the aspects of music dealt with in the Piṇḍotpattiprakaraṇa of the SR which, besides metaphysics, contains embryology, anatomy and Haṭhayogic anatomy. Only SHRINGY 1999 and FUNATSU 1991 are exceptions. SHRINGY 1999 treats this aspect in his commentary. FUNATSU 1991 also deals with it. The basic ideas of the metaphysics of music, which occur also in other musicological treatises, are explained by NIJENHUIS 1992, pp.1–22.

FUNATSU 1991 is a short essay on the first two sections (*prakaraṇa*) of the first chapter (*adhyāya*) of the SR, i.e., Padārthasaṅgrahaprakaraṇa and Piṇḍotpattiprakaraṇa. FUNATSU tries to identify the ideological background of some statements of the SR. Since this article is written in Japanese and might therefore not be easily accessible to the readers who do not know the language, I summarise the points relevant to my study:

According to FUNATSU (p.86ff), the verses describing Brahman (SR 1,2,4–5c) concur with the concept of *saccidānanda* ("existence, intellect, and joy") of the late Advaita Vedānta school. The verses SR 1,2,4–5c mention not only the traditional attributes of Brahman like *liṅga* ("mark"), *svayaṃjyotis* ("self-shining"), *advitīya* ("without a second"), *sarveśvara* ("lord of all") etc., but also the two qualifications "omnipotent" (*sarvaśakti*) and "omniscient" (*sarvajña*), which, according to FUNATSU, represent the position posterior to Rāmānuja who emphasised the personified aspect of Brahman.

Although FUNATSU (op. cit.) and SHRINGY (op. cit.) offer very worthy considerations on the SR, it is to be regretted that both scholars simply consider the SR to be a homogeneous text. They often start from the hypothesis that the whole text of the SR was composed by Śārṅgadeva himself.[18] This work is, however, most probably heterogeneous; it seems to be a patchwork consisting of parallels to earlier texts.[19] At least a considerably large part of the Piṇḍotpattiprakaraṇa is based on earlier texts, as I shall show in this study. Therefore, the statements of the SR do not necessarily represent the personal ideology of Śārṅgadeva. The SR sometimes contains contradictory views.

For instance, in contrast to the theological statement of Advaita Vedānta referred to in SR 1,2,4–5c, which is perhaps based on another text[20], Śārṅgadeva does not worship Viṣṇu but Śiva as Lord in his benedictory verse, SR 1,1,1. This verse might manifest Śārṅgadeva's own religious position, for benedictory verses are usually composed by authors themselves. It suggests Śārṅgadeva's ideological relation to the homeland of his ancestors, Kashmir.

§5. Meditation of sound

The Piṇḍotpattiprakaraṇa investigates the two aspects of the human body: the embryologico-anatomical aspect according to Indian medical science (SR 1,2,21–119) and the Haṭhayogic aspect of the body (SR 1,2,120–163).

18 It would, however, be untrue to say that SHRINGY ignores this problem, for he mentions parallels in Caraka and Suśruta (SHRINGY 1999, p.386ff). Nevertheless, he often uses expressions that might give the misleading impression that the whole text was composed by one and the same author.

19 Nearly forty verses in the introductory part of the seventh chapter on dancing (Nartanādhyāya) are the same as the introductory verses found in the Abhinayadarpaṇa ascribed to the legendary sage Nandikeśvara, cf. RAJA & BURNIER 1976, vi.
RAJA & BURNIER (ibid.) refers to the theory of Alain Danielou that Śārṅgadeva might have borrowed these verses from Nandikeśvara's text, but refutes this theory, considering the Abhinayadarpaṇa to be a forgery.

20 This is, however, my personal impression and can not be proven with the available material.

Śārṅgadeva's motive for treating the Haṭhayogic theory of *cakra*-s and respiratory tubes (*nāḍī*) (SR 1,2,120–163) is clearly related with the notion of *nāda-brahman* in the musical practices of certain groups of Yoga, as BECK 1993 (p.109ff) discusses. The notion of *nāda-brahman* is, according to BECK (ibid., p.107), of Yogic and Tantric origin, rather than Vedic or Upaniṣadic. The same thought as mentioned in the statement of the SR, that the union of fire and the vital wind (*prāṇa*) produces voice, is already found in texts on phonetics and grammar (BECK ibid., p.110).

According to the Haṭhayogic texts, the Yogin perceives various internal sounds during his meditation[21], which resemble to the sounds of crickets, a flute, thunder, bells, trumpets, drums etc.[22] Following these sounds (*nāda*), he perceives the "not-struck" i.e., primordial or unmanifest sound (*anāhata-nāda*).

The Haṭhayogapradīpikā, a treatise on Haṭhayoga practice, deals with the meditation of sound (*nāda*) called *nāda-upāsanā* "worship of sound". This work states that the various knots (*granthi*)[23] which are considered as obstacles situated along the route of the *cakra*-s in the Yogin's body are connected to the audition of various sounds (*nāda*).

BECK (ibid.), pp.111–118, mentions the present-day methods of the meditation of sound as being practiced by modern Hindu religious movements.

The SR's verses dealing with *cakra*-s and respiratory tubes (*nāḍī*) seem associated to this Haṭhayogic meditation of sound.[24]

However, we should not hastily jump to the conclusion that these verses of the SR dealing with the Haṭhayogic description of the body reflect the Haṭhayogic practice in the time of Śārṅgadeva. It is very likely that this text part is also a patchwork consisting of several textual layers. As I shall show in my textual analysis (cf. *Situating the text" §1.1.*), the verses treating the *cakra*-s (SR 1,2,120–145ab) and

21 Through the breathing exercise of the *kumbhaka* technique (BECK 1993, p.103).

22 Gheraṇḍasaṃhitā, 5,79–82 according to BECK 1993, p.103.

23 I.e. the *brahma-granthi* in the heart, the *viṣṇu-granthi* in the throat, and the *rudra-granthi*. The *rudra-granthi* deemed situated in the middle of the forehead, according to JAIDEVA SINH 2003, lv., plate 3.

24 For such a practice in the Nātha cult, see KIEHNLE 1997, pp.101–105.

the verses treating the respiratory tubes (SR 1,2,145cd–163ab) belong to layers different from each other. The latter verses are parallel to the Yogayājñavalkya (YY).[25] It is probable that the former verses are also parallel to, or based on, another text.[26] I do not think that these verses on *cakra*-s are Śārṅgadeva's own composition, as they contain many parallels to other Haṭhayogic texts on *cakra*-s, e.g. the Saḍdarśananirūpaṇa.[27]

In the Piṇḍotpattiprakaraṇa, we observe the juxtaposition of the topic of *cakra*-s and that of respiratory tubes (*nāḍī*) in the musicological frame. This juxtaposition does seem to show Śārṅgadeva's intention to incorporate the Haṭhayogic meditation of sound, but this juxtaposition, or patchwork, fails to construct a consistent theory. So, these verses might have merely served as "theoretical armament" to authorise a secular entertainment like music through sanctifying it as a means of meditation, rather than to describe the actual Haṭhayogic practice contemporary to the author.

Or, such a condition of the text perhaps shows that theory is not necessarily a prerequisite for practice. Suppose a person practices a certain method of meditation in which he sings. He usually does not need to describe it verbally to himself while practicing this method. Only if he wishes to transmit this method to others, e.g. his disciples, is he compelled to resort to verbal description. In such cases he could maybe adopt some ready-made theoretical system which could also be understood by others. If he relied on different already existing explanations for different topics, his presentation of the practice could refer to multiple systems inconsistent with one another; this theoretical inconsistency, however, bringing about no inconvenience to the practice itself.

The verses, SR 1,2, śl.140–145ab, give a unique statement in the SR, directly associating the Haṭhayoga practice with music. These verses deal with the theory that particular *cakra*-s and their particular petals

25 See my textual analysis in *Situating the text §2.3.*

26 This is my personal impression. Anyway, many of these verses have parallels in other works on *cakra*-s.

27 See my footnotes on these verses, in which I list their parallels in other texts, in the *English translation.*

work on one's cultivation of musical arts positively or negatively.[28] According to this theory, the self/soul (*ātman*) chooses one of the petals of a particular *cakra* to stay on. If that petal is a right one, the person becomes successful in music. If it is a wrong one, the person is never able to succeed in this art. However, the description in these verses is too concise to provide us with some concrete information on the real practice of Haṭhayoga.[29]

28 Summarising this theory, SHRINGY 1999 (p.393ff, Appendix II) gives a chart of the *cakra*-s' relationship to music.

29 One could, perhaps, suppose a method in which the Yogin contemplates particular *cakra*-s or petals, imagining that he leads his breath or voice from a *cakra* or petal to another one. Such a method might be suggested by SR 1,2,148ab–149 which states that the individual self (*jīva*) which rides on the vital wind (*prāṇa*) climbs up and down along the Suṣumnā tube. But according to my textual analysis, this statement belongs to a theory totally independent from the *cakra* theory. These verses (SR 1,2,145cd–163ab) are parallel to the Yogayājñavalkya (YY), and have nothing to do with the text part dealing with the *cakra*-s (SR 1,2,120–145ab). Thus, from a philological point of view, it would be problematic to forcibly link two matters belonging to different layers of text with each other.
On the other hand, the fact that both sets of verses are included could suggest that, irrespective of their actual origins, they were meant to be taken together *on the level of the SR*. Śārṅgadeva brings together the two texts which till then had nothing to do with each other, because *he* considered these two as related to each other. And he expects readers to grasp the same context as he has in mind.
SR 1,2,140–145ab, the statement about the positive or negative influence of the particular *cakra*-s and petals, is explained by the comm.S. This commentary does not consider that the individual self oscillates to and fro or up and down from one to another *cakra*, but that it remains stationary at one of the *cakra*-s or petals; the individual self is said to stand (*sthita*) at a particular *cakra* or petal. Not a single word is uttered on any migration of the individual self.
The term *jīva-sthiti* "the individual self's situation/being situated" is once mentioned by the comm.K on SR śl.120–122ab (Adyar edition, p.60, l.5 from the bottom). It runs, [...] *dala-catuṣṭaye janma-kāle jīva-sthityā*, "through the individual self's situation on [one of] the four petals *at the time of birth*". According to this commentary, the fruit of a certain petal of the *cakra*, which the individual self has chosen to stay on, is already fixed at the moment of birth. Thus, if accepting this opinion, one has to inevitably conclude that the theory of SR 1,2,140–145ab has nothing to do with the Haṭhayoga practice, or the meditation of music. Instead, this theory suggests something opposite, that one's success or failure in musical cultivation has already been determined since birth and is unchangeable afterwards. (This would not be the case if we

Some modern musicians claim that such a relationship between *cakra*-s and music really exists.

PESCH 1999, which is an introductory work to South Indian classical music (Karṇātaka music), discusses the correlation between human anatomy and the seven musical notes (ibid., pp.56–66).[30] He compares the theory of *cakra*-s with the method of voice culture used by Western singers, and points out some similarities between the Indian and Western theory (cf. ibid., p.65). He states, "As part of their training and regular exercise, many singers associate various places of the human anatomy, comparable to the seven *cakra*-s, with certain qualities. Some places [...] serve as focal points for specific sounds as well as qualities." According to him, the theory of Western voice culture associates 1. the lower spine with sensuality, 2. the navel with emotion, 3 the diaphragm with mood or radiance, 4. the heart with calmness or cordiality, 5. the palate, teeth, tongue, and throat with projection and flexibility, 6. the nose and eyebrows with modulation, 7. the crown of the head with refinement. He points out the similarity between these seven areas of the body of the West and the seven *cakra*-s of India. But he does not substantiate his argument. Thus, we get no information about which Western theorists or singers he means concretely.

Śāradātanaya, a contemporary of, but somewhat anterior to Śārṅgadeva, in his dramaturgical work, Bhāvaprakśana, associates the seven tones of the octave to seven different places located in the body.[31] These places are: base of the body, navel, heart, throat, root of the tongue, head, middle of the eyebrows.[32] SHRINGY 1999 (p.102) states, "these places *roughly* [italics added] correspond to the psychophysical centres", i.e., to the *cakra*-s. Based merely on this fact, he leaps to the conclusion that "the line of thought linking

---

take the Ānandāśrama edition into consideration. In this edition *janma-kāle* is missing.)

30 He refers to a work entitled Nāradapurāṇa (circa 10th century AD) which treats this topic.

31 SHRINGY 1999 (Vol. I, p.102; pp.116–7).

32 Bhāvaprakāśana, adhikāra 7 (pp.187–8): *ādhāragaḥ śukra-dhātur majjā-dhātus tu nābhigaḥ / hṛdayāśrayo 'sthi-dhātuḥ syān medo-dhātus tu kaṇṭhagaḥ // māṃsa-dhātus tālu-mūle rakta-dhātus tu mūrdhagaḥ / bhrū-madhyagaḥ syāt tvag-dhātuḥ kramād evaṃ sthitāḥ svarāḥ //*

musical sound (*anāhata nāda*) to the yogic experience of *anāhata nāda* already existed in his time".

But the matter actually does not turn out to be so simple, if we examine Śāradātanaya's statement in question more precisely. It is true that the seven places *very roughly* correspond to the *cakra*-s. But actually the theory here is quite different. According to Śāradātanaya, the seven tones of the octave originate in the seven elements (*dhātu*) of the body. These seven elements, in their turn, originate in the above-mentioned seven places of the body, no *cakra*-s being mentioned. Thus, the association of the seven tones with the seven places of the body is only secondary, and in a manner very different from what is explained by the *cakra* theory.

On the other hand some people practicing Yoga and some Indian classical musicians consider music to be a method of Yoga. NARUSE 1986, p.51, who is a Japanese practicing Haṭhayoga, states that there are many kinds of hazards connected with the Tantric methods like the ascension of the *kuṇḍalinī*. Instead of such dangerous Tantric exercises, he proposes overtone chanting as an alternative method with which laymen could experience something approximate to that of the Tantric exercises, but in safety. He states:

> I occasionally let [my pupils] practice the primary method of the overtone chanting of Tibetan Buddhism. In spite of the easiness to chant U A O E I in a group, many kinds of things can be experienced. Sometimes, one hears inner sounds which resemble those of synthesizer, piano, flute, gong etc. Christian hymns, Buddhist *sūtra* recitation, and Shintoistic hymns, too. Some people perceive the vibration of the *cakra*-s as a kind of brightness, or get visions. In some cases, one makes various mystical experiences like one's own spirit slipping out of the body, which is the feeling of the ascension of the *kuṇḍalinī*. Yet the experiences acquired through this overtone chanting and the experiences of the Tantric practice are not identical. The one is not necessarily better than the other.[33]

I do not possess the ability to remark on the validity of his statement. Whatever it might be, some modern Yogins are empirically acquainted with such a Haṭhayoga practice utilising music. However, it is difficult to judge whether the SR's *cakra* theory really deals with

33 Translated from the original Japanese.

this practice: As I have shown above, this text part in itself is merely a description of each *cakra* and offers nothing more than a few suggestions on the matter of actual practice.

§6. The validity and nature of the statements of the SR

In this connection, the question arises as to what truth a classical text like the SR is actually supposed to present. Do the statements of the SR present some reality, or do they rather draw theoretical sketches than reflect real practice?

As to the musical compositions in notes contained in Indian musicological texts, NIJENHUIS 1970 (p.186) makes a reserved remark that such notations do not guarantee that they reflect the real musical practice of that time; rather, they are faithful reproductions of foregoing old texts.

The SR seems to be a similar case. Besides the musical notations, another instance could be referred to. This instance could symbolise the nature of the data contained in the SR. The *khaṇḍameru* system of *svaraprastāra* in the first chapter of the SR[34] is an example of the importation of mathematics into scholarly musicology. This is a mathematical exercise in determining the number of possibilities in combinations of notes, and conversely determining the place in a series of a given combination. KATZ 1983 (p.72, note 25) doubts whether the scheme really had practical importance for musicians. These combinatory variations of musical notes do not necessarily correspond to reality, but provide us with "theoretical values", so to say.

The same thing as is observed for the technical data of music seems to apply in the case of the topics treated in the Piṇḍotpatti-prakaraṇa. The embryologico-anatomical and Haṭhayogic data in this section seem reproductions of older texts. Thus the section is a patchwork whose components originate from sources originally unconnected to each other, i.e. from contexts different from one

34 SHRINGY 1999 (Vol. I), p. 208ff (SR 1,4,63c and following verses).

another. This makes it doubtful whether the text of this section reflects the musical reality of the time of Śārṅgadeva. And if it indeed does not reflect this reality, then we might have something similar to the above-mentioned "theoretical values": the human body which Śārṅgadeva explains in this section is not that which *is* in reality, but which *should be* theoretically. What concerns him is rather to describe the theoretical boundaries which the human body could reach, than to describe the material reality of the body.

If this deduction be valid – and I do not see how else we can explain what we find – , then Śārṅgadeva in this section represents the human body in its three aspects: 1. the body as a cosmic manifestation, or a derivative of the supreme principle, Brahman; 2. the body in its embryologico-anatomical aspect; 3. the body in its Haṭhayogic aspect. The picture of the human body as a theoretical construction is here drawn from a threefold perspective.

Śārṅgadeva displays, so to speak, an anatomical chart, or model, something like an anatomical plastic figure we have in the lecture hall of a medical school. We can look into the inside of the figure in which various anatomical components like organs, intestines, nerves etc., colourfully painted, are packed. We can perhaps even detach each part from the bodily frame. Though such a figure utters no word, nor explains the functions of the organs verbally, students who observe it can get some rough information on what the human body is. Namely, they get the *image* of the body. However, the perception acquired through such a concrete object is something else than that acquired through a medical textbook. Such an object appeals to a sphere of the brain different from that of verbal comprehension. A student who sees the figure is perhaps not able to verbally define the body and its functions, but his image of the body could be far more vivid than the knowledge acquired from a book.

As for the SR's anatomical model, Śārṅgadeva does not resort to plastic from which the bodily parts are modeled. Instead, it is language that he resorts to as the material to model the individual anatomical details. In other words, he creates an *objet d'art*, or a collage in putting together scraps from various texts, and thus represents the body. He very often merely lists technical terms, appellations for various organs etc., without explaining or defining them. Each term by itself, like a part of the plastic figure, utters nothing. The perception acquired through such pedantic listing of specialised words resembles that acquired from a plastic figure. It might be vague, remains not exactly defined, but is vivid. Here language functions in a manner somehow different from the language of modern Western scientific writing. The language of the SR is not necessarily the means to define the thing, or to draw a distinct outline of the thing, but something equivalent or similar to the thing itself.

§7. On my translation method

I have discussed this matter at the beginning of my *English Translation.*

§8. Philosophical matters

Since this work focuses on the embryologico-anatomical contents of the text, philosophical matters are usually not considered.[35] Even when they are considered, the information is limited to a minimum requirement.

35 The commentaries seem to use Navyanyāya terminology in places. I must admit that my attempt to provide literal translation is not always appropriate when it comes to Navyanyāya technical terms. But I refrain from being involved in this field from the reason given above.

On the editions of the SR

For my study, I mainly used the Adyar edition (1943) of the SR, which contains the *mūla* text and its two commentaries, the Kalānidhi (comm. K.) by Kallinātha and the Sudhākara (comm. S) by Siṃhabhūpāla.

According to his preface (x), the editor of the Adyar edition consulted two preceding editions, the edition by Kālīvara Vedānta-vāgīśa and Śārada Prasāda Ghosha (sic), Calcutta 1879 and the Ānandāśrama edition 1896. The editor of the Adyar edition reports that, of the former one (Calcutta 1879), only the first volume containing the Svarādhyāya, to which the Piṇḍotpattiprakaraṇa also belongs, was published. It is also reported that the former (Calcutta 1879) also contained the commentary Sudhākara and the latter (Ānandāśrama 1896), the commentary Kalānidhi. This edition (Calcutta 1879) is unfortunately not available any more.

Besides, the Adyar edition includes the variants from the five manuscripts which are given in the footnotes of the Ānandāśrama edition; they are noted in Devanāgarī alphabets: *ka, kha, ga, gha* and *ṅa*. Further, the author of the Adyar edition consulted five other manuscripts; their variants are noted in Roman alphabets: A, B, C, D and E.

In the Adyar edition, the variants of a certain edition 'C. E.' are noted (e.g., p.36, variants 9 and 10 in SR śl.134). Strangely, the editor does not explain what 'C. E.' stands for, nor does he include 'C. E.' in the list of abbreviations (xl). I suppose that 'C. E.' is the abbreviation of something like "Calcutta edition", namely, the edition by Kālīvara Vedāntavāgīśa and Śārada Prasāda Ghoṣa, Calcutta 1879.

Besides the Adyar edition, I consulted the Ānandāśrama edition which contains the *mūla* text of the SR and the commentary Kalānidhi by Kallinātha. Mostly, the reading of this edition is worse than that of the Adyar edition. But, as the result of comparing the SR with its two parallels, the Śivagītā (ŚG) and the Yogayājñavalkya

(YY), I found several passages in which the reading of the Ānandāśrama edition is better.

The editor of the Ānandāśrama edition (1896) does not give sufficient information on the manuscripts. G.H. Tarlekar, the editor of the reprinted version of this edition (1985), states (reprinted version, p.2) that the editors of the first edition in 1896 (chap.1–5) and 1897 (chap.6–7) used three manuscripts containing the *mūla* text of the SR along with the commentary Kalānidhi, three manuscripts containing only the *mūla* text, one manuscript of the commentary Kalānidhi in Devanāgarī script and one manuscript in Telugu script. Tarlekar, giving some examples, also remarks (p.14) that the two commentaries on the SR, Kalānidhi and Sudhākara, adopt different readings in several cases.

As a summary, I list the editions and manuscript which are mentioned above.

I consulted the Adyar edition and the Ānandāśrama edition.

ka, kha, ga, gha, ṅa	Manuscript readings given as footnotes in the Ānandāśrama edition. These are also noted in the Adyar edition.
A, B, C, D, E	Reading of the manuscripts consulted by the editor of the Adyar edition.

I suppose that "C. E." indicates the edition by Kālīvara Vedāntavāgīśa and Śārada Prasāda Ghosha (sic) in Calcutta 1879.

Situating the text

§1. Introduction

§1.1. Saṅgītaratnākara and its Piṇḍotpattiprakaraṇa

As I have explained in the *Prologue* of this work, the musicological text Saṅgītaratnākara (SR), which was written by Śārṅgadeva in the 13th century, is considered the second most important musicological treatise after Bharata's Nāṭyaśāstra. Its author, Śārṅgadeva, came from a family of physicians which had its roots in Kashmir and was under the patronage of the Yādava dynasty in the Deccan. Śārṅgadeva himself was a minister of King Siṅghanadeva who belonged to that dynasty.

Curiously, this treatise on music contains a section which treats medical topics. The second section, Piṇḍotpattiprakaraṇa "the section of the arising/origination of the body", of the first chapter (SR 1,2) discusses, in its first half, embryology and anatomy, i.e. the same topics as contained in the chapter called śārīrasthāna of the classical medical texts like SU, AS etc. The second half of the Piṇḍotpattiprakaraṇa of the SR describes the Haṭhayogic theory of the *cakra*-s and respiratory tubes (*nāḍī*).

The Piṇḍotpattiprakaraṇa is presumably not homogenous, but constituted of a medley of quotations from several texts.[36] I found two parallel texts. One is the Śivagītā (ŚG), a Purāṇic text contained in the Gauḍīya version of the PadmaP,[37] and the other is the Yogayājñavalkya (YY), a Yogic text ascribed to the sage Yājñavalkya. The ŚG contains parallels to the embryologico-anatomical verses of the SR, while the YY contains parallels to the SR's verses

36 For the methodology of dealing with a hybrid text, cf. HACKER 1978.
37 ROCHER 1986, p.212ff.

on the Haṭhayogic theory of the respiratory tubes (*nāḍī*).[38] The remaining verses on the *cakra* theory probably have a parallel somewhere, but unfortunately I was not able to find it.[39]

§1.2. The sources of the SR

In the opening of the Piṇḍotpattiprakaraṇa, the author Śārṅgadeva briefly mentions the relationship between sound (*nāda*) and the human body (SR 1,2, śl.1–3). With this statement, he justifies the treating of embryologico-anatomical science in his musicological work. Then, he begins to describe the supreme self's (= Brahman's) manifestation in the human body (SR śl.4 and the following verses). At first he describes the process of Brahman's manifestation in the individual selves (*jīva*). Thereafter, he discusses embryology and anatomy up to SR śl.119a. He seems to consider embryology and anatomy to be a part of the description of Brahman's manifestation in the human body. In SR śl.119bcd, which is the conclusion to his discussion on embryology and anatomy, the author Śārṅgadeva states that, for more detail, the readers should consult an earlier work entitled Adhyātmaviveka[40] written by him, Śārṅgadeva.

Therefore I presume that Śārṅgadeva quoted the part treating embryologico-anatomical science (SR 1,2, śl.4–119a) from his own

38 The SR preserves a more complete version than the text of the YY. The medical theories of the SR are often very similar to those of the AS, but contain many deviations, too. The description of the vital winds (SR 1,2, śl.60cd–68ab) and the respiratory tubes (*nāḍī*) (SR 1,2, śl.145cd–163ab) is parallel (but not strictly identical) to that of the YY. This is an early theory of Yoga, mentioning only the navel cakra (*nābhi-cakra*), cf. *Situating the text §2.3.6.*

39 There should be a source for these verses, for Mataṅga's Bṛhaddeśī, a musicological work which is older than the SR, already suggests the existence of this theory. (According to WIDDESS 1995, p.125, the Bṛhaddeśī was compiled sometime during the latter half of the first millennium.) Bṛhaddeśī, anuccheda 29 (P.L. SHARMA 1992, p.44) associates the seven tones of the octave with the seven cakra-s and with the seven continents (dvīpa).

40 The title means "Investigation of the theme concerning the self [i.e. the relation between the supreme and the individual self]".

previous work, the Adhyātmaviveka.[41] Unfortunately, the Adhyātmaviveka is lost.[42]

Analysing this part of the SR, one gap, at least, is observed in the text. After listing the sixteen types of "bodies" (*vigraha, deha*)[43], the author Śārṅgadeva abruptly stops his discussion, saying, "afraid of [too great] an expansion of the text, we (= I) do not tell their characteristics" (SR 1,2, śl.74cd).[44] After that, he immediately changes the topic: he shifts to the discussion of the [main] limbs (*aṅga*) and secondary appendages (*pratyaṅga*) of the human body (*piṇḍa*), and begins to enumerate them (SR śl.75–119a). Thus, Śārṅgadeva seems to have made an omission here. Since he states, "we (= I) do not tell" (*na brūmaḥ*), the omitted part might have originally existed in the Adhyātmaviveka.[45] If this is correct, then the embryologico-anatomical text contained in the SR (śl.4–119a) would be based on an older, more complete version.

A similar problem arises with his list of the secondary appendages (*pratyaṅga*) (SR śl.75cd–119a), as he calls this part a "summary of [the list of] the secondary appendages" (SR śl.119a, *pratyaṅga-saṅkṣepa*), which means that this part is a summary of a longer version of the text discussing the secondary appendages. Two inferences are possible: The one is that Śārṅgadeva himself summarised or abbreviated the longer version in his Adhyātmaviveka. The other possible inference is that, in the Adhyātmaviveka, he had already quoted an even older text, and this text itself had already been summarised or abbreviated.[46]

Indeed, there is a trace of emendation in SR śl.100. The parallels to SR śl.100, which are found in the SU and AS, both contain a number of qualifiers to the term *pracchādikāḥ* ("coverings"). These qualifiers which might probably have once occurred between

41 Or, at least, this part is based on the Adhyātmaviveka.

42 SHRINGY 1999 (Vol.I), p.84, "But this book is probably not available today".

43 This notion corresponds to the *kāya* of SU śārīra., 5,81–98. It actually denotes the types of psyche. Cf. my footnote 425 on SR śl.74ab in the *English translation*.

44 SR 1,2, śl.74cd: *teṣāṃ lakṣmāṇi na brūmo grantha-vistara-kātarāḥ*.

45 Other arguments are also possible, though. E.g., the Adhyātmaviveka itself might have been a work of quotation, and the part in question could have already been omitted in the Adhyātmaviveka itself.

46 Did Śārṅgadeva use a medical text which was handed down by his ancestors?

śukrārtava-praveśinyas tisraḥ (SR śl.100, *pāda* b) and *pracchādikā matā* (SR śl.100, *pāda* d), are omitted in SR śl.100.[47] One more example of omission which could be surmised is SR śl.113. If we compare SR śl.113 with its parallels in the SU and AS, it seems that a long passage between *prabhavān* and *rasān* is omitted.[48]

As far as this part (SR 1,2, śl.23–19) is concerned, it does not seem that Śārṅgadeva was a creative author. Most of his verses have parallels in the AS or SU. I have the impression that Śārṅgadeva faithfully reproduced a source-text, maybe handed down in his family.

So, the embryologico-anatomical text in the SR (śl.23–19) deals with the same topics as the śārīrasthāna-s in SU and AS, and has a structure similar to them. Therefore, the source-text which the SR was based on seems to have been a part or chapter of a medical work, which corresponded to the śārīrasthāna-s of SU and AS.

§1.3. The Śivagītā and its parallels to the SR's embryologico-anatomical verses

Parallels to the embryological and anatomical verses of the SR are found in the eighth and ninth chapter of the ŚG respectively. These parallels are exactly identical to each other in wording, except for a few variants. The ŚG has handed down parallels to almost all of the SR's verses on embryology, while the parallels to the SR's verses on anatomy are only sporadically found there.

§1.4. Common source text of SR and ŚG?

As noted in §1.3., most of the verses on embryology (SR śl.23–43) and most of the verses in the beginning part of anatomy (SR śl.44–70) have parallels in the ŚG. The parallels to the former ones are contained in the eighth chapter of the ŚG, and the parallels to the

47 For further detail, see my discussion in the footnote 554 on SR śl.100cd (coverings) in the *English translation.*

48 For further detail, see my discussion in the footnote 597 on SR śl.113 (*rasa*-s) in the *English translation.*

latter ones, in the ninth chapter.[49] But the parallels to the verses thereafter (SR śl.1–119) are very few. (Cf. §2.1.1. and §2.2.1.)

I take it for relevant to assume a common source, say source text, of the SR and ŚG, for the part before SR śl.71, in which most of the SR's verses have a parallel in the ŚG. In contrast, it is difficult to assume such a common source for the part after SR śl.71, in which only five verses[50] scattered in the text have a parallel in the ŚG.

However, I am not able to go ahead with this problem, as the available material is too limited to allow an attempt at reconstructing a source text. So I have to break off this argument here.

§1.5. Comparison with further medical and non-medical texts

I compared the SR's embryologico-anatomical description with the śārīrasthāna-s contained in the classical medical texts such as SU, AS, AH and CA. I compared it also with the embryologico-anatomical descriptions contained in non-medical texts such as the YS and Purāṇa-s like AgniP, ViṣṇudhP, GaruḍaP etc. The results of this comparative study are given in the footnotes on my *English translation* of the SR. In the following, I summarise them:

In general, the theories of the SR are close to those of the two classical medical texts, AS and SU.[51] Of AS and SU, the AS contains more expressions which are similar[52] to the SR than the SU does.[53]

In SR śl.25[54], terminological elements from both SU and AS are contained. In this verse, the elements which are found in SU śārīra., 3,18 and in AS śārīra., 2,12 supplement each other. The expressions

49 This fact might legitimate that, in my analysis, I divide the SR's embryologico-anatomical text in two parts, i.e. the part before SR śl.43, on embryology, and the part after SR śl.44, on anatomy.

50 More correctly, five passages, or cases, i.e. SR śl.79ab; 90cd–91; 92cd–94ab; 114cd; 116–118. Cf. §2.2.1 and §2.2.2.

51 The actual matter is of course a bit more complicated.

52 The SR contains sometimes even terminology identical to that of the AS.

53 I have analysed the matter in my respective footnotes on SR śl.76cd–78; śl.94cd. SR śl.96 and śl.98. SR śl.104–105ab; śl.110; ;sl.111–112a. Also see SR śl.33ab and śl.34b which belong to the SR's passages parallel to ŚG. Their counterparts in the ŚG are analysed by COMBA 1981.

54 The ŚG's parallel to SR śl.25 is analysed by COMBA (ibid.).

of SU and AS partly overlap, but the AS seems to have also adopted an expression which is parallel to, or maybe originated from the CA (*sarvāṅgāvayavendriya*).[55] Besides, the AS adds, to this statement, another theory on the secondary appendages appearing after birth.[56] In SR śl.25, the SR integrates elements from both SU and AS which have, from the outset, some expressions in common.

In SR śl.47, the elements found in SU śārīra., 4,33 and the elements found in AS śārīra., 5,22 compensate each other. In this case, the terminology of AS śārīra., 5,22 has nothing to do with the SU's terminology, but accords with CA śārīra., 3,12.[57] Thus, the SR's embryologico-anatomical text integrates different traditions, the SU's terminology, the AS's terminology, and sometimes, the CA's terminology.[58]

To SR śl.33cd–34a, only AS and AH have parallels. To SR śl.38cd and śl.40cd–41, only the AS has a parallel. To SR śl.101, only AS and YS have parallels.[59]

The CA's parallels do not correspond to the SR as literally as the SU or the AS does. Although their contents correspond with the SR, they are different in wording; in contrast to that, some parallels in SU and AS show striking closeness to the SR both in contents and wording. There is no case in which only the CA has a parallel to the SR. Therefore, I presume that the SR's theories according with the

55 Of course, the AH and AS explicitely amalgamate elements from the SU (as well as CA and other texts), but here the mentioned parts from the AS are not found in the SU, and vice versa. On the sources of the AS, cf. MEULENBELD 1999, 1A, p. 621ff.

56 Cf. my footnote 545 on SR śl.25 in the *English translation*.

57 Still, I am not completely sure whether AS śārīra., 5,22, is directly based on CA śārīra., 3,12, for the AS contains some expressions different from the CA's, and mentions *alaulya*, which the CA does not. The AS might be based on some other medical work belonging to the same tradition as the CA.

58 For details, cf. my footnote 687 on SR śl.47 in the *English translation*.
I quote the original texts of these parallels (the expressions parallel to the SR are underlined): SU śārīra., 4,33: *śarīropacayo balaṃ varṇaḥ sthitir hāniś ca rasajāni.*
CA śārīra., 3,12 (on *rasaja*): *śarīrasyābhinirvṛttir abhivṛddhiḥ prāṇānubandhas tṛptiḥ puṣṭir utsāhāś ceti /12/.*
AS śārīra., 5,22: *rasajāni, kṛtsnasya dehasya sambhavo vṛttir vṛddhis tṛptir alaulyam puṣṭir utsāhaś ca.*

59 SR śl.101.

CA are indirectly adopted into the text of the SR through the intermediary of the AS.

Therefore, for the correct situating of the SR, it seems crucial to investigate the textual situation of the AS, especially the relationship between SU and AS. MEULENBELD 1999, IA, pp.623–626, discusses this matter. In p.626, he concludes that the AS is later than Dṛḍhabala's revised and complete version of the CA, and very probably, also posterior to the revised and completed version of the SU.[60]

In the cases where both, the SU and AS, have parallels to the SR, I compared their parallels with each other, in the footnotes on my *English translation* of the text of the SR. There, I discussed which text, AS or SU, contains the expression which is closer to the SR. This is a complicated matter to elucidate, for the AS itself is mostly

60 On the other hand, my observation of the text of the śārīrasthāna of the AS does not agree with what MEULENBELD 1999, IA, p.623, remarks. He states, "In general, the *Saṃgraha* [= AS] deals much more freely with material also occurring in the *Suśrutasaṃhitā* than with that found in the *Carakasaṃhitā*. The former is less often quoted literally, while very numerous verses of *Saṃgraha* and *Carakasaṃhitā* are identical." As far as the śārīrasthāna of the AS is concerned, the SU is far more often quoted than the CA, and the AS's manner of quoting is considerably literal and faithful to the SU.
In some cases, the AS and the SU are parallel to each other, but these parallels are not literally identical, i.e. deviate from each other in wording. Even in such cases, I have the impression that it is rather the SU that deals with the material freely. The versions in the AS have very brief and compact forms, while the versions in the SU give the impression that they are extended with secondary insertions; the SU's versions are verbose, containing more explaining words than the AS's versions do. It looks as though the version of the SU as we know it today is a secondarily expanded form of an older, more compact text.
Indeed, MEULENBELD (ibid., p.623) mentions P.V. Sharma's opinion that the revised and completed version of the SU as now known to us was not yet available to the author of the AS, though MEULENBELD does not agree with P.V. Sharma.
If we agree with P.V. Sharma, then some parts of the text of the AS (śārīrasthāna) might preserve a more archaic form than the SU. The AS would contain an older version of the text, which dates to a time before the revision of the SU. Intriguingly, the embryologico-anatomical text in the SR resembles the AS in its compactness. Unfortunately, researching this topic further would lead us away from our actual discussion.

based on the SU.[61] In addition, we have previously seen that at least two cases (SR śl.25 and śl.47) suggest that the embryologico-anatomical text of the SR contains both elements from SU and AS. Even in the cases where the SR seems closer to the SU than to the AS, it is possible to imagine that the SR has actually mingled both elements from SU and AS; but the result of this blending might by chance look closer to the SU.

On the other hand, the embryologico-anatomical text of the SR shows, in some points, similarity to the embryologico-anatomical descriptions contained in the non-medical (Purāṇic) works, as we shall see later. The SR and these works share some topics which are not dealt with in the classical medical works.[62]

So, the possibility cannot be denied that the embryologico-anatomical text in the SR presents an independent branch of the textual tradition of a school which deviates from the branches of SU and AS.[63]

Therefore, I would like to call attention to the embryologico-anatomical description in the SR, which might throw more light on the development and mutual relationship of the medical texts.[64]

61 According to MEULENBELD 1999, IA, p.625, the AS considers Gayadāsa's version of the SU authoritative. He cites the AS's passages which prove this fact, for example, in his notes 243, 247, and 248 on AS śārīrasthāna (MEULENBELD IB, p.629).

62 E.g., the topics in SR 1,2, śl.34cd–36ab (the foetus' practice in the uterus) and śl.42–43 (birth).

63 It is certainly also possible that the author of this text, though being mostly based on SU and AS, adopted elements from various other works, but I prefer to see the matter in this way for the moment.

64 The SR uses a special appellation, *kaṇḍarā*, for the kind of cord (*snāyu*) which is called *vṛtta* by the SU. Intriguingly, the SU considers *kaṇḍarā* to be a terminology used by surgeons (cf. my footnote 905 on SR śl.95, *kaṇḍarā* in the *English translation*). This might suggest that Śārṅgadeva's ancestors in Kashmir belonged to the surgical school mentioned by the SU.

§1.6. Anatomical theory of vocal manifestation by ancient Indian musicologists

The authors of Indian classical musicological texts seem to have been interested in the mechanism of the human body since the period before the SR.

Mataṅga's Bṛhaddeśī, an important musicological text of the time between Bharata's Nāṭyaśāstra and the SR, explains the anatomical aspect of the vocal manifestation in the body: sound is produced through the union of wind and fire in the human body.[65] This text also informs us of the theory which is ascribed to an ancient scholar,[66] that the quality of the human voice is determined by the three morbific entities (*doṣa*) of the Indian classical medical theory.[67]

The Bṛhaddeśī informs us of the musical theory of its predecessor, Kohala, that the seven tones of the octave arise from the seven elements (*dhātu*) of the human body, i.e. skin (*tvac*), blood, flesh, fat, bone, marrow and semen.[68] According to the Indian classical medical theory, these seven elements – but with *rasa* in the place of *tvac* – are considered to be the gradual results of metabolism which is promoted by the digestive fire.[69] A dramaturgical text,

65 Bṛhaddeśī verses 20 (P.L. SHARMA 1992, pp.8–9).

66 The sage Tumburu.

67 Bṛhaddeśī, 3, anuccheda 4. Cf. P.L. SHARMA 1992, p.11: *apare tu vāta-pitta-kapha-sannipāta-bheda-bhinnāṃ catur-vidhāṃ śrutiṃ pratipedire* etc.

68 Bṛhaddeśī anuccheda 29 (P.L. SHARMA 1992, pp.44–45): *nanu kathaṃ sapta svarā iti niyamaḥ? ucyate, yathā sapta-dhātv-āśritatvena saptaiva dhātavo rasādayo jñeyāḥ; tathā cāha suśrutaḥ; tvag-asṛṅ-māṃsa-medosthi-majjā śukrāṇi dhātavaḥ iti; tathā sapta-cakrāśritatvena sapta-dvīpāśritatvena vā saptaiva svarā iti.*

69 In the Indian classical medical theory, the first of the seven elements (*dhātu*) is usually not skin (*tvac*), but nutrient fluid (*rasa*). This tradition of *tvac* in the metabolic chain is very old, cf., e.g., JAMISON 1986. However, "the medical texts also refer to an undifferentiated entity *tvac-* which is a *dhātu-*, and which often seems to be, and in a few cases actually is expressly said to be, part of the chain of metabolic transformation of food, taking the place of *rasa-*" (DAS 2003A, p.547, on *tvac-*). DAS ibid., §10,7ff, discusses this problem. In §10,8 (ibid., p.276), he draws attention to "the confusion, in Tantric texts, caused by two different lists, the one being the known list of the seven *dhātu*-s, but with skin in the place of *rasa-*, the other a list of substances, which are the same as in the list of the seven *dhātu*-s (with skin as the first element), but without the last

Śāradātanaya's Bhāvaprakāśana, which is contemporary with the SR, handed down and even developed this old theory.[70]

§1.7. Embryology in musicological works

As for the topic of embryology, the SR is not the only musicological text that treats this topic. For instance, the Bhāvaprakāśana, the above-mentioned dramaturgical text, in its seventh adhikāra, treats it in relationship with music. Also, the Saṅgītasāroddhāra, a musicological text which is chronologically later than the SR, briefly mentions embryology.[71] The reason for this is concisely explained by Śārṅgadeva, the author of the SR. He states that it is necessary because sound (*nāda*) is produced in the human body (*śarīra*).[72] The Sudhākara commentary on SR 3 (= Prakīrṇakādhyāya), śl.82 states that not only sound (*nāda*), but also melody (*rāga*) is manifested in the human body.

§1.8. Embryology and music in the YS

It is remarkable that the Yājñavalkyasmṛti (YS) mentions music after its embryologico-anatomical description. It states that one can attain liberation (*mokṣa*) through the practice of music. In this context, it lists musicological terms and several genres of songs which are mentioned by the Nāṭyaśāstra. This statement of the YS seems to

---

element, namely semen; according to the texts, the constituents of the list of the six substances are found in six "sheaths" or "repositories" (*kośa*)." Like the Bhāvaprakāśana, the list of the seven *dhātu*-s in SR śl.79ab mentions skin as the first element. Remarkably, the SR (śl.76c and 80c) states that these *dhātu*-s result from the maturation through the respective fires/heats of the *kośa*-s (*svasvakośāgninā*). Thus, if DAS' observation is right, this fact might suggest the SR's relationship to the Tantric tradition. Cf. my argument in §2.2.2 and §2.2.3 of the *Situating the text.*

70 Cf. SHRINGY 1999 (= vol.I), p.117. Śāradātanaya's Bhāvaprakāśana (G.O.S. No.XLV), p.186, ll.5–6.

71 The Saṅgītopaniṣatsāroddhāra (G.O.S. No.133), 1,11–24 deals with embryology.

72 SR 1,2,3.

have been well-known to the Indian musicologists, as it is quoted by Abhinavagupta in his commentary on the Nāṭyaśāstra and by Kallinātha in his commentary on the SR. Therefore, it is very likely that the author of the SR was also aware of the above-mentioned verses of the YS. If so, he could have inherited this tradition of the embryologico-anatomical argument in a musicological context from the YS. (Cf. my discussion on the topic in §4., *Embryology, Asceticism and Music: YS and SR.*)

Both YS and SR describe the whole process of human manifestation from the beginning until the end: the individual self (*jīva*) which derives from the supreme self (Brahman) enters into the uterus of the mother; the embryo is gradually formed month by month, and is at last born into the world. This is the way from the supreme self downward to earthly existence.

The way back is also described, namely, how to liberate the individual self from the human body, i.e. the method of how to return to the supreme self, which is Haṭhayoga. The YS considers music to be the substitute for Haṭhayoga. Thus, liberation through Haṭhayoga or music is a process which points to the inverted process of human birth.

§1.9. Influence of the pregnant woman's auditory perception on the foetus

On the other hand, the śārīrasthāna-s of the SU and CA inform us of the influence of the pregnant woman's auditory perception on her foetus.[73] According to these passages, not only the nourishment which the pregnant woman takes in, but also all the five kinds of perception,[74] including the sense of hearing, could influence the foetus positively or negatively. The pregnant woman should therefore avoid uncomfortable noises, and listen to comfortable sounds like

73 The statements that the mother's auditory perception influences the foetus are found e.g., in CA, śārīra., 8,9; *8,16*; 8,21; 8,24; also in SU śārīra., 2,25; *10,3*.

74 I.e., the senses of touching, tasting, smelling, seeing and hearing, cf. CA śārīra., 7,7.

religious tales.[75] According to the SU, the heart of the mother and that of the foetus are connected to each other through the umbilical cord from the fourth month onward, and the mother and the foetus are mentally synchronised, too.[76] But the classical medical texts never explicitly mention music in relationship to embryology. Therefore, I can give no evidence for the use of music as a treatment for the pregnant woman and her foetus.

§1.10. Cakra-s and respiratory tubes

As already mentioned, the section, Piṇḍotpattiprakaraṇa, of the SR, in its second half, deals with the *cakra*-s and the respiratory tubes (*nāḍī*). Its verses on the ten *cakra*-s seem to have a source, but this source is unknown to us. On the other hand, the verses on the respiratory tubes are parallel to those mentioned in the YY, the treatise on Haṭhayoga, whose theory seems to be archaic, as it mentions only one *cakra*, i.e., the *cakra* of the navel (*nābhi-cakra*).[77] In the corresponding verses, the SR's wording is more compact than the YY's. The YY makes considerably more repetitions. The SR seems to preserve a version more original than the YY, as will be discussed later.

§1.11. Conclusion

Thus, the Piṇḍotpattiprakaraṇa of the SR seems to consist merely of quotations from various texts. The author seems to have quoted passages from other texts, without taking great troubles to revise them. He did not make much effort at putting the quotations into a

75 The manuscript D of the SR (1,2,36), to which the ŚG has an identical verse, is the only text that mentions the foetus' acquisition of the ability of hearing in the eighth month. Except for it, there are neither medical nor non-medical texts that make such a statement.

76 Cf. SU, śārīra., 3,55: *niḥśvāsocchvāsa-saṅkṣobha-svapnān garbho 'dhigacchati / mātur niśvasitocchvāsa-saṅkṣobha-svapna-sambhavān //*. Cf. CA, śārīra., 4,24.

77 A segment of the fourth chapter of YY are also found in SR 1,2, śl.59–68ab, i.e., amid the verses treating embryologico-anatomical science.

context and correlating them to one another. The quotations are often awkwardly adjoined to one another.[78]

Nevertheless, this section is quite interesting, for it presumably preserves versions of the texts more original than the other texts available to us, and it can surely offer some hints on the lost textual traditions.

Besides, some embryologico-anatomical statements in the SR, being put into the Haṭhayogic or musicological context, might assume totally new connotations. For instance, for the classical medical texts, the comparisons of the human body with a tree (SR śl.77b; śl.103a) or with a ship (SR śl.96) are no novel topics at all.[79] However, the same comparisons occur also in other contexts, e.g., in the Haṭhayogic texts like the Caryāpada-s, or in their medieval and modern developments like the Baul songs.[80] I wonder if there could have been some relationship between the musicological tradition of the SR and that of the Haṭhayoga. As a matter of fact, the SR, in its third section, deals with the topic of the correspondence between micro- and macrocosm, which is a typical topic in the Indian medieval mystic poetry.[81]

78 The "Juxtaposition" of HACKER 1978.

79 For the parallels in the classical medical works, see my footnotes 820, 926 and 907 on these verses in the *English translation.*

80 For the concept of the human body as a tree in the Caryāpada-s and the tradition of the Bauls, cf. DAS 1992, p.410, note 154. The concept of the body as a boat is found in the Caryāpada, song no.38 by Sarahapā, which is discussed by HARDER 2011 (under the key term, "boat journeys") in relationship to the Baul songs.

81 For example, SR 1,3, śl.55cd–56ab: *jambū-śāka-kuśa-krauñca-śālmalī-śveta-nāmasu // dvīpeṣu puṣkare caite jātāḥ ṣaḍjādayaḥ kramāt.* To explain this verse, the commentary S quotes from Mataṅga's Bṛhaddeśī: *nanu katham sapta svarā iti niyamaḥ? ucyate, yathā sapta-dhātv-āśritatvena saptaiva dhātavo rasādayo jñeyāḥ, tathā cāha suśrutaḥ, "tvag-asṛn-māṃsa-medo-sthi-majjā-śuklāni dhātavaḥ" iti. tathā sapta-cakrāśritatvena sapta-dvīpāśritatvena vā saptaiva svarāḥ.*

For microcosmic and macrocosmic correspondences in the tradition of the Bauls, cf. DAS 1992, p.389, note 7 and note 208 (on the seven divisions of the earth).

§2. Comparison with the two parallel texts

The embryologico-anatomical verses of the SR have parallels in ŚG and YY. Charts of the corresponding verses are given below. A precise comparative analysis of each verse can be found in my footnotes on the *English translation* of the text of the SR.

§2.1. Verses in ŚG adhyāya 8, which are identical to the embryological verses of SR

§2.1.1. The list of the identical verses in the two texts

ŚG 8,15 (perhaps 8,13, too) up to 8,38a are parallels to the SR's verses on embryology, with the exception of SR 1,2, śl.26–27 (the relation between the foetus' position in the uterus and its sex) and SR 1,2, śl.28–32 (the agony which the foetus suffers in the uterus). The next verse, ŚG 8,38b, also has a counterpart in the SR which has a similar content, but is not parallel in wording.[82]

The ŚG's verses on embryology (ŚG 8,15–38a) are not only parallel, but even identical[83] to SR 1,2, śl.24b–40ab. The wording (i.e. vocabulary and syntax) of both texts is identical, except for a few trivial deviations.

The chart of the correspondences of the verses in the two texts is given below, with the SR's verse numbers first, then those of the ŚG.

82 With the term "parallel" I mean cases in which two verses share the same vocabulary, even if their syntactic structure might differ from each other. These two verses are presumed to be derived from one and the same source. With "similar" here I mean a case in which one verse adopts vocabulary different from another verse's, though both verses deals with the same topic. In this case, the two verses are not necessarily derived from the same source.

83 With "identical" I mean a case in which two verses share the same vocabulary *and* the same syntactic structure. Namely, it is a case in which one and the same verse is handed down in two different texts.

The ŚG's verses which contain the same topic as the SR, but are not identical[84] to the SR's, and those which have no parallelism but are considered noteworthy for the purpose of a better text analysis, are marked with an asterisk [*].

* SR 1,2, śl.16. (Four kinds of birth)
ŚG 8,3 deals with the same topic. A few, but not all, words contained in it are parallel to those in SR 1,2, śl.16. The syntax is different. So, this verse cannot be called identical to SR 1,2, śl.16.[85]

SR 1,2, śl.22. (Union of semen and blood)
ŚG 8,13 seems to have a trace of parallelism, containing parallel wording, though not completely identical.[86] I have the impression that the two verses are derivatives of one and the same verse, that means, either of SR or ŚG altered this original verse. ŚG 8,13 mentions the first stage of the embryo, *drava*. The term *drava* is found in SR śl.23a.

* SR 1,2, śl.23–24ab. (Embryonic conditions in the first and second month)
ŚG 8,14 describes a totally different type of development, i.e. *budbuda, kalala, peśī, ghana* and *piṇḍa*.[87] These stages are subsequent, and the three, i.e. *peśī, ghana* and *piṇḍa*, have nothing to do with the difference of the child's sex, in contrast to the newer theory of the SR. The type of development described in the ŚG resembles that of the Purāṇa-s[88], and differs from that of the classical

84 That means, the syntax contained in one verse differs from that in another.

85 ŚG 8,3: *jarāyujo 'ṇḍajaś caiva svedajaś codbhijjas tathā / evaṃ catur vidhaḥ prokto deho 'yaṃ pāñcabhautikaḥ //.*

86 ŚG 8,13: *janma-karma-vaśād eva niṣiktaṃ smara-mandire / śukraṃ rajaḥ-samāyuktam prathame māsi tad dravaṃ //.*
Cf. SR 1,2,21cd: *śuddhārtavāyā yoṣāyā niṣiktaṃ smara-mandire.* SR 1,2,22cd: *jīva-karma-preritaṃ tad garbham ārabhate tadā.* SR 1,2,23ab: *dravatvam prathame māsi kalalākhyam prajāyate.*
Compare the underlined expressions with each other. For details, cf. my footnote 522 on SR śl.22 in the *English translation.*

87 ŚG 8,14: *kalalam budbudaṃ tasmāt tataḥ peśī bhaved idam / peśī ghanaṃ dvitīye tu māsi piṇḍaḥ prajāyate //.*

88 For the Purāṇic theory of embryonic development, cf. SUNESON 1991. According to his study, the Purāṇic theory is more archaic than that of the classical medical texts. The theory of the ŚG accords with the series of the terms, *kalala, arbuda/budbuda, peśī,* and *ghana*, which is mentioned by the Mahābhārata and the Mārkaṇḍeyapurāṇa (SUNESON, ibid., p.111). Intriguingly, the embryological description in the Taittirīyopaniṣadbhāṣyavārttika, from which the ŚG has many quotations (cf. COMBA 1981, p.217), contains the same series of the embryonic terms (SUNESON ibid., p.112).

medical texts.[89] ŚG 8,16–17 states that the embryo's position, i.e. right, middle or left, in the uterus decides its sex.

* SR 1,2, śl.24cd (Sprouting of the hands, feet and head)
ŚG 8,15ab has the same content, but it is not an identical verse.[90]

SR 1,2, śl.25. (The simultaneous manifestation and the primitive and minute condition of the main and subsidiary parts of the body)
= ŚG 8,18.

* SR 1,2, śl.26ab. (Normality and abnormality)
ŚG has no parallel to SR śl. 26ab.

SR 1,2, śl. 26cd. (Fourth month)
= ŚG 8,19ab.SR 1,2, śl.27ab. (Characteristics of the three sexes)
= ŚG 8,19cd.

* SR 1,2, śl.27cd. (Character of the third sex)
The same topic is dealt with in ŚG 8,20ab, but in a different manner.[91]

SR 1,2, śl.28–30ab. (*Dohada*)
= ŚG 8,20cd–22.* SR 1,2, śl.30cd–32. (Relation between the mother's *dohada* and the future child)
These verses find no counterpart in the ŚG. The contents of SU, śārīra., 3,22 correspond with those of the SR.[92]

SR 1,2, śl.33ab. (Fifth month)
= ŚG 8,23ab.

According to the chart "Embryonic development in Other Postvedic Texts" of SHIVARAM 2001, pp.83–84, the series of the embryonic terms contained in other Purāṇa-s like AgniP, BhaviṣyaP, BhāgavataP, ViṣṇudhP, YS, etc. deviate from the above-mentioned one.
This fact might suggest that this verse (ŚG 8,14) is also based on the Taittirīyabhāṣyavārttika, like the other verses which are not parallel to the SR. But unlike the ŚG, the series in the Mahābhārata, MārkaṇḍeyaP and Taittirīyopaniṣadbhāṣyavārttika do not mention *piṇḍa*. In addition to that, according to SHIVARAM ibid., pp.83–84, none of the non-medical texts mentions *piṇḍa* in the second month.

89 For details of the classical medical theory, cf. my footnote 535 on SR śl.23 in the *English translation.*

90 ŚG 8,15ab: *karāṅghri-śīrṣakādīni tṛtīye sambhavanti hi /15ab/.*

91 ŚG 8,20ab: *napuṃsake ca te miśrā bhavanti raghunandana /20ab/.*

92 Cf. my footnote 587 on SR śl.32 in the *English translation.*

SR 1,2, śl.33cd–34ab. (Sixth and seventh month)
= ŚG 8,23cd–24ab.

SR 1,2, śl.34cd–35ab. (Position of the foetus in the uterus)
= ŚG 8,24cd–25ab (Verse 25 consists merely of two *pāda*-s, *ab*). The following verses (ŚG 8,25cd–32), which are not contained in the SR, describe the agonies which the foetus suffers in the uterus.

SR 1,2, śl.35cd–36ab. (*Jātismara*. Meditation of the foetus)
= ŚG 8,33.

SR 1,2, śl.36cd–37ab. (Eighth month. *Ojas*)
= ŚG 8,34.

SR 1,2, śl.37cd–38ab. (The *ojas* wavers between the mother and the foetus; consequently, the child born in the eighth month does not survive.)
= ŚG 8,35.

SR 1,2, śl.38cd–39ab. (Temporary subsistence because of *saṃskāra*. Proper birth after the ninth month.)
= ŚG 8,36.

SR 1,2, śl.39cd–40ab. (Umbilical cord)
= ŚG 8,37abcd. (8,37 consisting of six *pāda*-s)

§2.1.2. Analysis

The SR and ŚG are identical in their verses describing fertilisation until birth, except for a few deviations. In some passages the SR preserves more non-Āyurvedic materials than the ŚG, while in other passages the ŚG is more non-Āyurvedic, as shown below. This shows that the matter is far more complicated than one text simply quoting the other. The verses which are identical to each other might have been taken from a common source, but I will not go further into this matter, because the available materials (SR and ŚG) are insufficient to discuss this point.

Instead, I make, in the following, the divergences of the two texts clear. Doing so might help elucidate the relationship of the two texts to each other and their textual development.

For the various embryonic stages in the first and second month, the ŚG (8,14) adopts the older theory of embryonic development which is handed down in the Purāṇa-s. In contrast, the SR (1,2, śl.23–24ab) adopts the theory which is dealt with in the classical medical texts[93]; according to this theory, the three embryonic states, i.e. *peśī, ghaṇa* and *piṇḍa*, are associated with the determination of the embryo's sex. On the other hand, the SR does not have a parallel to ŚG 8, śl.16–17, which mentions the relation between the embryo's position in the uterus and its sex (determination of its sex).[94]

Although SR śl.23 and ŚG 8,13d–14 present different theories, i.e. the classical medical theory and the Purāṇic theory respectively, these two verses nevertheless show partial parallelism.[95]

Intriguingly, two *pāda*-s from different *śloka*-s in the SR (SR śl.21d, *niṣiktaṃ smara-mandire* and SR śl.22c, *jīva-karma-preritaṃ*) are found together in the ŚG as belonging to one and the same *śloka* (ŚG 8,13ab, *janma-karma-vaśād eva niṣiktaṃ smara-mandire*).[96]

But the part preceding this (just before SR śl.22) is problematic. Though the SR and ŚG both describe the formation of the embryo, the manner of description differs.

The SR (śl.18–22) describes transmigration, during which the individual self (*kṣetrajña, jīva*) travels through space, wind, smoke, cloud, rain-cloud, rain, plants and food into semen, which is ejacu-

93 I presume that the SR substituted the Purāṇic theory of the common source text with the classical medical theory, while the ŚG preserved the Purāṇic theory. The other possible interpretation is, of course, that the common source text did not contain this part, and that both the SR and ŚG, independently from each other, adopted different theories, the Purāṇic one and the classical medical one respectively.

94 The correlation between the foetus' position in the uterus and the determination of its sex is, however, mentioned later on in SR 1,2, śl.41. But it is not a verse identical to ŚG 8,16–17.

95 SR śl.23: *dravatvam prathame māsi kalalākhyam prajāyate / dvitīye tu ghanaḥ piṇḍaḥ peśīṣad-ghanam arbudam /23/.*
ŚG 8,13d–14: *prathame māsi tad-dravam /13c/ kalalam budbudaṃ tasmāt tataḥ peśī bhaved idam / peśī ghanaṃ dvitīye tu māsi piṇḍaḥ prajāyate /14/.*
Besides the names of the embryonic states, the expression *prajāyate* is also shared by the two texts.

96 If it is relevant to consider *jīva-karma-preritam* to be a variant of *janma-karma-vaśād.*

lated into the vagina. The SR's description clearly resembles that of the Chāndogyopaniṣad.[97]

In contrast, the ŚG (8,5–13) begins its description of embryonic development only from the moment of ejaculation and conception,[98] without dealing with the transmigration of the individual self. The statements in this part of the ŚG often agree with the classical medical theory, especially with the SU. ŚG 8,6cd–7ab is parallel to the SU (śārīra., 3,5) and AS (śārīra., 2,4).[99] ŚG 8,8 is parallel to SU śārīra., 3,12.[100] ŚG 8,10 is parallel to SU śārīra., 2,26.[101]

Thus, we have here the case opposite to the case of SR śl.23 and ŚG 8,13d–14. It is the SR (śl.18–22) that adopts a non-Āyurvedic theory, i.e. an Upaniṣadic theory.

Although the SR (śl.16–17) and ŚG (8,3 and 8,12) both deal with the types of birth like *aṇḍaja, svedaja, jarāyuja* etc., they are not identical in wording. The SR's verses and the ŚG's verses seem to have respectively originated in different sources.

The verses proceding this (SR śl.4–15cd), which describe the process of Brahman's manifestation as the individual self and the creation of the world, function as the prologue to the embryologico-anatomical description. The ŚG does not have parallels to this part of the SR.

After mentioning *dohada* (SR 1,2, śl.28–30ab; ŚG 8, śl.21–23a), the SR (śl.30cd–32) enumerates the various types of the mother's

97 Cf., my footnote 506 on SR śl.18, *abhra* in the *English translation.*

98 ŚG 8,5: *śukra-soṇita-sambhūtā vṛttir eva jarāyujaḥ / strīṇāṃ garbhāśaye śukram ṛtu-kāle viśed yadā /5/ yoṣito rajasā yuktaṃ tad eva syāj jarāyujam /6/.*

99 ŚG 8,6cd–7ab: *bāhulyād rajasā strī syāc chukrādhikye pumān bhavet /6/ śukra-śoṇitayoḥ sāmye jāyate ca napuṃsakaḥ /7ab/.*
AS śārīra., 2,4: *tata eva ca śukrasya bāhulyāt pumān ārtavasya bāhulyāt strī, tayoḥ sāmye napuṃsakam.* The wording of SU śārīra., 3,5 is almost the same as in the AS.

100 ŚG 8,8: *tatrāyugma-dine strī syāt pumān yugma-dine bhavet.*
SU śārīra., 3,12ab: *yugmeṣu tu pumān prokto divaseṣv anyathābalā.*

101 ŚG 8,10: *ṛtu-snātā yasya puṃsaḥ sākāṅkṣam mukham īkṣate / tad-ākṛtir bhaved garbhas tat paśyet svāmino mukham /10/.*
SU śārīra., 2,26: *pūrvam paśyed ṛtu-snātā yādṛśaṃ naram aṅganā / tādṛśaṃ janayet putram bhartāram paśyed ataḥ.*

dohada which have influence on the character of the future child. The ŚG lacks a parallel or corresponding statement.

This part seems to be of an origin different from the text surrounding it. I have come to this assumption, because, in the SR, the description of the embryonic condition in each month is usually limited to a half-verse (i.e. two *pāda*-s) or two half-verses (i.e. four *pāda*-s = a *śloka*), except for the fourth month which is in question. For instance, the first (SR śl.23ab) and fifth month (SR śl.33ab) are respectively described in two *pāda*-s, and the second (SR śl.23cd–24ab) and sixth month (SR śl.33cd–34a) are described in four *pāda*-s, respectively. Six *pāda*-s, or one and a half verse, are allotted to the description of the eighth month (SR śl.36cd–37). In contrast to these, the description of the fourth month, including that of *dohada*, contains twenty-six *pāda*-s (= six and a half verses) and is thus disproportionately detailed. Therefore, the statement on the relationship between the mother's *dohada* and the future child's character (SR śl.30cd–32) is probably a secondary insertion. Indeed, this statement (SR śl.30cd–32) has a striking parallelism to SU, śārīra., 3,22. This theory might have been adopted from the classical medical texts.[102]

The ŚG (8, śl.25–33) describes the agony which the foetus suffers in the uterus, i.e., the topic commonly discussed by the religious texts dealing with embryology, such as the Purāṇa-s. In contrast to that, the SR shares only the first and last verse of this part (SR śl.34cd–35ab and 35cd–36ab) with the ŚG, while the verses in-between are missing. So the SR touches only slightly on the topic of the foetus' agony in the uterus.

Actually, it is already proven that this part (ŚG 8, śl.26–34) is an insertion: COMBA 1981 (pp.203–205, notes 29–35) remarks that the verses between ŚG 8, śl.25 and śl.33 (parallel to SR śl.34cd–35ab and SR śl.35cd–36ab, respectively) are quotations from the Taittirīyopaniṣadbhāṣyavārttika by Sureśvara[103], a direct pupil of Śaṅkara.

102 That means either that the SR is based on the SU, or that both, the SR and SU, have a common source.

103 For the information about this text, see COMBA 1981, p.177, note 12.

In other words, the ŚG strengthened its ethico-religious character through quoting verses which express disgust for the human body and birth.

The verses at the end of the embryological description (SR śl. 40cd–43)[104] do not have parallels in the ŚG. Instead, the ŚG deals with ethico-religious topics. In the part next to ŚG śl.37–38a (parallel to SR śl.39cd–40ab), the ŚG (from ŚG 8,38b–40 onward) describes the traumatic moment of birth in more detail than the SR does, and thereafter begins the description of various kinds of agonies which a human being has to suffer during his life. Almost all the verses after ŚG 8,38b–40[105] are quotations from the Taittirīyopaniṣadbhāṣyavārttika (cf. COMBA 1981, p.217, *appendice*, which lists the quotations from this Vārttika). In contrast to that, the SR does not mention the agony suffered by the foetus/baby at birth. The SR describes the moment of birth briefly in two verses (SR 1,2, śl.41 and 42); however, some expressions of the SR imply such an agony, i.e. *saṅkucad-gātra* "whose limbs are contracting" (SR śl.41a), *rujad-gātra* "whose limbs are paining" (SR śl.42c) and *yantra* "torturing instrument" (SR śl.42d).[106]

In short, the SR is more closely related to the classical medical texts than the ŚG.[107] In contrast, the ŚG has strengthened its ethico-religious character through adopting quotations from the Taittirīyopaniṣadbhāṣyavārttika.[108]

Maybe we could try to reconstruct the common source of the SR and ŚG, through analysing which parts the two texts have in common, and in which parts they deviate from each other. But I have chosen not to discuss this any further at this point, as momentarily I

104 After that, from SR śl.44 onward, the anatomical description begins. For this part again, parallels are found in the ninth chapter of the ŚG (see below §2.2.1.).

105 That means the verses thereafter upto the end of the eighth chapter of the ŚG.

106 Usually, the descriptions of the moment of birth in the classical medical works are brief, e.g., the CA and SU, in contrast to the Purāṇic texts like BhāgP, ViṣṇudhP, AgniP and YS (cf. COMBA 1981, p.209, note 42).

107 The exception is ŚG 8,6cd–8, which accords with the SU and AS. In the place corresponding to this, SR śl.18–22 contains an Upaniṣadic theory.

108 The Taittirīyopaniṣadbhāṣyavārttika itself deals with embryology, according to SUNESON 1991, p.112.

consider the material available as insufficient. Because of these insufficiencies, I am not able to differentiate whether the SR replaced Purāṇic theories in the source text with the classical medical theories, or if the source-text did not contain these verses at all from the start. The two works, the SR and ŚG, might have independently from each other supplemented different verses in these places. For these places, as we have seen above, it is difficult to judge which of the SR and ŚG preserves a more original version, or if these verses were perhaps their respective secondary supplements, and originally not contained in the common source-text at all. As this is quite an interesting but complex matter, one can only hope that future researchers will be able to elucidate the problem.

A similar problem will be encountered in the part dealing with anatomy (SR śl.44–119cd), where the matter is even more complicated (cf. §2.2. "*Verses in ŚG adhyāya 9, which are Identical to the Anatomical Verses of SR*").

§2.1.3. Comparison of SR śl.24b–40ab with other texts, according to COMBA (1981)'s study

In her study, COMBA 1981 compares the ŚG with other texts, i.e. the classical medical texts (SU, CA, AH), the Purāṇa-s (BhāgP, ViṣṇudhP, AgniP), the Garbhopaniṣad, the YS, and the Jaina text on embryology entitled Taṇḍulaveyāliya (cf. COMBA 1981, p.176, note 5).[109]

109 For the information of the ŚG, cf. ROCHER 1986, p.212. Among the studies on the ŚG mentioned by COMBA 1981, p.175 (footnote 2), I was able to obtain only VALLAURI 1942.
It is only with reservation that ROCHER (ibid.) and VALLAURI (ibid.) admit that the ŚG belongs to the Padmapurāṇa. ROCHER (ibid.) states that the ŚG is not included in the Ānandāśrama edition of the Uttarakhaṇḍa of the Padmapurāṇa. ROCHER (ibid., p.207) quotes Wilson's notice that "the different portions of the Padmapurāṇa are in all probability as many different works". ROCHER (ibid.) remarks that "there are number of sections in the Uttarakhaṇḍa which also had an existence of their own".
VALLAURI (ibid., p.3) reports that the ŚG belongs to the Padmapurāṇa, according to the majority of his manuscripts, but that a scholar, Rājendralāla Mitra, mentions a source, according to which it belongs to the Matsyapurāṇa. Nevertheless, he did not find it in any edition of the two Purāṇa-s.
COMBA (1981, p.175, footnote 2) reports as follows: the majority of the manuscripts claim that the ŚG belongs to the Padmapurāṇa; but the other

Regrettably, her study does not consult Vāgbhaṭa's Aṣṭāṅgasaṅgraha (AS) which actually contains the highest amount of verses parallel to the ŚG and SR, as I have discussed in my footnotes on the *English translation* of the SR.

For convenience, I have summarised some points of COMBA's analysis of the ŚG, to make its textual position clear. I feel that this is helpful because COMBA's article is not accessible to all readers, as it is not easily available and the language of her discussion is Italian.

I have given the number of the SR verses which are identical to these ŚG verses in brackets.

— ŚG 8, śl.15 locates the manifestation of the five limbs (*aṅga*) in the third month.[110] COMBA (ibid., note 18) reports that all the texts she consulted are in concordance with the ŚG, except for the Garbhopaniṣad, BhāgP and Taṇḍulaveyāliya.
— ŚG 8, śl.18 (= SR 1,2, śl.25) mentions the simultaneous manifestation of the minute limbs (*aṅga*) and secondary appendages (*pratyaṅga*).[111] This is identical to the theory of the CA and SU (cf. COMBA 1981, note 22).
— ŚG 8, śl.21cd–22 (= SR 1,2, śl.28–30ab) locates the double-heartedness (*dohada*) at the same time as the full manifestation of the limbs (*aṅga*) and secondary appendages (*pratyaṅga*) in the

manuscripts state that it belongs to the Bhāgavata-, Viṣṇudharmottara-, Śiva-, or Kūrmapurāṇa. Nevertheless, the ŚG does not occur in the southern version of the Padmapurāṇa which was published in Poona and Bombay; however, the ŚG is not found in the edited texts of the other Purāṇa-s, although these Purāṇa-s contain numerous other Gītā-s. (Summary of COMBA's report.)

COMBA (ibid., in the same footnote) suggests two possibilities. The first possibility is: the ŚG is said to belong to a certain Purāṇa, so that it might acquire authenticity, but actually it does not belong to that Purāṇa. The second possibility is: the ŚG is inserted into a certain Purāṇa in the form of a manuscript.

110 ŚG 8,15: *karāṅghri-śīrṣakādīni tṛtīye sambhavanti hi / avibhaktiś ca jīvasya caturthe māsi jāyate.*

SR 1,2,24cd: *tṛtīye tv aṅkurā pañca karāṅghri-śiraso matāḥ /24cd/*

SR 1,2, śl.24cd is not an identical verse, but contains similar wording. The contents of the verses are actually the same in both texts.

111 ŚG śl.18abcd and SR 1,2,25cd–26ab are identical: *aṅga-pratyaṅga-bhāgāś ca sūkṣmāḥ syur yugapat tadā / vihāya śmaśru-dantādīn janmānantara-sambhavān.*

fourth month.[112] Among the texts she consulted, only the SU agrees with this (cf. COMBA ibid., note 18).[113]

— ŚG 8, śl.23ab (= SR 1,2, śl.33ab) mentions the awakening of consciousness (*citta*) and the abundance of flesh and blood.[114] COMBA's comments may be summarised (COMBA ibid., note 26) as follows: For the abundance of flesh and blood, similar descriptions are found in the YS and Taṇḍulaveyāliya, but the nearest description is made by the CA, although the CA does not mention consciousness (*citta*). In contrast to the CA, the SU mentions it. The Purāṇa-s which are consulted locate the development of consciousness in the sixth month.

— ŚG śl.23cd–24ab (= SR śl.33cd–34ab) locates the clear division of the bones, cords (*snāyu*), etc. in the sixth month, and the fullness of the limbs in the seventh month.[115] COMBA (ibid., note 27) writes that the CA and AH have the same as the ŚG. The CA however, similarly to the SU, states that all the body parts appear simultaneously, i.e. the two classical medical texts do not set particular parts such as the bones, cords etc. apart. The AH allots the cords and nails to the sixth month. COMBA considers, therefore, the AH as the nearest to the ŚG (i.e. also to the SR) on this point, among the texts she consulted.[116]

— YS 3,71 has a very similar wording to ŚG śl.35 (SR śl.37cd–38ab)[117] describing the oscillation of the *ojas* (cf. COMBA ibid., note 38).

112 ŚG śl.21–23a: *tāṃ ca dvihṛdayāṃ nārīṃ āhur dauhṛdinīṃ tataḥ /21cd/ adānād dohṛdānāṃ syur garbhasya vyaṅgatādayaḥ / mātur yad-viṣaye lobhas tad-ārto jāyate sutaḥ /22/* For SR śl.28–30ab's variants, cf. my footnotes 572, 577 and 580 in the *English translation.*

113 Also see COMBA 1981, note 25, stating that other texts often locate it in the third month.

114 ŚG 8, śl.23 (= SR 1,2, śl.33ab): *prabuddham pañcame cittam māṃsa-śoṇita-puṣṭatā.*

115 ŚG śl.23cd–24ab (SR śl.33cd–34ab): *ṣaṣṭhe 'sthi-snāyu-nakha-keśa-loma-viviktatā / bala-varṇau-copacitau saptame tv aṅga-pūrṇatā //.* The SR contains the variant *roma* (for *loma*).

116 Actually, the AS, which COMBA 1981 does not consult, is the nearest to the SR, cf. my footnote 602 on SR śl.34b in *English translation.*

117 ŚG śl.35: *mātaraṃ ca punar garbhaṃ cañcalaṃ tat pradhāvati / tato jāto 'ṣṭame garbho na jīvaty ojasojjhitaḥ //.* The identical verse, SR śl.37cd–38ab, contains a few variants, cf. my footnote 624 in the *English translation.*

— For the topic of the umbilical cord mentioned in ŚG śl.37abcd (= SR śl.39cd–40ab),[118] COMBA (1981, note 40) states that, for the chronological scheme of development of the embryo's nutrition, the ŚG adopts that of the SU, though in a fragmentary way, without mentioning the period when the embryo is nourished through other means than the umbilical cord.

Śr̥ṅgeri Abhinava Nr̥siṃha Bhāratī[119], one of the commentators on the ŚG, might have been aware of the parallelism between the SR and ŚG, as suggested by COMBA's statement (1981, p.185) that his commentary (called "edition A" by COMBA) quotes from the SR. But COMBA herself seems not to be aware of this parallelism, as she makes no further comment on this matter.

COMBA (1981, pp.175–177) analyses the position of the embryological description in the ŚG in comparison with the other texts as follows:

It shows the relative independence from the three classical medical texts, though it does not contradict them in a direct manner. Sometimes it presents a remarkable similarity with the SU.[120] At other times, it follows the general principles common to the CA and SU, while sometimes it keeps itself apart from them and stands near to the AH.[121]

For example, according to the CA and SU, the limbs (*aṅga*) of the embryo develop simultaneously. The ŚG states the same in 8, śl.

118 ŚG śl.37abcdef: *mātur asra-vahāṃ nāḍīm āśrityānvavatāritā / nābhistha-nāḍī garbhasya mātr-āhāra-rasāvahā / tena jīvati garbho 'pi mātr-āhāreṇa poṣitaḥ //*. The fifth and sixth *pāda*-s (ef) have no equivalents in the SR. The identical verse, SR śl.39cd–40ab, contains variants, cf. my footnote 631 in *English translation*.

119 He was, according to COMBA (1981, p.184, note 55), the head of the Śāradā Pīṭha of Śr̥ṅgeri from 1599 until 1622.

120 E.g., ŚG 8,6 (without any parallel in the SR) and SU śārīra., 3,5 (cf. COMBA 1981, p.175, note 3). ŚG 8,6: *yoṣito rajasā yuktaṃ tad eva syāj jarāyujam / bāhulyād rajasā strī syāc chukrādhikye pumān bhavet.*

121 In such cases, the AS, which COMBA 1981 does not consult, actually stands nearer to the SR, than the AH does (cf. my footnote 114, above in *Situating the text*).

18[122] (= SR 1,2, śl.25). In spite of that, in ŚG 8, śl.23–24ab[123] (= SR 1,2, śl.33–34ab) and ŚG 8, śl.34[124] (= SR 1,2, śl.36cd–37ab), there are precise references to the differentiation of the individual parts of the embryonic body from the sixth till eighth month, in contradiction to the statement on the simultaneous development of the limbs.

In other cases, the ŚG has similarity to the Purāṇic texts, because of the vocabulary it has in common with these. This kind of vocabulary does not occur in the medical texts.[125]

122 ŚG 8,18: *aṅga-vibhāgāś ca sūkṣmāḥ syur yugapat tadā / vihāya śmaśru-dantādīn janmānantara-sambhavān.*
COMBA (1981, p.176) writes śl.23, which is obviously a mistake. SR śl.34cd–36ab (=ŚG śl.25 and 33) deal with the posture of the foetus, and its remembrance of the previous lives (*jātismara*) and its practice (*abhyāsa*) and meditation; these are not precisely discussed in the medical texts, neither in the SU, AS and AH, nor CA. These topics are only dealt with in the Purāṇa-s.
However, *jātismara* and *abhyāsa* are mentioned in the SU. SU śārīra., 2,57 states that those who have been imprinted/educated (*bhāvita*) by means of *śāstra*-s in their previous life (*pūrva-deha*) or have predominantly *sattva-guṇa*, acquire the remembrance of their previous lives (*pūrvajātismara*); the next verse 2,58 states that one receives the same qualities (*guṇa*) which one has practiced or been habituated to (*abhyasta*) in one's previous body (*pūrva-deha*). Although the preceding verse 2,55 mentions the foetus' movements which are simultaneous with the mother's, verses 2,57–58 generally have nothing to do with 2,55. Thus, the manner in which the SU deals with *jātismara* and *abhyāsa* is different from that of the Purāṇa-s; the two topics in the SU have nothing to do with the foetus' posture or agony in the uterus. It might be possible to guess, though difficult to prove, that the order in which the two topics are situated after the topic of the foetus' movement might be a vestige suggesting that the source which the SU is based on originally had a character closer to the Purāṇa-s; the SU altered the source-text, retaining the order.

123 ŚG 8,23–24ab: *prabuddham pañcame cittam māṃsa-śoṇīta(sic.)-puṣṭatā / ṣaṣṭhe 'sthi-snāyu-nakha-keśa-loma-viviktatā /23/ bala-varṇau copacitau saptame tv aṅga-pūrṇatā /24ab/.*

124 ŚG 8,34: *aṣṭame tvak-chrutī syātām ojas tejaś ca hṛdbhavam / śuddham āpīta-raktaṃ ca nimittaṃ jīvitam matam.*

125 E.g., ŚG śl.14, which enumerates the early stages of the embryonic development, i.e. *budbuda, kalala, peśī, ghana* and *piṇḍa*. This verse is *not* parallel to SR śl.23–24ab, which accords with the classical medical texts.
ŚG 8,14: *kalalam budbudaṃ tasmāt tataḥ peśī bhaved idam / peśī ghanaṃ dvitīye tu māsi piṇḍaḥ prajāyate //.*

Sometimes, the ŚG accords with the contents of the SU, but not with that of the CA or AH.[126]

At other times, it synthesises elements from all three classical medical texts.[127]

Some elements in the ŚG do not appear in the corresponding parts of the embryological descriptions of the CA, SU or AH.[128] Another passage is in concordance with the YS.[129]

COMBA draws the conclusion that the ŚG presupposes and follows the theories of the three classical medical texts, but sometimes integrates other elements, or condensates the principal data from various texts.

COMBA states that the passages with ethico-religious character in the ŚG generally consist of the quotations from the Taittirīyopaniṣadbhāṣyavārttika of Sureśvara, a direct pupil of Śaṅkara. COMBA's observation is indeed reinforced by the fact that

---

SR śl.23–24ab: *dravatvam prathame māsi kalalākhyam prajāyate / dvitīye tu ghanaḥ piṇḍaḥ peśīṣad-ghanam arbudam /23/ puṃ-strī-napuṃsakānāṃ syuḥ prāg-avasthāḥ kramād imāḥ /24ab/.*

126 Cf. ŚG 8, śl.15–23a; cf. COMBA 1981, note 7.
These verses are verses identical to SR 1,2, śl.24cd–30ab, except for ŚG 8,15cd, ŚG 8,16 and ŚG 8,17. I quote these verses which are not contained in the SR in the following:
ŚG 8, 15cd–17: *avibhaktiś ca jīvasya caturthe māsi jāyate /15cd/ tataś calati garbho 'pi janmanā jaṭhare svataḥ / putraś ced dakṣiṇe pārśve kanyā vāme ca tiṣṭhati /16/ napuṃsakas tūdarasya bhāge tiṣṭhati madhyataḥ / ato dakṣiṇa-pārśve śete mātā pumān yadi /17/.*

127 Cf. ŚG 8,23ab (identical to SR 1,2, śl.33ab): *prabuddham pañcame cittam māṃsa-śoṇīta(sic.)-puṣṭatā.*

128 COMBA (1981, p. 176, note 9) refers to ŚG 8,34 (parallel to SR 1,2, śl.36cd–37ab) which mentions the colour of the *ojas.* ŚG 8,34: *aṣṭame tvak-chrutī syātām ojas tejaś ca hṛdbhavam / śuddham āpīta-raktaṃ ca nimittaṃ jīvitam matam.*
Against her opinion, however, a similar statement is found in CA, sūtra., 17,14, although it does not belong to the chapter on embryology (śārīrasthāna), but to the chapter sūtrasthāna (cf. my footnote 616 on SR 1,2, śl.37ab, "red" in *English translation*).

129 Cf. ŚG 8,35 (parallel to SR 1,2, śl.37cd–38ab) and YS 3,71. ŚG 8,35: *mātaraṃ ca punar garbhaṃ cañcalaṃ tat pradhāvati / tato jāto 'ṣṭame garbho na jīvaty-ojasojjhitaḥ.*
YS 3,71: *punar dhātrīm punar garbham ojas tasya pradhāvati / aṣṭame māsyato garbho jātaḥ prāṇair viyujyate.*

the verses of the ŚG (8,26–32; 8,38b and the following verses) parallel to the Taittirīyopaniṣadbhāṣyavārttika, which describe the foetus' agony in the uterus, have no parallel in the SR, as examined above.

Except for the few deviations which have been elucidated, including the quotations from the Taittirīyopaniṣadbhāṣyavārttika, the texts of the SR and ŚG are identical in the part mentioned above (SR śl.25–40ab = ŚG 8,18–37abcd). Thus, what COMBA remarks on the ŚG, as I summarised above, applies to the SR, too. I have investigated into the relation or parallelism between the SR and AS, on which COMBA has not done any research,[130] and given the results in the footnotes on my *English Translation* of the SR's text.

§2.2. Verses in ŚG adhyāya 9, which are identical to the anatomical verses of SR

For the SR's verses on anatomy, there are identical ones in the ŚG. They are contained in the ninth chapter (*adhyāya*) of the ŚG, titled Śarīranirūpaṇa, "the Investigation of the Body", following the embryological description in the eighth chapter.

§2.2.1. List of the identical verses in ŚG adhyāya 9 and SR

The chart of the identical parallels[131] between the SR's and the ŚG's verses is given in the following. These parallels are identical to the letter.

SR 1,2, śl.44 = ŚG 9,10cdfg (six *bhāva*-s, i.e. components) derived from the mother, father etc.)
SR 1,2, śl.45 = ŚG 9,11 (*mātṛbhāva*, i.e. components derived from the mother)
SR 1,2, śl.46 = ŚG 9,12 (*pitṛsamudbhava*, i.e. those derived from the father)
SR 1,2, śl.47 = ŚG 9,13 (*rasaja*, i.e. those derived from *rasa*)
SR 1,2, śl.48 = ŚG 9,14 (*ātmaja*, those derived from the self)

130 Which is unfortunately a substantial defect, because the AS is very often the closest parallel to the SR (i.e. ŚG). (Cf. my footnote 114 above in *Situating the text.*)

131 "Identical parallels", in this case, mean the parallel verses which are strictly identical in wording, both in vocabulary and syntax.

SR 1,2, śl.49 = ŚG 9,15 (*jñānendriya*)[132]
SR 1,2, śl.50 = ŚG 9,16 (*gocara, karmendriya*)
SR 1,2, śl.51 = ŚG 9,17 (*kriyā* and pair of *antaḥkaraṇa*, i.e. *manas* and *buddhi*)
SR 1,2, śl.52 = ŚG 9,19 [133] (*viṣaya* and *kriyā* of *manas*)

SR 1,2, śl.53ab (two different theories on the origin of the *indriya*-s) has no parallel in ŚG.

SR 1,2, śl.53cd–55 = ŚG 9,20–22ab (*sattvaja*, those derived from the *sattva*)
SR 1,2, śl.56ab = ŚG 9,22cd (*sātmyaja*, those derived from suitableness)
SR 1,2, śl.56cd = ŚG 9,23ab (the body receives qualities from the gross elements[134]).
SR 1,2, śl.57-60ab = ŚG 9,23cd–26 (qualities received from the wind)
SR 1,2, śl.60cd–68ab = ŚG 9,27–34 (ten kinds of vital wind, i.e. *Prāṇa* etc.[135])
SR 1,2, śl.68cd–70 = ŚG 9,35–37ab (the qualities received from fire and water)

* SR 1,2, śl.71ab (the hard substances, such as beard, hair, nails etc., including the list of the derivatives of earth) has no identical-parallel in the ninth chapter of the ŚG.[136] Still, SR 1,2, śl.71c, in which wind etc. are called *dhātu-prakṛti*, has no identical parallel in the ŚG, but shows similarity to ŚG 9,43ab, in which *vāta, pitta* and *kapha* are called *dhātu*-s.[137]

SR 1,2, śl.79ab = ŚG 9,37cd (metabolic chain of the seven *dhātu*-s)[138]

SR 1,2, śl.82cd–83ab shows similarity to ŚG 10,24cd–25, though it is not identical[139].

132 The description in SR śl.49–53ab serves to introduce the notion of *antaḥkaraṇa* and *sattva*.
133 To the preceding and succeeding verses of the ŚG (ŚG 9,18ab and 9,20ab), there are no identical parallels in the SR.
134 The gross elements (*mahābhūta*) are usually five, but here (SR śl.57–70) only three, wind, fire and water, are mentioned.
135 The ten kinds of vital winds are included in the qualities belonging to the gross element wind.
136 Namely, SR śl.71cd is the sequel to the foregoing verses. It seems that the common source-text once contained the verse (SR śl.71cd), and that the ŚG does not quote it. Thus this case is an example for the SR preserving a more complete version of the text than the ŚG.
137 ŚG 9,43ab: *vāta-pitta-kaphās tatra dhātavaḥ parikīrtitāḥ.*
138 ŚG 9,38–42 describes metabolic change in a manner different from that in SR śl.79cd–80.

SR śl.84 mentions *svapna* and *suṣupti*,[140] which are mentioned in ŚG 10,58ab and 12,34.[141]

SR 1,2, śl.90cd–91 = ŚG 9,46abcdef (360 bones and their names according to their forms)

SR 1,2, śl.92cd–94ab = ŚG 9,47–48 (*asthisandhi*, i.e. junctures of bones)

SR 1,2, śl.114cd = ŚG 9,49ab (35,000,000 beard-hairs and hairs)[142]

SR 1,2, śl.116–118 = ŚG 9,43cd–45 (amounts of various body fluids)[143]

§2.2.2. Analysis

Now I shall analyse the above-mentioned condition of the texts.

To put it in a nutshell, for the first half of the anatomical description of the SR (from SR 1,2, śl.44 up to SR1,2, śl.70), the SR and ŚG show as clear identical parallelism[144] as in the case of the foregoing embryological description (SR śl.24–40ab).

But in the rest of the anatomical description, it is not so. The verses appearing after SR 1,2, śl.71 only sporadically have identical parallels in the ŚG.

That the ŚG has no verse identical to SR 1,2, śl.71ab (see the list above) seems to indicate the following: SR 1,2, śl.71ab mentions nails, hairs, bones etc. as the derivatives of earth, namely, it continues the foregoing list of the derivatives of earth (SR 1,2, śl.70cd). Therefore SR 1,2, śl.71ab must have also belonged to the common source-

139 ŚG 9,24cd–25: *tasya madhye 'sthi hṛdayaṃ sanālam padma-kośavat /24cd/ adho-mukhaṃ ca tatrāsti sūkṣmaṃ suṣiram uttamam / daharākāśām ity uktaṃ tatra jīvo 'vatiṣṭhati /25/.*

140 SR śl.86cd explains the etymology of the word *svapiti*, according to ChāndogyaUp 6,8,1, as the commentary S points out.

141 ŚG 10,58ab and 12,34ab. 10,58ab: *jāgrat-svapna-prasuṣupty-ākhyam evehāmutra lokayoḥ*. 12,34: *jāgrat-svapna-suṣupty-ādi-prapañco yaḥ prakāśate / tad-brahmāham iti jñātvā sarva-bandhaiḥ pramucyate.*

142 This is remarkable, as this statement does not accord with any classical medical texts, but with the YS.

143 This theory is treated in the classical medical texts, but seems to be old, because the YS and other Purāṇa-s also hand down the same theory.

144 Cf. my footnote 129 above, in *Situating the text.*

text.[145] My assumption is supported by the fact that the parallels[146] in the AS and VişnudhP also mention nails, hairs and bones as derivative of the earth-element. Also, the parallel in the AgniP mentions nails and hairs.[147] Thus the ŚG abruptly breaks off the text which seems to have been originally longer.

As already remarked, the second half of the anatomical descript-tion (SR 1,2, śl.71–118) has only sporadic parallels in the ŚG. Actually, in the SR itself, there is a gap between SR 1,2, śl.74 and śl.75. In SR śl.74cd, the author Śārṅgadeva says that he breaks off the text to avoid prolixity.[148] In the verses between SR śl.71ab and 74cd, another topic is dealt with, namely various "bodies" derivative of *sattva, rajas* and *tamas*, or, the "bodies" (*vigraha, deha*) named after various deities (*brahman, indra, yama* etc.) and species (*piśāca*-s, *asura*-s, birds, snakes, fish, trees etc.) The first verse of this part, SR śl. 71cd, mentioning *prakṛti*, is so fragmentary that we fail to grasp what it concretely means.[149] Actually, the parallel contained in the SU (śārīra., 4,62–80) is a considerably longer text.[150] Without the help of this parallel in the SU, it would be impossible to understand SR

145 If we take this case on its own, it would be possible to suppose that the ŚG is based on the SR. It would then be possible to presume that the ŚG quoted the foregoing verses directly from the SR. In this case, it would be unnecessary to assume a common source-text. But the actual matter is more complicated, because the ŚG, in another case, seems to preserve an older and more original version of the text than the SR. On account of this, I consider it convenient to assume a common source text. This matter will extensively be discussed below and in the next section.

146 These parallels are not identical to the SR in vocabulary and syntax, although they deal with the same topics as those in the SR, and sometimes contain similar vocabulary.

147 Cf. my footnote on SR śl.71ab.

148 SR śl.74cd: *teṣāṃ lakṣmāṇi na brūmo grantha-vistara-kātarāḥ.*

149 SR śl.71ab: *vātādi-dhātu-prakṛtir vyomādi prakṛtis tathā.*

150 My reason to consider this part of the SU to be parallel to the very fragmentary statement of SR śl.71cd is as follows. The SU here associates the three morbific entities (wind etc.) with the five gross elements (*mahābhūta*-s). SU śārīra., 4,62 mentions seven *prakṛti*-s derived from combinations of the three morbific entities. SU śārīra., 4,80 mentions the *prakṛti*-s pertaining to the gross elements (*bhautikī prakṛti*). Immediately after treating the topic of *prakṛti*, it lists the bodies (*kāya*) derivative of three *guṇa*-s, namely, it has the same order of the same topics. Also see my footnote 785 on SR śl.71cd ("*wind etc.*") in the *English translation*.

śl.71cd. This very brief statement, SR śl.71cd, presumably seems to have been a summary of an older, longer text which might have contained a detailed description similar to that of the SU. Thus this case forces us to assume a source-text.

Unfortunately, we have no information to judge whether Śārṅgadeva himself summarised the source-text in order to avoid prolixity, or the source-text itself was already summarised.[151] Or else, it might also be possible that the source-text did not contain the part SR śl.71cd–74, and Śārṅgadeva inserted it from another medical work which was close to the SU.[152] The matter thus seems to be quite complicated.

In the following part from SR śl.75 onward, he opens a long text listing the limbs (*aṅga*) and secondary appendages (*pratyaṅga*) of the body.[153] This list is concluded by SR śl.119.[154] The anatomical topics treated in this part are the same ones as in the SU and AS. They are listed in an order similar to those in the SU and AS.

Śārṅgadeva calls this part, SR śl.75–118, "a summary/shortening of the secondary appendages" (śl.119a, *pratyaṅga-saṅkṣepa*). So the text, SR śl.75–118, is a shortened version of a more complete text. Indeed, many topics in the SU und AS are missing here. Perhaps the older version of this text before being summarised might have contained such topics, too, though there is no evidence. Anyway, Śārṅgadeva instructs the reader to consult his other work, Adhyātmaviveka, for more detail (SR śl.119). This work which is now lost could also have contained the topics omitted here in the SR.

In this part (SR 1,2, śl.75–119), verses identical to the ŚG are rare. They are five in total, as shown below.

151 I mean that already in the medical text, which would have been handed down in Śārṅgadeva's family, this statement could have taken the summarised form.

152 I suspect SR śl.30cd–32cd, to which the ŚG has no identical-parallels, of having similar circumstances. Only the SU (śārīra., 3,22–27) has a parallel to this part, whose statement is very close to that of the SR (cf. my footnote 587 on SR śl.32 in the *English translation*).

153 SR śl.75: *piṇḍasyāhuḥ ṣaḍ-aṅgāni, śiraḥ pādau karau tathā / madhyaṃ cety. atha vakṣyante pratyaṅgākhilāny api /75/.*

154 SR śl.119: *iti pratyaṅga-saṅkṣepo vistaras tv iha tattvataḥ / asmad-viracite 'dhyātmaviveke vīkṣyatām budhaiḥ /119/.*

SR śl.79ab[155] = ŚG 9,37cd (metabolic chain of the seven elements (*dhātu*))
SR śl.90cd–91[156] = ŚG 9,46abcdef (360 bones and their names according to their forms)
SR śl.92cd–94ab[157] = ŚG 9,47–48 (junctures of bones)
SR śl.114cd[158] = ŚG 9,49ab (35,000,000 beard-hair and hair)
SR śl.116–118[159] = ŚG 9,43cd–45 (amounts of various body liquids)

To SR śl.114cd (= ŚG 9,49ab), on the amount of beard-hairs and hairs, there is no parallel in the medical texts. The unique parallel is the YS.[160] To SR śl.116–118 (= ŚG 9,43cd–45), on the amounts of various liquids, not only the medical texts but also the non-medical texts, the YS and Purāṇa-s, contain parallels.[161] The metabolic chain of the seven elements (*dhātu*) mentioned by SR śl.79ab (= ŚG 9,37cd), too, is a topic which occurs not only in the medical texts, but also in the non-medical texts. In addition, we should remark that, in the SR and ŚG, the metabolic chain begins with skin, unlike in the classical medical texts, in which the chain usually begins with the nutrient fluid (*rasa*). The theory of the metabolic chain beginning with skin seems to be more archaic than the theory of the chain beginning with the nutrient fluid.[162] Only the statements on the bones and the junctures of bones belong to the genuine classical medical theory according to the SU, and are not dealt with in the non-medical texts.[163]

155 *tvag-asṛg-māṃsa-medo-'sthi-majja-śuklāni dhātavaḥ.*

156 *asthnāṃ śarīre saṅkhyā syāt ṣaṣṭi-yuktaṃ śatatrayam /90cd/ valayāni kapālāni rucakās taruṇāni ca / nalakānīti tāny āhuḥ pañcadhā 'sthīni ca /91/.*

157 *dve śate tv asthi-sandhīnāṃ syātām atra daśottare /92cd/ korakāḥ pratarās tunnāḥ sīvanyaḥ syur ulūkhalāḥ / sāmudgā maṇḍalāḥ śaṅkhāvartā vāyasatuṇḍakāḥ /93/.*

158 *sārdha-koṭi-trayaṃ romṇāṃ śmaśru-keśās tri-lakṣakāḥ /114cd/ lakṣāṇāṃ* [...].

159 *(lakṣāṇāṃ) saṃhitā-mānaṃ jalāder adhunocyate / daśāñjali jalaṃ jñeyaṃ rasasyāñjalayo nava /116/ raktasyāṣṭau sapta syuḥ śleṣmaṇas tu ṣaṭ / pittasya pañca catvāro mūtrasyāñjalayas trayaḥ /117/ vasāyā medaso dvau tu majja eko 'ñjalir mataḥ / ardhāñjaliḥ śiromajjā śleṣma-sāro balaṃ tathā /118/.*

160 YS 3,102: *trayo lakṣās tu vijñeyāḥ śmaśru-keśāḥ śarīriṇām* /102ab/ [...] /102/ *romṇāṃ koṭyas tu pañcāśac catasraḥ koṭya eva ca sapta-ṣaṣṭis tathā lakṣāḥ sārdhāḥ svedāyanaiḥ saha* /103/. For details, see my footnote 967 in the *English translation.*

161 For details, see my footnote 973 in the *English translation.*

162 Cf. JAMISON 1986.

163 This list of bones is unique, contained only in the SU among the medical classical texts (cf. my footnote 883 on SR śl.91 in the *English translation*),

Thus, I am not sure if it is apt to assume a common source-text for this part, SR śl.75–119. The identical verses in the SR and ŚG are too scarce.[164]

Furthermore, the matter is made more complicated by the fact that the ninth section of the ŚG contains several verses on anatomy which are not found in the SR. I discuss these verses in the following.

ŚG 9,9[165] states that six cases/sheaths (*koṣa*)[166] arise from the food eaten by the parents; three of them, i.e. cords, bones and marrow, are derived from the father; and the other three, skin, flesh and blood, from the mother. This theory partly overlaps with that of the *mātr̥ja* substances and *pitr̥ja* substances treated by SR śl.45[167] (= ŚG 9,11) and SR śl.46[168] (= ŚG 9,12). So, at first sight, it gives the

---

though the Hastyāyurveda also contains it (cf. my footnotes 879–881on SR śl.91, the terms, *valaya, kapāla* and *taruṇa* in the *English translation*).
On the other hand, the science of bones seems to have a very old tradition which began developing in a considerably earlier period, i.e. in the Atharva Veda (cf. HOERNLE 1907, pp.8–9; p.109ff). Although the list of bones in the SU has no parallel in any other texts (except for the SR and the Hastyāyurveda), the terms, *kapāla* and *nalaka*, occur in the non-medical texts like the YS, too, if the presumption of HOERNLE is right (see the following). HOERNLE 1907, p.58, presumes that the names of bones, *kapola,* and *phalaka*, contained in the YS are ancient misreadings of *kapāla* and *nalaka*, respectively.

164 I give a possible argument in the following, but I cannot guarantee the aptness of my speculation: The common source-text, which might have been closely related to the Purāṇa-s, contained only these five verses which the SR and ŚG are now sharing. The ŚG preserves the original version of the text. But the version of the SR has been enriched with many secondary supplements originating in the classical medical texts.
Of course, the opposite argument is possible too: the ŚG might have omitted many verses from the common source-text, preserving only the five verses which the two works still have in common.

165 *pitr̥bhyām aśitād annāt ṣaṭ-koṣaṃ jāyate vapuḥ / snāyavo 'sthīni majjā ca jāyante pitr̥s (=pitr̥tas?) tathā /9/ tvaṅ-māṃsaṃ śoṇitam iti mātr̥taś ca bhavanti hi /10/.*

166 Cf. my footnote 69 of the *Situating the text*.

167 SR 1,2,45: *mr̥davaḥ śoṇītaṃ (sic.) medo majjā plīhā yakr̥d gudam / hr̥n-nābhīty evam ādyās tu bhāvā mātr̥bhavā matāḥ /45/.*

168 SR 1,2,46: *śmaśru-keśa-loma-kacāḥ snāyu-sirā-dhamanayo nakhāḥ / daśanāḥ śuklam ity ādyāḥ sthirāḥ pitr̥-samudbhavāḥ /46/.*

impression that it merely mentions exactly the same topics as in ŚG 9,11 and ŚG 9,12 in advance.

But let us think of the fact that the elements (*dhātu*) of the metabolic chain are only twice called "sheath" (*koṣa*) by the SR (śl.76c and 80c), and the fact that the metabolic chain in the SR begins with skin, deviating from the classical medical texts. The term *koṣa* never occurs in the parallels of the classical medical texts, the SU or AS, which in other points contain almost the same vocabulary as that of the SR. ŚG 9,9 seems to elucidate what *koṣa*, mentioned in SR śl.76c and śl.80c, actually denotes. If my hypothesis is right and it is in accordance with the SR's theory, ŚG 9,9 might also originate from the same source as SR śl.76c and 80c, namely from the source I have up to now called the "common source-text". But proving this is a task which goes beyond the topic of this thesis.

ŚG 9,38[169] states that the food eaten by human beings becomes threefold through the fire of the belly[170]. ŚG 9,39[171] mentions the three kinds of results derived from water in the same manner. ŚG 9,40[172] mentions the three results of fiery energy (*tejas*). ŚG 9,41–42[173] lists the metabolic chain whose seven members arise one after another.

But peculiarly, although ŚG 9,43ab has no parallel in the SR, this verse calls wind, bile and phlegm *"dhātu"*[174], which seems to be in concordance with the statement of SR śl.71c that wind etc. are *dhātu-prakṛti*. Here again, the ŚG seems to compensate for the SR's unclear, fragmentary statement, like in the above-mentioned case of

169 *annam puṃsāśritaṃ tredhā jāyate jaṭharāgninā.*

170 The fire of the belly (*jaṭharāgni*) is mentioned by ŚG 8,30b, too. SR śl.79d mentions it (*jāṭharavahni*) in a similar context, in which the metabolic change is dealt with.

171 *apāṃ sthaviṣṭho mūtraṃ syān madhyamo rudhiram bhavet / prāṇaḥ kaniṣṭho bhāgaḥ syāt tasmāt prāṇo jalātmakaḥ /9,39/.*

172 *tejaso 'sthi sthaviṣṭhaḥ syān majjā madhyama-sambhavaḥ / kaniṣṭhā vāṅ matā tasmāt tejo-'b-annātmakaṃ jagat /9,40/.*

173 *lohitāj jāyate māṃsam medo māṃsa-samudbhavam / medaso 'sthīni jāyante majjā cāsthi-samudbhavā /9,41/ nāḍyo 'pi māṃsa-saṅghātāc chukram majjā-samudbhavam /9,42/.*

174 ŚG 9,43ab: *vāta-pitta-kaphās tatra dhātavaḥ parikīrtitāḥ.* For this problem, cf. DAS 2003A, p.553ff.

ŚG 9,9. But I am not in a position to wander into the maze of this matter.

§2.2.3. Comparative study of SR 1,2, śl.40bc–119 with medical and non-medical texts

For the verses SR 1,2, śl.40bc–119 which sporadically have identical parallel verses in the ŚG as demonstrated above, I made a comparative study of other medical and non-medical texts.

For this study, I consulted the following texts: Suśrutasaṃhitā (SU), Aṣṭaṅgasaṅgraha (AS), Aṣṭāṅgahṛdayasaṃhitā (AH), Carakasaṃhitā (CA), Yājñavalkyasmṛti (YS), Agnipurāṇa (AgniP) and Viṣṇudharmottarapurāṇa (ViṣṇudhP). In addition, I sometimes referred to the Garuḍapurāṇa (GaruḍaP) and the Hastyāyurveda in the several cases where these two texts seemed relevant. In the following, I summarise the results of my research which are given in detail in the respective footnotes on the verses in the *English translation*.

The theories of the SR accord with those of the medical texts, SU, AS, AH and CA, when all of the four present an identical or similar theory.[175]

The SR very often accords with the SU and AS.[176] The SR is often closer to the AS than to the SU.[177] The opposite case is, how-

175 Cf. SR śl.44–46; śl.75.

176 Cf. SR śl.51abc; śl.76. SR śl.81–82ab and śl.82cd–83ab, to which the AgniP and the ViṣṇudhP, too, are parallel. Also cf. the most verses in SR śl.86cd–114ab. A comparison with the SU and AS suggests that the SR, in śl.100cd and śl.113, might have made a broad omission of an older text (for detail, cf. my footnotes 918 and 961 in the *English translation*). Also see SR śl.28ab and COMBA's study (1981) on ŚG 8,20cd, the identical verse to SR śl.28ab.

177 Cf. SR śl.76cd–78; śl.94cd. SR śl.96 and śl.98. SR śl.104–105ab; śl.110; śl.111–112a.
Also see SR śl.33ab and śl.34b. These two verses belong to the verses identical to the ŚG which are studied by COMBA 1981. To SR śl.33cd–34a, only the AS and AH have parallels. To SR śl.38cd and śl.40cd–41, only the AS has parallels.

ever, observed, too.[178] Sometimes the SU and AS compensate each other: the elements of both texts are found together in the SR.[179] In contrast to this, as regards the theories dealt with in SR śl.71cd–74ab, the AS does not mention them, while the SU and CA do.

The AH often uses expressions briefer than the AS. Most of the AH's passages parallel to the SR have their counterpart in the AS.[180] The AH very often lacks the passages which the SR and AS share. But there is one case in which the AH is closer to the SR than the other medical texts.[181]

The theory dealt with in SR śl.102ab accords only with the AS and AH, but differs from the SU.[182]

In comparison with the other medical texts, the CA contains few concordances with SR śl.86cd–119, i.e. the verses chiefly dealing with the enumeration of various components of the body and their quantities.[183] Although the CA assigns the seventh chapter (adhyāya) of the śārīrasthāna to the enumeration of the components of the body and their quantities, the system presented there has very little to do with that of the SR.

But in one case, the AS and CA share the same theory, and are in the closest relation to the SR.[184] As for the theory mentioned in SR śl.116–118, the AS, CA and the other texts accord with the SR, except for the SU.

178 Cf. SR śl.83d–84ab, which is parallel only to the SU. SR śl.101 accords only with the AS and YS. Also cf. SR śl.39cd–40ab, and COMBA's study (1981) on ŚG 8,37abcd, which are verses identical to them.

179 Cf. SR śl.47. Cf. SR śl.25 and COMBA's study (1981) on the identical verse, ŚG 8,18.

180 Cf. SR śl.33cd–34ab which is included in the passages which are identical to the ŚG. As for the topic mentioned in SR śl.33cd–34a, only the AS and AH accord with the SR, while the SU and CA do not mention it.

181 Cf. SR śl.54–55, which accords with the AgniP and the ViṣṇudhP, too. Only the AH has a counterpart to SR śl.79d–80, which is the list of the seven *dhātu*-s, but I am not sure if it should be called parallel, or is a coincidence.

182 Also see SR śl.104–105, to which the AS and AH are more closely related than the SU is.

183 Most of the statements accords with the SU and AS, and very often with the AH.

184 Cf. SR śl.48, śl.56ab, śl.112ab. Cf. SR śl.27bcd and 28a, which belongs to among the verses identical to the ŚG's verses analysed by COMBA 1981.

Some theories of the SR accord only with the YS, although such cases are few in number.[185]

There is no case where the SR only accords with the AgniP and the ViṣṇudhP. When the SR accords with the two Purāṇa-s, the SU and AS already accord with the SR.[186] An exceptional case is SR śl.54–55, with which only these two Purāṇa-s and the AH accord.

In one case, the SR contains a unique theory.[187]

SR śl.60cd–68ab is obviously an insertion, quoted from an old text which the SR has in common with the Yogayājñavalkya (YY), a Haṭhayogic text.

SR śl.79 mentioning the metabolic chain of the seven elements (*dhātu*) beginning with skin is unique. The theory about the chain of the seven elements (*dhātu*) is mentioned in the classical medical texts. But there, the chain usually begins with the nutrient fluid (*rasa*), although cases where the chain is mentioned as beginning with skin are sporadically found in the classical medical texts, too. This variant chain is extensively analysed by DAS 2003A, p.237ff (§10.7.).

The statement of SR śl.84cd–86cd on the two states of sleep, i.e. *svapna* and *suṣupti*, is also unique. This is an Upaniṣadic theory. These two states of sleep are described in Bṛhadāraṇyakopaniṣad 2,1,18–19[188], mentioned in Kaivalyopaniṣad 17 and Brahmabindu-

185 Cf. SR śl.60–68, which shows a close parallelism to the YY. Also cf. SR śl.114–116a, which partly shows some similarity with the GaruḍaP and the Tandulaveyāliya. SR śl.101 accords not only with the YS, but also with the AS. If the reading *tvaksmṛti* in SR śl.36 is correct, it would accord to YS 3,81cd, too, cf. my footnote 612 in the *English translation*.

186 SR śl.71ab accords with the AS and the two Purāṇa-s. Also see applicable verses in SR śl.86cd–118.

187 Cf. SR śl.51cd.

188 Bṛhadāraṇyakopaniṣad 2,1,19 (RADHAKRISHNAN 1953, p.190) mentions the 7200 tubes (*nāḍī*) radiating from the heart, through which the self (*ātman*) enters the heart during deep sleep. RADHAKRISHNAN, explaining this paragraph, points to Aitareyopaniṣad 1,3,12, which suggests that one of these tubes leads to the opening of Brahman (*brahmarandhra*) at the top of the head. This is obviously an archetype of Haṭhayogic theory.
The 7200 tubes around the heart are mentioned by SR śl.65, which has a parallel in the YY, a Haṭhayogic text (for further information, see my footnote 754 and 756 on SR śl.65, *nāḍī*, in the *English translation*).

upaniṣad 11, and also dealt with by Śaṅkara in his commentary on Brahmasūtra 3,2,1–9[189]. The etymological explanation of the term *svapiti* as meaning "he sleeps" in SR śl.86ab is according to Chāndogyopaniṣad 6,8,1, which is quoted by the commentary S (Adyar ed., p.55, the last line).

In brief, SR śl.40bc–119 accords with the SU and AS more often than with the other texts.

§2.3. Parallelism of SR and YY

§2.3.1. Parallelism of SR 1,2, śl.145cd–163ab and 1,2, śl.60cd–68ab to YY

After discussing the *cakra* theory (SR śl.120–145cd), the SR discusses a further theory (SR śl.145ab–163ab), in which it deals with breaths or vital winds (*prāṇa*) and the fourteen respiratory tubes (*nāḍī*), such as the *Suṣumnā* etc. This part is parallel to the first half of the fourth chapter titled "Śarīravyavacchedavidyā" ("the Science of the Division of the Body") of the Yogayājñavalkya (YY), a Haṭhayogic text.[190]

In addition to this, SR śl.60cd–68ab is also parallel to the YY.

The next paragraph, Bṛhadāraṇyakopaniṣad 2,1,20, compares the self's moving to and fro through the tubes to a spider moving on threads (*yathorṇanābhis tantunoccaret*), which looks like small sparks coming from the fire (*yathāgneḥ kṣudrā visphuliṅgā*), though, of course, the original Upaniṣadic thought might be different from the later Haṭhayogic theory. This work states that this is the manner in which all vital winds, all worlds, all divinities and all beings emanate from the supreme self. Intriguingly, the simile of a spider is mentioned by SR śl.148, and the simile of sparks by SR śl.6b (*yathāgner visphuliṅgakā*).

189 DEUSSEN 1883, pp.369–380.

190 I referred to the edition of the Trivandrum Sanskrit Series (noted under Yogayājñavalkya 1938 in my bibliography). There is a French translation, GEENENS 2000. For the information of the YY, cf. GEENENS 2000, pp.9–16. GEENENS used the critical edition by Sri Prasad Divanji, Bombay 1954 (see my bibliography under Yogayājñavalkya). Unfortunately this edition was not available to me. GEENENS' translation suggests that this edition contains considerable deviations from the readings of the Trivandrum Sanskrit Series edition.

SR śl.59–68ab mentions the ten kinds of vital winds, i.e. the *Prāṇa*[191] etc. including the subsidiary ones such as the *Nāga, Kūrma* etc. This theory is different from the theory of vital winds presented by the classical medical texts. The five subsidiary vital winds like the *Nāga* etc. are not mentioned at all by the classical medical texts, but only in the Purāṇa-s. In SR śl.60cd–68ab, the nouns *Prāṇa, Apāna*, etc. are nominative, despite the context demanding accusatives.[192] This theory of vital winds seems to be an insertion from a text other than a classical medical one. The number of 72,000 tubes (*nāḍī*) mentioned by SR śl.65a, too, exposes the Yogic origin of this theory.

This part is parallel[193] to the second half of the fourth chapter of the YY.

In summary, it appears that the YY's fourth chapter on the theory of the vital winds and the respiratory tubes was split into two parts and thereafter was separately inserted into the SR. The SR, however, preserves a more complete version than the YY. So, there could once have been a common source-text which the SR and YY are based on. Besides, the YY also contains verses which are not parallel to the SR. They might not have been originally handed down by the hypothetical common source-text.

I give the chart of the parallel verses of the SR and YY below.

191 There are two kinds of *prāṇa*: one is the generic name for the vital winds (*Prāṇa, Apāna, Udāna* etc.), and the other is the proper noun for the first member of these winds. To distinguish them, I mark the former with an initial small letter, *prāṇa*, and the latter with an initial capital letter, *Prāṇa*.

192 SR śl.59–60ab: *prāṇāpānau tathā vyāna-samānodāna-sañjñakān / daśeti vāyu vikṛtīs tathā gṛhṇāti lāghavam /60ab/.* The human body obtains (*gṛhṇāti*) the ten kinds of vital winds from the gross element of wind. So, the names of the ten vital winds are accusative in SR śl.59. SR śl.60cd–68ab, in which these names are nominative, is an embedded part which explains each of these ten vital winds. I.g. śl.60cd: *teṣām mukhyatamaḥ prāṇo nābhi-kandād adhaḥ sthitaḥ /60cd/.* From śl.68cd again, the substances obtained from the gross element of fire are enumerated in accusative case.

193 "Parallel" means that the two texts contain wording close to one another, i.e. that they partly have vocabulary in common, but they are not identical in syntax.

§2.3.2. *List of YY verses parallel to SR1,2, śl.145cd–163ab*

The wording of these verses is parallel, but not completely identical. It seems that the two texts are based on the same source, but the word arrangement or syntax has been changed and some expressions have been substituted with their synonyms. Therefore I do not use = but ≈, meaning "comparable".

YY 4,1–10 have no parallels in the SR.
YY 4,1–5 briefly introduce the contents dealt with in the fourth chapter. YY 4,6–8 mention the measure of the *Prāṇa*. YY 4,9 mentions the fire situated in the subtle body (*sūkṣma-śarīra*), and YY 4,10 the efficacy of Yoga.

SR śl.145cd (2 *aṅgula*-s above the *ādhāra*, 2 *aṅgula*-s below the penis) ≈YY 4,14ab.[194]
SR śl.146ab (*deha-madhya*) ≈ YY 4,11cd.[195]
SR śl.146c (flame of fire) ≈ YY 4,13ab.[196]
SR śl.147ab (bulb of the body and its size) ≈ YY 4,16.[197]
SR śl.147cd (*Brahmagranthi*, i.e. another name of the bulb) has no parallel in the YY.
SR śl.148ab (*nābhi-cakra* with twelve spokes) ≈ YY 4,18.[198]
SR śl.148cd (the *jīva* wanders there like a spider) ≈ YY 4,19.[199]

194 YY 4,14ab states that the middle of the body (*dehamadhya*) is above the anus by two fingers, and below the genitals by two fingers. (YY 4,14: *gudāt tu dvy-aṅgulād ūrdhvam adho meḍhrāc ca dvy-aṅgulāt / deha-madhyaṃ tayor madhyaṃ [...] //.*)
GEENENS 2000 (p.202, note on 1–3) states that the *Mulādhāracakra* is not mentioned at all in YY, but the anus and genitals are often considered places (*sthāna*) important for the *kuṇḍalinī*.

195 YY 4,11cd, *dehamadhye śikhi-sthānaṃ tapta-jāmbūnada-prabham.*
The following verse, YY 4,12, gives information on the form of the *dehamadhya*: slender in birds and reptiles, square in quadrupeds, triangular in human beings. (YY 4,12: *tri-koṇam manujānāṃ tu caturaśraṃ catuṣpadām / varttulaṃ tat tiraścāṃ tu satyam etad bravīmi te.*) Also, YY 4,17 differentiates the forms of the *kanda* according to species.

196 YY 4,13ab *tan-madhye tu śikhā tanvī śivā tiṣṭhati pāvakī.*

197 YY 4,16: *kanda-sthānam manuṣyāṇāṃ deha-madhyān navāṅgulam / caturaṅgulam utsedham āyāmaṃ ca tathāvidham //.*
YY 4,14–15 states that the position of the body varies from one species to another, like in the case of the form of the *dehamadhya*.

198 YY 4,18: *tan-madhyaṃ nābhir ity uktaṃ nābhau kanda-samudbhavaḥ / dvādaśāra-yutaṃ tac ca tena dehaḥ pratiṣṭhitaḥ //.*

199 YY 4,19: *cakre 'smin bhramate jīvaḥ puṇya-pāpa-pracoditaḥ / tantu-piñjara-madhya-stho yathā bhramati lūtikaḥ //.*

SR śl.149 (the individual self (*jīva*) rides on the vital wind (*prāṇa*), through the *Suṣumnā* tube, up to the *Brahmarandhra*); the YY contains no parallel, although a similar topic is dealt with in YY 4,20–24.[200]

SR śl.150 (tubes stretching from the bulb) has no parallel in the YY, although YY 4,25 deals with a similar topic.[201]

SR śl.151–153a (list of the names of the fourteen prominent tubes) ≈ YY 4, 26–28ab.[202]
SR śl.153ab (three main tubes) ≈ YY 4,28cd.[203]
SR śl.153cd (*Suṣumnā*, way to liberation) ≈ YY 4,29.[204]
SR śl.154 (*Iḍā* and *Piṅgalā*, the moon and the sun) ≈ YY 4,31cd–32.[205]

SR śl.155ab (*Suṣumnā* as the destroyer of time) has no parallel, though YY 4,33–35ab deal with a topic similar to SR śl.154cd–155ab.

SR śl.155cd (positions of the *Sarasvatī* and *Kuhū*) ≈ YY 4,35cd.[206]
SR śl.156ab (positions of the *Gandhārī* and *Hastijihvā*) ≈ YY 4,36ab.[207]

200 YY 4,20: *jīvasya mūle cakre 'sminn adhaḥ prāṇaś caraty asau / prāṇārūḍho bhavej jīvaḥ sarva-bhūteṣu sarvadā /20/* 4,23: *mukhenaiva samāveṣṭya brahmarandhramukhaṃ tathā / yogakāle tv apānena prabodhaṃ yāti sāgninā /23/ sphurantī hṛdayākāśe nāgarūpā mahojjvalā / vāyur vāyu-sakhenaiva tato yāti suṣumnayā /24/.*
SR śl.149 and YY 4,20 do not seem to be *directly* based on the same source-text. Therefore I do not call this relation "parallel", but "similar".

201 YY 4,25: *kanda-madhye sthitā nāḍī suṣumneti prakīrtitā / tiṣṭhanti paritaḥ sarvāś cakre 'smin nāḍi-sañjñakā.* Though the three underlined expressions are shared by the SR and YY, it does not seem that these two verses are *directly* based on the common source text. Therefore I do not call this textual relation "parallel" but, "similar".

202 YY 4,26: *nāḍīnām api sarvāsām mukhyā gārgi caturdaśa / iḍā ca piṅgalā caiva suṣumnā ca sarasvatī.* YY 4,27: *vāraṇā caiva pūṣā ca hasti-jihvā yaśasvinī / viśvodarā kuhūś caiva śaṅkhinī ca tapasvinī //* YY 4,28ab: *alambusā ca gāndhārī mukhyāś caitāś caturdaśa.*

203 YY 4,28cd: *āsām mukhyatamās tisras tisṛṣv ekottamā matā.*

204 YY 4,29: *mukti-mārgeti sā proktā suṣumnā viśva-dhāriṇī / kandasya madhyame gārgi suṣumnā supratiṣṭhitā.*

205 YY 4,31cd: *iḍā ca piṅgalā caiva tasyāḥ savye ca dakṣiṇe.* YY 4,32: *iḍā tasyāḥ sthitā savye dakṣiṇe piṅgalā sthitā / iḍāyām piṅgalāyāṃ ca carataś candra-bhāskarau.*

206 YY 4,35cd: *sarasvatī kuhūś caiva suṣumnā-pārśvayoḥ sthite.*

207 YY 4,36ab: *gāndhārī hasti-jihvā ca iḍāyāḥ pṛṣṭha-pārśvayoḥ.*

SR śl.156cd (positions of the *Pūṣā* and *Yaśasvinī*) has no parallel, although the extensions of the *Yaśasvinī* and *Pūṣā* are mentioned in YY 4,40 and 41 respectively.[208]

SR śl.157ab (position of the *Viśvodarā*) ≈ YY 4,36cd.[209]
SR śl.157cd (position of the *Vāruṇī*) ≈ YY 4,37ab.[210]
SR śl.158ab (position of the *PayasvinI*) ≈ YY 4,37cd.[211]
SR śl.158cd (position of the *Śaṅkhinī*) ≈ YY 4,38ab.[212]
SR śl.159a (position of the *Alambusā*) ≈ YY 4,38cd.[213]

Between SR sl.159a and śl.159b, there is a gap. From SR śl.159b on, the extensions of the tubes (*nāḍī*) are listed. The following verses (SR śl.159bcd–163ab) are parallel to YY 4, 39–45, but the order of these parallel verses is different in the SR and YY, as shown below.[214]

SR śl.159bc (extensions of the *Iḍā* and *Piṅgalā*); YY 4,40cd mentions the extension of the *Piṅgalā*. YY 4,43cd mentions that of the *Iḍā*.[215]

SR śl.159d (extension of the *Kuhū*) ≈ YY 4,39ab.[216]
SR śl.160a (extension of the *Sarasvatī*) ≈ YY 4,42ab.[217]
SR śl.160bc (extension of the *Gāndhārī*) ≈ YY 4,43ab.[218]
SR śl.160cd–161a (extension of the *Hastijihvā*) ≈ YY 4,44ab.[219]
SR śl.161ab (Extension of the *Vāruṇī*) ≈ YY 4,39d.[220]
SR śl.161bc (Extension of the *Yaśasvinī*) ≈ YY 4,40ab.[221]

208 YY 4,40–41: *yaśasvinī ca yāmyasya pādāṅguṣṭhāntam iṣyate / piṅgalā cordhvagā yāmye nāsāntaṃ viddhi me priye /40/ yāmye pūṣā ca netrāntam piṅgalāyāṃ tu pṛṣṭhataḥ / tapasvinī tathā gārgi yāmya-karṇāntam iṣyate /41/.*
209 YY 4,36cd: *kuhoś ca hastijihvāyā madhye viśvodarā sthitā.*
210 YY 4,37ab: *yaśasvinyāḥ kuhor madhye vāruṇā ca pratiṣṭhitā.*
211 YY 4,37cd: *pūṣāyāś ca sarasvatyāḥ sthitā madhye tapasvinī.*
212 YY 4,38ab: *gāndhāryāḥ sarasvatyāḥ sthitā madhye ca śaṅkhinī.*
213 YY 4,38cd: *alambusā ca viprendre kanda-madhyād avasthitā.*
214 The order of the quarters (*pāda*) in SR śl.161abc is, however, the same as that in YY 4,39cd–40ab.
215 YY 4,40cd, *piṅgalā cordhvagā yāmye nāsāntaṃ viddhi me priye;* YY 4,43cd, *iḍā ca savya-nāsāntaṃ savya-bhāge vyavasthitā.*
216 SR śl.159d is parallel to YY 4,39ab, *pūrva-bhāge suṣumnāyās tv āmeḍhrāntaṃ kuhūḥ sthitā.*
217 YY 4,42ab: *sarasvatī tathā cordhvam ājihvāyāḥ pratiṣṭhitā.*
218 YY 4,43ab: *gāndhārā savya-netrāntā iḍāyāḥ pṛṣṭhataḥ sthitā.*
219 YY 4,44ab: *hastijihvā tathā savya-pādāṅguṣṭhāntam iṣyate.*
220 YY 4,39cd: *adhaś cordhvaṃ ca vijñeyā vāraṇā sarva-gāminī.*

*SR śl.161d (Extension of the *Viśvodarā*); YY 4,44cd mentions it, but unlike the SR, states that it is situated in the middle of the belly.[222]

SR śl.162a (extension of the *Śaṅkhinī*) ≈ YY 4,42cd.[223]
SR śl.162bcd (extensions of the *Pūṣā* and *Payasvinī*) ≈ YY 4,41.[224]
SR śl.163ab (Extension of the *Alambusā*) ≈ YY 4,45ab.[225]

YY 4,45cd and 4,46 conclude the list of the tubes (*nāḍī*). The next verse (YY 4,47) begins the description of the ten vital winds. This list of the ten vital winds, *Prāṇa* etc. (YY 4, 47–71), seems to be parallel to SR śl.59–68ab (see the next chapter).

The SR at the same time mentions the respective actions and places of the ten vital winds, but the YY passage does not. The YY separately deals with their functions in YY 4,67ff.

§2.3.3. List of YY verses parallel to SR 1,2 śl.59–68ab

The list below shows the YY's verses parallel to SR 1,2, śl.59–68ab, which deal with the ten vital winds. Although the parallelism is obvious, the expressions of the two texts often deviate from each other. We have previously (§ 2.1.1.) seen that these verses of the SR (śl.59–68ab) are identical to ŚG 9,25cd–34.

These verses on the ten vital winds are the only ones that the ŚG shares with the YY. The ŚG does not contain a counterpart to the SR's description of Haṭhayoga (SR śl.145cd–163ab) which is parallel to the YY. The ten vital winds originally belong to Haṭhayogic theory, but, in the SR (śl.59–68ab) and YY (9,25cd–34), they are inserted into the embryological frame.

List:

= means that the parallels between the SR and ŚG are exactly identical to each other;

221 YY 4,40ab: *yaśasvinī ca yāmyasya pādāṅguṣṭhāntam isyate.*

222 YY 4,44cd: *viśvodarā tu nāḍī tundamadhye vyavasthitā.*

223 YY 4,42cd: *ā savya-karṇād viprendre śaṅkhinī cordhvagā matā.*

224 YY 4,41: *yāmye pūṣā ca netrāntam piṅgalāyāṃ tu pṣṣṭhataḥ / tapasvinī tathā gārgi yāmya-karṇā(ntam iṣya)te.*

225 YY 4,45ab: *alambusā mahābhāge pāyu-mūlād adhogatā.*

≈ means that the parallels are not perfectly identical but comparable, with changes of arrangement or use of synonyms, between the SR and ŚG on one side, and the YY on the other.

SR 1,2, śl.59 = ŚG 9,25cd–26ab (names of the ten vital winds)
≈ YY 4,47.[226]

SR 1,2, śl.60cd–61ab (*Prāṇa* and its places) = ŚG 9,27
≈ YY 4, 49cd–50ab.[227]
SR 1,2, śl.61cd (its actions) = ŚG 9,28 (consisting of two *pāda*-s) ≈ YY 4,66ab.[228]

SR 1,2, śl.62 (*Apāna* and its places) = ŚG 9,29abcd
≈ YY 4,53.[229]
SR 1,2, śl.63ab (its actions) = ŚG 9,29ef ≈ YY 4,66cd.[230]

SR 1,2, śl.63cd (*Vyāna* and its places) = ŚG 9,30ab ≈ YY 4,52.[231]
SR 1,2, śl.64ab (its actions) = ŚG 9,30cd ≈ YY 4,67ab.[232]

SR 1,2, śl.64cd–65ab (*Samāna* and its places) = ŚG 9,31
≈ YY 4, 54cd–56.[233]
SR 1,2, śl.65cd (its actions) = ŚG 9,32ab ≈ YY 4,68ab.[234]

SR 1,2, śl.66ab (*Udāna* and its places) = ŚG 9,32cd
≈ YY 4,54ab.[235]

226 YY 4,47: *prāṇo 'pānaḥ samānaś ca udāno vyāna eva ca / nāgaḥ kūrmo 'tha kṛkaro devadatto dhanañjayaḥ //.*
227 YY 4,49cd–50ab: *prāṇa evaitayor mukhyaḥ sa sarva-prāṇa-bhṛt sadā /49/ āsya-nāsikayor madhye hṛn-madhye nābhi-madhyame /50ab/.*
228 YY 4,66ab: *niḥśvāsocchvāsa-kāsāś ca prāṇa-karmeti kīrtitāḥ.*
229 YY 4,53: *apāna-nilaye kecid guda-meḍhroru-jānuṣu / udare vaṅkṣaṇe kaṭyāṃ jaghane tau vadanti hi //.*
230 YY 4,66cd: *apāna-vāyoḥ karmaitad viṇ-mūtrādi-visarjanam.*
231 YY 4,52: *vyānaḥ (śrotrā)kṣi-madhye ca kṛkaṭyāṃ gulphayor api / ghrāṇe gale sphijoddeśe tiṣṭhaty atra na saṃśayaḥ //.*
232 YY 4,67ab: *hānopādāna-ceṣṭādi vyāna-karmeti ceṣyate.*
233 The order of the *Samāna* and *Udāna* is the reverse of that mentioned in YY. YY 4,54cd–56: *samānaḥ sarva-gātreṣu sarvam prāpya vyavasthitaḥ /54cd/ bhukta-sarva-rasaṃ gātre vyāpayan vahninā saha / dvisaptati-sahasreṣu nāḍī-mārgeṣu sañcaran /55/ samāna-vāyur evaikaḥ sāgnir vyāpya vyavasthitaḥ / agnibhiḥ saha sarvatra sāṅgopāṅa-kalebare /56/.*
234 YY 4,68ab: *poṣaṇādi samānasya śarīre karma kīrtitam.*

SR 1,2, śl.66cd (its actions) = ŚG 9,33ab ≈ YY 4,67cd.[236]

SR 1,2, śl.67ab (five subsidiary vital winds, i.e. *Nāga* etc., which are situated in the skin, bones etc.) = ŚG 9,33cd ≈ YY 4,57ab.[237]
SR 1,2, śl.67cd–68ab (their actions) = ŚG 9,34
≈ YY 4, 68cd–70ab.[238]

YY 4, 58–65 describe the process of the digestion and assimilation of nourishment, and excretion. In these, the vital winds play important roles.[239]

§2.3.4. Relation between the two texts

Summarised, the manner of the parallelism between the SR and YY is as follows: except for the introductory verses (YY 4,1–10) and the verses describing the digestive process (YY 4,58–65), the first and second half of the YY's fourth chapter are respectively parallel to SR śl.145cd–163ab and SR śl.59–68ab.

It does not seem that the SR directly borrowed these passages from the YY, but probably from an unknown source common to the SR and YY. This seems to be shown by the textual condition, in which the two texts only partly overlap each other; one often compensates the other. For example, SR 1,2, śl.61cd (= ŚG 9,28) contains a more complete list for the actions (*karman*) of the *Prāṇa* than its parallel, YY 4,66ab.[240] On the other hand, the YY more completely lists the places of the vital winds (compare, for example, SR 1,2, śl.64ab with its parallel, YY 4,67ab[241]).

---

235 YY 4,54ab: *udānaḥ sarva-sandhi-sthaḥ pādayor hastayor api.*

236 YY 4,67cd: *udāna-karma tat proktaṃ dehasyonnayanādi yat.*

237 YY 4,57ab: *nāgādi-vāyavaḥ pañca tvag-asthy-ādiṣu saṃsthitāḥ.*

238 YY 4,68cd–70ab: *udgārādi-guṇo yas tu nāga-karmeti cocyate /68cd/ nimīlanādi kūrmasya kṣutaṃ vai kṛkarasya ca / devadattasya viprendre tandrī karmeti kīrtitam /69/ dhanañjayasya śophādi sarva-karma prakīrtitam /70ab/.*

239 These verses state that nourishment is cooked/digested in the water which is heated by the fire in the belly. The flame of this fire is fanned by the vital winds. This flame might be the same as the one which is situated in the *deha-madhya* mentioned in the SR śl.146c (parallel to YY 4,13).

240 YY 4,66ab: *niḥśvāsocchvāsa-kāsāś ca prāṇa-karmeti kīrtitāḥ.*

241 YY 4,67ab: *hānopādāna-ceṣṭādi vyāna-karmeti ceṣyate.*

The YY's passages which are not parallel to the SR seem to be the YY's own inventions. In order to demonstrate this, I show two examples from the YY in the following.

The first example is YY 4,11–13.

YY 4,11: *ātmastham mātariśvānaṃ yogābhyāsena nirjaya /*
dehamadhye śikhisthānaṃ taptajāmbūnadaprabham //

Conquer the fire situated in the self through the practice of Yoga.
In the middle of the body, [there is] a place of/for the flame [which has] the luster of heated/melted gold.[242]

YY 4,12: *trikoṇam manujānāṃ tu caturaśraṃ catuṣpadām /*
vartulaṃ tat tiraścāṃ tu satyam etad bravīmi te //

[The place of/for the flame is] triangular for human beings, but square for quadrupeds,
but (*tu*) it is round for [those who are] oblique (= crawling animals without legs). I tell you, this is true.[243]

YY 4,13: *tanmadhye tu śikhā tanvī śikhā tiṣṭhati pāvakī /*
dehamadhyaṃ ca kutreti śrotum icchasi cec chr̥ṇu //

But in the middle of that (= the place of/for the flame) a slender flame, a purifying (*pāvakī*)[244] flame, is situated.
If you want to hear where the "middle of the body" is, listen![245]

242 GEENENS 2000, p.64, translates, "C'est par la méditation propre au *yoga* que l'on conquiert le feu intérieur, / qui habite le corps subtil, et qui demeure, / étincelant, au centre du corps. Il a l'éclat de l'or pur / qui provient de la Jambū." *Qui demeure, étincelant, au centre du corps* seems to correspond to *dehamadhye śikhisthānaṃ*. It is unclear whether he translates it freely, or whether the edition he consulted contains different readings.

243 GEENENS (ibid.)'s translation looks a little different: "Chez lez oiseaux et chez les reptiles, / oblongue est la constitution ignée. / Chez les quadrupèdes, ell'est comme un carrée, / et triangulaire chez l'homme, je te le dis en vérité." *Lez oiseaux et chez les reptiles* seems to correspond *tiraścāṃ*. But I am skeptic concerning the aptness of his interpretation, as reptiles also have four legs, except for snakes. The categorization here rather seems to be concerned with the mode of movement, or the number of legs.

244 The term *pāvakī* may mean "fire", too.

YY 4,11cd is parallel[246] to SR śl.146ab. YY 4,13ab is again parallel to SR śl.146cd. But YY 4,12abcd and YY 4,13cd do not have parallels in the SR. So it is the question whether YY 4,12abcd, mentioning the respective forms of the *dehamadhya* of various species, is an original verse handed down from a common source-text, or a secondary composition.

I interpret this in the following manner.

YY 4,12abcd is a secondary composition. YY 4,13cd asks about the place of the thin flame, although the slender flame has been already mentioned in YY 4,13ab (*tanmadhye tu śikhā* "but, in the middle of that, the flame"); in the same verse, it is said to be situated in the *dehamadhya*. Supposing that there were no YY 4,12abcd between YY 4,11cd and YY 4,13ab, such a roundabout way of statement would not be necessary. The role of YY 4,13ab seems to be only to supply continuity to verse 4,11cd, which was interrupted by the insertion of YY 4,12abcd.

As a matter of fact, the SR lacks a parallel to YY 4,12abcd. SR śl.146ab and SR śl.146cd, which are the counterparts of YY 4,11cd and YY 4,13ab respectively, are put in conjunction. The linking of the two verses produces a feeling of continuity in reading.

Therefore, I presume that the SR preserves a more compact and more original version of the text than the YY does, and that the YY contains secondary insertions. An alternative would be to assure that different traditions have been combined in the YY passage.[247]

---

245 GEENENS translates it quite differently, "La flamme mince, ou le feu, est toujours là / au milieu, chez tous ceux-là. / L'endroit où il se trouve, est-ce là ce que tu veux savoir? / Il est au centre du corps!" The version of the YY he consulted might contain different readings.

246 The term "parallel", again, means that it is "not an identical verse", but contains terminology similar to that of the SR.

247 Cf. HACKER 1978, pp.489–490: "Aber es gibt Fälle, wo in längeren Stücken derselbe Stoff mehrmals in verschiedener Darstellung behandelt wird. Wenn man mehrere solche Fälle beobachtet hat, so ist der Schluß unabweisbar, daß solche Juxtapositionen eine primitive Redaktionsmethode sind, die mehreren Überlieferungen in gleicher Weise gerecht werden will. Von da aus wird man dann auch solche Fälle beurteilen, wo einzelne Verse oder kürzere Versreihen denselben Gegenstand in verschiedener und manchmal mehr oder weniger divergierender Weise behandeln."

The other example is YY 4,67–69cd. YY 4,67–69cd deals with the respective actions (*karma*) of the principal vital winds such as *Prāṇa, Apāna, Samāna, Udāna* and *Vyāna*.[248] For each wind, only one or two actions are mentioned. In contrast to this, SR (śl.62–66cd) contains a more complete list of their actions.

On the other hand, for the respective actions of the subsidiary vital winds, i.e. *Nāga* etc., the YY (4,68cd–70ab) contains a more detailed list than the SR (śl.67cd–68ab).[249]

Thus the two texts complement each other. In this case, it is easily assumed a common source-text on which the two texts are both based. This common source-text is assumed to have contained the pieces of information in a complete condition. An alternative explanation would be that the common source-text contained no list or a smaller list, and that the SR and YY added the other items afterwards, independently from each other.[250]

The statements of the SR and those of the YY are always in concordance, except for a unique case. The only case in which the

248 These verses of the YY are quoted in the list of §2.3.3. *"List of YY Verses Parallel to SR 1,2 śl.59–68ab"*.

249 YY 4,68cd–70ab: *udgārādi-guṇo yas tu nāga-karmeti cocyate /68cd/ nimīlanādi kūrmasya kṣutaṃ vai kr̥karasya ca / devadattasya viprendre tandrī karmeti kīrtitam /69/ dhanañjayasya śophādi sarva-karma prakīrtitam /70ab/.*

250 E.g. SR śl.61cd might have secondarily added *śabdoccāraṇa*, which the parallel YY 4,66ab does not share.
But in the case where a verse has the term *ādi* ("etc.") at the end of the listed items, it would appear more likely that it is an abbreviated version of an original list which was longer. In the case where both texts contain *ādi*, the original list would then have been still longer than both. Thus, YY 4,67ab contains *hānopānādi*, while the parallel, SR śl.64ab, *prāṇāpānādi-dhr̥tityāga-grahaṇādy*. YY 4,67cd contains *dehasyonnayanādi*, while the parallel, SR śl.66cd, *dehonnayanotkramaṇādi*. In these two cases, the SR's list contains one more item than the YY's, but the original list would have contained still more items.
Conversely, YY 4,57ab contains *tvag-asthy-ādiṣu*, while the parallel, SR śl.67ab, contains *tvag-ādi-dhātūn āśritya*. In this case, we know that the *dhātu*-s are actually seven, beginning with skin, according to the theory of SR śl.79ab.
For the expressions like *udgārādi, nimeṣādi, kṣuta-prabhr̥ti* (SR śl.67cd), *tandrā-prabhr̥ti* and *śophādi* (SR śl.68a), both, the SR and YY, contain the same word with *ādi* or *prabhr̥ti* at the end of the compound.

two texts are inconsistent with each other is the following. The extension of the tube *Viśvodarā* is mentioned by both SR śl.161d and YY 4,44cd. But, in these verses, the two texts allot it to different areas, "all over the body" (SR) and "in the middle of the belly" (YY).

As the two examples above clearly show, for most cases, the SR preserves a version of text which seems closer to the postulated common source-text. But the YY, too, sometimes offers supplementary pieces of information. [251]

As already mentioned above in the beginning of this section (§2.3.4.), except for the introductory verses (YY 4,1–10) and the verses describing the digestive process (YY 4,58–65), the first and second half of the YY's fourth chapter are respectively parallel to SR śl.145cd–163ab and SR śl.59–68ab. Roughly speaking, the YY's fourth chapter is split in two, and each of the two parts has a parallel in the SR; but the order is reversed, namely, SR śl.59–68ab, i.e. the parallel to the second half of the YY's fourth chapter, comes before SR śl.145cd–163ab, i.e. the parallel to the first half. Here I am inclined to assume that these two parts originally formed one sequence or one chapter in a common source-text.[252]

But the two problems which I am going to describe in the following make the matter appear more complicated.

Firstly, the respective conditions of the texts, SR śl.145cd–159a (on the tubes), SR śl.159bcd–163ab (on the extensions of the tubes) and SR śl.59–68ab (on the vital winds) are different from one another. As already observed in the lists of the parallelisms between the SR and YY, the order of the verses also is parallel between SR śl.145cd–159a and their counterparts in the YY. In contrast, the order

251 A theory of nourishment is dealt with in YY 4,59–66. This theory is not presented by the SR. But the SR's description of the *Samāna* (SR śl.64cd–65) is based on this theory. These verses (SR śl.64cd–65) are parallel to YY 4,55cd and 4,56.
On the other hand, SR śl.80 presents the theory of nourishment according to the classical medical tradition.

252 This "common source-text" does not necessarily have to be the one identical to the source-texts of the SR's other verses, which I mentioned in my previous arguments.

in SR śl.159bcd–163ab is completely different from their counterparts in the YY. The order of verses in SR śl.59–68ab and that of their counterparts in the YY is also different. Thus, for each of the latter two parts (SR śl.159bcd–163ab and SR śl.59–68ab), a textual confusion or reorganisation seemed to have taken place in either of the textual traditions of the SR or the YY, as the result of which the order of verses was totally changed. Probably we should rather allot each of these three parts to different layers of the textual tradition, than to consider them as belonging to one homogeneous source-text, i.e. they might have respectively derived from three different sources.

Secondly, there is the problem of SR śl.44–70. Into SR śl.44–70 which is the part parallel to the YY, SR śl.59–68ab has been inserted, as seen above. On the other hand, SR śl.44–70 is exactly identical to ŚG 9,10cdfg–37.[253] So, the common source-text of the SR and ŚG must have already contained these parallels to the YY. In contrast, the ŚG lacks counterparts to SR śl.145cd–163ab which are parallel to the YY (YY 4,11cd and following). The question is if the common source of the SR and ŚG contained only the former part (SR śl.59–68ab), or both parts (SR śl.59–68ab and SR śl.145cd–163ab). In the first case, it would suggest that the tradition of the SR supplemented the passages, SR śl.145cd–163ab, which originally came from another textual tradition. This case would then contradict my assumption above, that these two parts originally formed one sequence.

Nevertheless, this does not totally contradict my assumption that the two parts (SR śl.59–68ab and SR śl.145cd–163ab) both belonged to the same sequence or chapter in the source text, because the SR's statement on the action of the *Samāna* (SR śl.64cd–65)[254] presupposes the theory of nourishment dealt with in YY 4,59–66.[255]

Thus, the actual matter seems very complicated, but I possess too little information and material to elucidate the problem of the source-text sufficiently. In order to reach a definite conclusion, it is necessary to obtain and research more parallel texts. This, however, is not the purpose of this work.

253 Except for SR śl.53ab.

254 Parallel to YY 4,55cd and 4,56, respectively.

255 An opposing argument is possible, though. It could be assumed that YY 4,59–66, which have no parallel in the SR, were developed in the YY; the YY then secondarily added this part dealing with the theory of nourishment.

§2.3.5. *On the YY*[256]

In the following, I give some information about the YY, as it might give some clues to the comprehension of its parallel in the SR.

GEENENS 2000 presents this work as follows:

The origin of the YY is supposed to be in the south-west of India, because all of its manuscripts are from Malabar. It describes the manner of Yoga practiced about the 10th century. It contains all the elements that Patañjali did not analyse, namely, *prāṇāyāma, pratyāhāra, dhāraṇā* and *dhyāna*. (GEENENS 2000, p.9) But this work preserves the ancient theory, in which only the navel is called "*cakra*", like in Patañjali's Yogasūtra.[257]

The fourth chapter describes the human body according to the Yoga theory, i.e., the vital fire, tubes (*nāḍī*) and vital winds (*prāṇa*) (ibid., p.10).

Although the YY is entitled after Yājñavalkya, i.e. the principal figure of the Bṛhadāraṇyakopaniṣad, its author could not be the same Yājñavalkya. It cannot be the same author of the code titled Yājñavalkyasmṛti, either. Supposedly, the unknown, genuine author[258] of the YY wanted to give authority to his book through attributing it to the ancient sage. According to its contents, it was written in the milieu of Vaiṣṇava-s and is permeated with devotion (*bhakti*) (ibid., p.11).

The YY has many parallels to other texts.[259] These texts are the commentary of Śaṅkara on the Śvetāśvataropaniṣad, the Sarvadarśanasaṇgraha of Mādhava, and the Laghuyogavāśiṣṭhavṛtti of Mummaḍideva. The YY seems to have influenced certain Yoga Upaniṣad-s, especially the Jābaladarśanopaniṣad and the Śāṇḍilyaupaniṣad (ibid., pp.12–13).

256 About the YY, also see KANE 1990 (vol.I, part 1, pp.434–435).

257 Cf. GEENENS 2000, p.217 and my discussion §2.3.6.

258 It seems to me that there could have been several authors.

259 GEENENS 2000 states that the YY contains many parallels to a text which he calls "Yogavasiṣṭha". GEENENS' Annex I (ibid., p.214) contains a list of the topics shared by the YY and this so-called "Yogavasiṣṭha". But actually, the Yogavāsiṣṭha which is commonly known to us is not a Haṭhayogic text, but a philosophical one. I wonder which text he actually means with "Yogavasiṣṭha".

Although GEENENS discusses other texts' parallels to the YY, he does not further investigate the fourth chapter which contains parallels to the SR. From his study, we cannot obtain any further information on the fourth chapter. I hope for further investigation into this matter in the future.

On the other hand, the topics, i.e. respiratory tubes (*nāḍī*) and vital winds (*prāṇa*), are mentioned in many other Haṭhayogic texts.[260] Intriguingly, the Śāradātilaka, a Tantric text, treats the tubes (1,41–43ab) and vital winds (1,44cd–45ab) in the relationship with embryology (1,48cd–50). It seems that these topics were handed down to the later Haṭhayogic tradition.

This suggests that the SR's verses which are parallel to the YY are related to this Haṭhayogic tradition.

§2.3.6. "Cakra" in YY and its Parallel in SR

GEENENS' study on the YY's fourth chapter helps us better understand some points of the theory contained in SR 1,2, śl.145cd–163ab.

In this theory, the area *ādhāra*, which literally means "basis", is not called "*cakra*" (SR śl.145c), while the navel is called *cakra* (SR śl.148a).

According to the theory of the YY, the navel (*nābhi-cakra*) is the only area which is called *cakra*. Though the YY mentions the centres or points resembling *cakra*-s, through which the *kuṇḍalinī* ascends, it does not call them "*cakra*-s". This fact suggests that the theory contained in the YY is an archaic one. As a matter of fact, the navel *cakra* (*nābhi-cakra*) is only once mentioned in Patañjali's Yogasūtra (3,29). Except for 3,29, the Yogasūtra does not mention *cakra*-s (cf. GEENENS, ibid., p.217).[261]

260 For the parallels, cf. my footnote 1127 on SR śl.151.

261 On the other hand, the SU compares the shape of the navel, from which many vessels radiate, to a wheel (*cakra*). Cf. SU śārīra. 5,5: *nābhisthāḥ prāṇinām prāṇāḥ prāṇān nābhir vyupāśritāḥ / sirābhir āvṛtā nābhiś cakra-nābhir ivārakaiḥ*. Also cf. SU śārīra. 7,4–5: *yāvaty astu sirāḥ kāye sambhavanti śarīriṇām / nābhyāṃ sarvā nibaddhas tāḥ pratanvanti samantataḥ // nābhisthāḥ prāṇinām prāṇāḥ prāṇān nābhir vyupāśritāḥ / sirābhir āvṛtā nābhiś cakra-nābhir ivārakaiḥ //*. Compare it with SR śl.106, which states that ducts (*dhamanī*) are situated around the navel like the spokes of a wheel (*cakra-nābhāvarā*).

Thus the archaic theory in the fourth chapter of the YY, in this matter, is comparable to that of the Yogasūtra. On the other hand, we have formerly seen GEENENS state that the YY contains Haṭhayogic theory. GEENENS does not make it clear to which degree the YY's theory could be called "archaic".

However, that may be, the *cakra* in SR śl.146d (*cakrāt tasmān*) seems to be nothing else but the navel *cakra* (*nābhi-cakra*) mentioned by the YY and the Yogasūtra.

§3. The body and music

§3.1. The notion of nāda as suggested by the text and its commentaries

As to the reason why he deals with embryologico-anatomical science in his musicological work SR, the author Śārṅgadeva briefly states as follows:

> SR 1,2, śl.3cd: *so 'yam prakāśate piṇḍe tasmāt piṇḍo 'bhidhīyate.*
>
> This [sound (*nāda*)] manifests/illuminates [itself] in the [human] body (*piṇḍa*), therefore, the body (*piṇḍa*) is explained.[262]

The author states also,

> SR 1,2, śl.17cd: *tatra nādopayogitvān mānuṣaṃ deham ucyate.*

The Dhyānabindūpaniṣad (in Yogopaniṣad) v.54 (p.199), describes the shape of the *kanda* ("bulb") as a circle/wheel (*cakra*) consisting of tubes (*nāḍīmayaṃ cakraṃ*).

262 The expression *piṇḍo 'bhidhīyate* might be interpreted differently: "[the *nāda*] is named body, [when it is manifested]". It might then contain a philosophical notion such as that the whole worldly manifestation is constituted of sound (*nāda*). This expression of the SR might then suggest the idea that the body can be considered as a worldly manifestation constituted of sound.

With regard to that (= the four kinds of bodies (*śarīra*) of the living beings such as *svedaja, udbhedaja, jarāyuja* and *aṇḍaja*), the human body is mentioned, because of being suitable for sound.

The commentator Kallinātha (comm. K) explains this statement as follows:

Among the mentioned fourfold bodies, [there is no sound] in three kinds, [i.e.] the bodies of lice etc., because of the unconsciousness of creeping plants etc., because of the extreme minuteness of lice etc., [and] because of the lack of complete element[s] (*dhātu*), tube[s] (*nāḍī*) etc. of birds etc. And (*ca*) [there is no sound] in those whose birth is [from] the foetal envelope (*jarāyu*) either (*api*), because of the incapability of pronunciation (*uccāraṇa*) of the body of the animals etc., due to [their] being beasts (*tiryañc*), despite the existence of completeness of element[s] (*dhātu*), tube[s] (*nāḍī*) etc. By elimination (*pāriśeṣyāt*), the human body alone is suitable for sound.[263]

From this commentary, we can infer the characteristics of the notion *nāda*, although there is, of course, no guarantee that the author and the commentator share the same view on the notion. According to the commentator Kallinātha, *nāda* ("sound") is something which is produced with consciousness (*cetana*), through the completed organic components such as *dhātu*, *nāḍī* etc.[264]; it should consist of pronun-

263 Comm. K on SR 1,2, śl.17 (Adyar ed., p.35): *ukta-caturvidha-deha-madhye tri-vidheṣu yūkādi-deheṣu latāder acetanatvād, yūkādeḥ atisūkṣmatvāt, vihagādeḥ sampūrṇa-dhātu-nāḍy-ādy-abhāvāt, jarāyujeṣv api paśv-ādi-dehasya dhātu-nāḍy-ādi-sampūrṇatā-sadbhāve 'pi tiryaktvenoccāraṇāśakteś ca, pāriśeṣyāt mānuṣaṃ deham eva nādopayogi.*

264 It is easily understood that the respiratory tubes (*nāḍī*) are necessary for vocalization. The seven elements (*dhātu*) of the body are the origin of the seven tones of the octave according to a musical theory, cf. Bhāvaprakāśa adhikāra 7 (G.O.S. No.XLV, p.186, ll.5–8): *anye dhātubhya utpannāḥ svarā ity eva jānate / dhātavaḥ sapta bhūtānām antaḥ saptāgnayaḥ sthitāḥ // kecid agnaya ity evaṃ kecid ūṣmeti manvate / tvag-asṛṅ-māṃsa-medo-'sthi-majjā-śuklāni dhātavaḥ //* This theory is explained in more detail in p.187, l.14 up to p.188, l.7. Verses p.186, ll.9–21 explain that the ducts (*dhamanī*) based on (*āśrita*) the semen, marrow, bone, fat, flesh, blood (I read *asra* instead of *asru*) and skin are respectively concentrated in the bulb (*kanda*), navel, heart, throat, the root of the palate (*tālu-mūla*), head and the middle of the eyebrows. The self (*ātman*), situated in the cave of the heart (*daharākāśa*), sets the vital winds in motion; the vital winds make the *dhātu*-s penetrate the ducts, and produce fires in them; from these fires, *nāda* arises (p.187, ll.1–8).

ciation (*uccāraṇa*). The sole being that fulfils these requirements is the human being. [265]

The requirement that the *nāda* must be produced with consciousness and in the form of pronunciation (*uccāraṇa*) suggests that this *nāda*, the primordial form of music, is conceived as something similar to the pronunciation or articulation of a language, that is also an emanation of consciousness.

Another hint as to what the notion *nāda* actually is, is given by the commentator Kallinātha, in a remote part of the book (in his commentary on SR's 3rd adhyāya (Prakīrṇakādhyāya), śl.82).

This part of the work describes the qualities or talents of a musician. In his commentary to this, Kallinātha explains the term *śārīra* which means in-born talent of singing, as follows: *yathā dhvaniḥ śarīreṇa sahodbhavati, tathā tasya rāgābhivyakti-śaktatvam api śarīreṇa sahodbhavati. na hy abhyāsenāgantukam ity arthaṇ* "The meaning [of the text] is that (*iti*), like resonance (*dhvani*) arises together with the body (*śarīra*), so does his (= the musician's) representative ability to express (*abhivyakti-śaktatva*) melodies (*rāga*) arise together with the body; indeed, it is not acquirable through [postnatal] training". If we consider *dhvani* here as a synonym for *nāda* – like Siṃhabhūpāla, the other commentator on SR 1,2, śl.1–3, does, glossing the term *nāda* through *dhvani* –, this statement would suggest that the musical representation (*rāga* etc.) itself is derived from the resonance or the *nāda* which is originated in

265 This statement by Kallinātha contradicts SR 1,3, śl.46cd–47ab which states that the peacock, *cātaka*-bird, goat, *krauñca*-bird, cuckoo, frog, elephant pronounce/utter (*uccārayanti*) the seven musical tones, respectively.
The idea that animals and birds create the tones of the octave also occurs in other musicological works. For example, in Nāradīyaśikṣā 1,5,4 quoted by Abhinavagupta (on Nāṭyaśāstra 28,21. GOS ed. p.12, l.8), it is stated that the cuckoo sings the fifth tone of the octave in the season of flowers (*puṣpa-sādhāraṇe kāle kokilo vakti pañcamam*).
Mataṅga's Bṛhaddeśī quotes his predecessor Kohala's statement which is parallel to SR 1,3, śl.46cd–47ab. It runs: *Saḍjaṃ vadati mayūra ṛsabhaṃ cātako vadet / ajā vadati gāndhāraṃ krauñco vadati madhyamam // puṣpa-sādhāraṇe kāle kokilaḥ pañcamaṃ vadet / prāvṛṭ-kāle tu samprāpte dhaivataṃ darduro vadet // sarvadā ca tathā, devi, niṣādaṃ vadate gajaḥ //*. It has to be noted that the first half of the second verse is parallel to the above-mentioned Nāradīyaśikṣā 1,5,4 !

and very closely connected to the organism of the body.[266] This means, in other words, that musical representation is similar to the linguistic one, as suggested by SR 1,2, śl.2. That is to say, it goes through an articulative or analytical process like language; musical representation is, nevertheless, an organic process.

§3.2. Vocal process as described in the 1st adhyāya, 3rd prakaraṇa of SR

The process of the manifestation of *nāda*, or sound/voice, is described in SR adhyāya 1, prakaraṇa 3. I shall explain these statements on the vocal process in the following.

SR 1,3, śl.3–4 describes the manifestation of resonance (*dhvani*) in the human body:

> This self (*ātman*), being desirous to speak/express_itself (*vivakṣamāṇo*), urges the mind. The mind strikes (*āhanti*) the fire situated in the body. It (= the fire) urges the wind. (śl.3)
> Then it (= the wind), situated in the knot of Brahman, moving to the upper region, manifests resonance (*dhvani*) in the navel, heart, throat, head and mouth, in turn. (śl.4)
> Sound (*nāda*), situated in the five places, takes five names: the super-minute [sound], the minute [sound], the ample [sound], the not ample [sound] and the artificial [sound], respectively.[267] (śl.5)
> [The wise ones] know [the syllable] *na* [of the word *nāda*] as the name of the vital wind (*prāṇa*), [the syllable] *da* as [the name of] fire.
> Therefore, [that] born from the union of *prāṇa* and fire is called sound (*nāda*). (śl.6)[268]

266 A statement of the Bṛhaddeśī (Verse 54, anuccheda 15; P.L. SHARMA, 1992, p.29) is comparable with this. It runs: *ucyate, rāga-janako dhvaniḥ svara iti* "It is said that the resonance which produces the *rāga* is the musical tone (*svara*)".

267 Cf. Kubjikāmatatantra 11,80b: *sūkṣmaś caiva susūkṣmaś ca, vyaktāvyakto 'tha kṛtrimaḥ*. The five-fold *nāda* is counted among the sixteen fruits of the Viśuddhi-cakra (HEILIJGERS-SEELEN 1990, p.57).

268 *ātmā vivakṣamāno 'yam manaḥ prerayate manaḥ / dehasthaṃ vahnim āhanti sa prerayati mārutam /3/ brahmagranthisthitaḥ so 'tha kramād ūrdhvapathe caran / nābhi-hṛt-kaṇṭha-mūrdhāsyeṣv āvir bhāvayati dhvanim /4/ nādo 'tisūkṣmaḥ sūkṣmaś ca puṣṭo 'puṣṭaś ca kṛtrimaḥ / iti pañca-bhidādhatte pañca-*

The self sets the mind in motion, and the mind produces fire. This fire seems to be identical with the flame of fire (*agniśikhā*) situated in the middle of the body (*dehamadhya*), mentioned in SR 1,2, śl.146. The fire in its turn urges the wind which is considered to be situated in the knot of Brahman. Sound (*nāda*) is produced through the union of the fire and the vital wind (*prāṇa*).

According to SR 1,2, śl.148–147, the individual self (*jīva*) dwells in the knot of Brahman, and sometimes moves to and fro over the net of the tubes (*nāḍī*) spreading from the knot of Brahman, like a spider moves over its net; the individual self climbs up and down through the *Suṣumnā* tube between the knot of Brahman and the aperture of Brahman, riding on the vital wind (*prāṇa*), or breath.

This seems to mean the following: In vocalisation, the vital wind or breath is led upward, and, together with it, the individual self is also brought up to the aperture of Brahman, i.e. the path out of the body into bliss.

This theory is also intimated by the opening verse of the SR (1,1,1), dedicated to the god Śiva in the form of sound (*nāda*). It states that sound manifests itself in the heart-lotus, during the contemplation, in which the mind (*citta*) follows the movement of the vital wind originating from the knot of Brahman.[269]

SHRINGY 1999 (vol.I, p.396) expresses his personal opinion that the situation of the three registers of the human voice corresponds to that of the three *cakra*-s, i.e. *Anāhata*, *Viśuddhi* and *Lalanā*.

The same theory of the vocal process as described in SR 1,3, śl.3–6, is dealt with in other musicological and linguistic texts, too.[270]

---

sthānasthitaḥ kramāt /5/ na-kāram prāṇa-nāmānaṃ da-kāram analaṃ viduḥ / jātaḥ prāṇāgni-saṃyogāt tena nādo 'bhidhīyate /6/.

269 SR 1,1,1:
brahmagranthi-mārutānugatinā cittena hṛtpaṅkaje
sūrīṇām anu rañjakaḥ śrutipadaṃ yo 'yaṃ svayaṃ rājate /
yasmād grāmavibhāgavarṇaracanālaṅkārajātikramo
vande nādatanuṃ tam uddhurajagadgītam mude śaṅkaram /1/.

270 On the musicological texts dealing with it, see the following. As an example of the linguistical texts, see the Pāṇinīyaśikṣā.
SINHA, K.P. 1993, p.94, informs us of the mention of this theory in the Yogaśikhopaniṣad. The original text quoted by him (p.98, note 10) runs as follows: Yogaśikhopaniṣad 5,13–4: *vāyunā vahninā sārdham brahmagranthim*

The Haṭhayogapradīpikā (3,66–69) also contains a similar theory:

> Fließt der *apāṇa* einmal nach oben, dann erreicht er den Platz des Feuers. Von diesem Wind gefächert, erstreckt sich die Spitze der Flamme. Wenn aber Feuer und *apāṇa* sich zusammen dem *prāṇa* nähern, dessen Wesen die Hitze ist (*uṣṇasvarūpa*), so lodert die Flamme innerhalb des Körpers brennend auf. Dann wird die *Kuṇḍalinī* aus ihrem Schlaf gerissen. Wie eine mit einem Stock geschlagene Schlange, bäumt sie sich zischend auf. Als schlüpfte sie in ein Erdloch, schleicht sie sich in die *brahmanāḍī* (die *suṣumnā*) hinein.[271]
> (Tr. by M. HULIN 1999, pp.197–198)

Here, the union of the vital wind (*prāṇa*) and fire is utilised to wake up the sleeping power, *kuṇḍalinī*. The vocalisation is here not directly dealt with, but the statement is closely associated with it. As a matter of fact, the *anāhata-nāda*, "sound not-struck" or "unmanifest sound", perceived by Yogins played a great role in the practice of Haṭhayoga.

Besides Brahmānanda's commentary on the verse quoted above, Haṭhayogapradīpikā 3,66, explains *agniśikhā* ("the flame of fire") through *jaṭharāgniśikhā* ("the flame of fire in the belly"), namely the fire of digestion. M. HULIN 1999 (p.198, footnote 7) comments on this as follows: "Das Feuer, von dem hier die Rede ist, ist offensichtlich das der Verdauung, das zur gleichen Zeit die Rolle eines Opferfeuers spielt, sofern das Kochen/die Verdauung als ein inneres Agnihotra verstanden wird."[272]

---

bhinatti sā, viṣṇugranthiṃ tato bhittvā rudragranthau ca tiṣṭhati. He also reports upon a parallel statement from the Yogakuṇḍaly-upaniṣad (1,67,85–86), "The *kuṇḍalinī*-power risen from slumber by the stroke of fire and air first crosses the knot (*sic.*) of Brahmā" (SINHA ibid., p.94).

271 Haṭhayogapradīpikā 3,66–69:
apāna ūrdhvage jāte prayāte vahni-maṇḍalam / tadānalaśikhā dīrghā jāyate vāyunāhatā /66/ tato yāto vahny-apānau prāṇam uṣṇasvarūpakam / tenātyanta-pradīptas tu jvalano dehajas tathā /37/ tena kuṇḍalinī suptā santaptā samprabudhyate / daṇḍāhatā bhujaṅgīva niśvasya ṛjutāṃ vrajet /38/ bilam praviṣṭeva tato brahmanāḍyantaraṃ vrajet / (tasmān nityam mūlabandhaḥ kartavyo yogibhiḥ sadā) /69/.
Brahmānanda's commentary on śl.66 quotes Yogayājñavalkya 4,11 (which is parallel to SR 1,2,146) in order to explain the term "circle of fire" (*vahnimaṇḍala*).

272 AS śārīra., 6,92 mentions the five vital winds (*vāyu*) which cause various kinds of abnormalities like giddiness (*mūrchanā*) etc. These abnormalities are

Indeed, Śāradātanaya, the author of the musicological text Bhāvaprakāśana, who is contemporary to Śārṅgadeva, presents the theory of a preceding musical school that the seven tones of the octave are produced from the seven elements (*dhātu*) of the human body, i.e. skin, blood, flesh, fat, bone, marrow and semen.[273] In this regard, we should remind ourselves that according to Indian classical medical theory, these seven elements are considered as the results of metabolic evolution, for which the digestive fire or heat plays a great role.

The following verses, SR śl.7–8, explain the process, in which the twenty-two microtones (*śruti*) of the octave, namely the base of music, are produced.

> SR 1,3, śl.7–8
> Meanwhile, in practice (*vyavahāra*), that (= *nāda*, or sound) is called in three manners: 'low' (*mandra*) in the heart, 'middle' (*madhya*) in the throat, 'high'[274] (*tāra*) in the head. And [they become] double one after another. (śl.7)
> Its twelve differences are deemed microtones (*śruti*) due to hearing.
> The oblique/horizontal tubes (*nāḍī*) attached to the [two] upper tubes (*nāḍī*) [situated in] the heart are deemed twenty-two.
> In those [twenty-two tubes], as many microtones arise one after another, united with the state of [becoming] higher and higher, due to the stroke (*āhati*) of the vital wind. (śl.8–9)

In śl.8, the body frame is compared to that of a harp[275], to which twenty-two strings are attached. The vital wind strikes the horizontal tubes like a finger picking the gut-strings of a harp. The comparison

---

sometimes accompanied by sound (*śabdavān īṣac-chabdaḥ*), due to the excess or shortage of fire and food. The term, *śabda*, here denotes the noise which emanates from the body.

273 Bhāvaprakāśana adhikāra 7 (G.O.S. No.XLV, p.186, l.5–8). Cf. SHRINGY 1999, p.117: The first member, skin, deviates from the classical medical theory, which has *rasa* instead. See on this note 828 on SR śl.79 (*dhātu*) in my *English translation.*

274 I.e. high in pitch.

275 According to LATH 1978, pp.199–200, this is a harp-type *vīṇā* shaped like a bow. LATH states: It is placed in a way that the bow-tip points away from the ground; such a *vīṇā* is found as a stone-relief on a Sānchī gateway (2nd century B.C.).

of the human body to a harp has a long tradition in Indian literature as I shall show in *§5 "the Comparison of the Human Body with the Musical Instrument"*. The topic, the three vocal registers inside the human body[276], is described by other musicological texts too, in association with the frame of a harp.[277]
According to P.L. SHARMA 1992, pp.154–155, the description of the process of vocalisation in SR 1,3, śl.1–10ab is based on a preceding literary tradition such as the Bṛhaddeśī, a musicological text, and the Pāṇinīyaśikṣā, a grammatical text. In the following, I discuss P.L. SHARMA's study of these two texts. The verses from the two texts, quoted below, are already analysed by P.L. SHARMA 1992, pp.154–155. I nevertheless discuss them here again, with some new remarks added, because they are crucial to understanding the context of the SR's Piṇḍotpattiprakaraṇa.

In the Bṛhaddeśī, the author Mataṅga quotes his predecessors' statements:

> Bṛhaddeśī, 20
> *yad uktam brahmaṇaḥ sthānam brahma-granthiś ca yaḥ smṛtaḥ /*
> *tan-madhye saṃsthitaḥ prāṇaḥ prāṇād vahni-samudgamaḥ //*
> *vahni-māruta-saṃyogān nādaḥ samupajāyate /*
> *nādād utpadyate bindur nādāt sarvaṃ ca vāṅmayam //*
> *iti kecit.*
> *kanda-sthāna-samuttho hi samīraḥ sañcarann adhaḥ /*
> *ūrdhvaṃ ca kurute sarvāṃ nāda-paddhatim uddhatām //*
> *iti anye vadanti.*

276 The topic of the positions of the three registers in the human body is also dealt with by the Nāradīyaśikṣā, the text on Vedic music, cf. LATH 1997. Nāradīyaśikṣā, 1st prapāṭhaka, śl.7: *uraḥ kaṇṭhaḥ śiraś caiva sthānāni trīṇi vāṅmaye / savanāny āhur etāni sāmni cāpy adharottare /7/.*
The Nāradīyaśikṣā, however, does not mention the process of vocal manifestation, while SR (śl.2, śl.3, śl.3–4) does. For the chronology of the Nāradīyaśikṣā, fifth century A.D., cf. SCHARFE 1977, pp.176–177 (on the Śikṣā-s).

277 Cf. Dattila śl.8–9: *nṛṇām urasi mandras tu dvāviṃśatividho dhvaniḥ / sa eva kaṇṭhe madhyaḥ syāt, tāraḥ śirasi gīyate /8/ uttarottaratāras tu vīṇāyāṃ tv adharottaraḥ / iti dhvaniviśeṣās te śravaṇāc chruti sañjñitaḥ /9/* (cf. M. LATH 1997, p.196).

na-kāraḥ prāṇa ity āhur da-kāraś cānalo mataḥ /
nādasya dvipadārtho 'yaṃ samīcīno mayoditaḥ //20//

"Inside that (*tan-madhye*) which is spoken of (*yad uktam*) as the place (*sthāna*) of Brahman and which is remembered (*yaḥ smṛtaḥ*) as the knot of Brahman (*brahmagranthi*), the vital wind (*prāṇa*) is situated. Fire arises from the vital wind; sound (*nāda*) is born from the contact of fire and [the vital] wind (*māruta*).
From sound (*nāda*), the drop (*bindu*) arises, from sound (*nāda*) all of speech (*vāṅmaya*, lit. that which consists of speech or language)." So (*iti*) [say] some.
"The [vital] wind (*samīraḥ*) which has arisen from the place of the bulb (*kanda*), moving about up and down, produces the whole intense course of sound (*nāda*)."[278]
So say others.
"They call the syllable *'na'* the vital wind (*prāṇa*), and the syllable *'da'* is known as fire; this is mentioned as the right meaning of the two words (*pada*) of *nāda* by me."
(The original text is quoted from SHARMA, P.L. 1992, p.7)[279]

In Bṛhaddeśī śl.21, sound (*nāda*) is classified as five: minute (*sūkṣma*), super minute (*atisūkṣma*), manifest (*vyakta*), unmanifest (*avyakta*) and artificial (*kṛtrima*).

The statement in Bṛhaddeśī śl.22–23ab deals with the relation between sound and the body. In this statement, the five kinds of sound (*nāda*) are allotted to various areas of the body. The minute sound (*sūkṣma nāda*) arises in the cave (*guhā*)[280], the super minute one (*atisūkṣma*) in the heart, the manifest one (*vyakta*) in the throat,

278 I follow P.L.SHARMA's (1992, p.7) interpretation which sees *sañcaran, adhaḥ* and *ūrdhvaṃ* as being in the same context. This is supported by SR 1, 2, śl.149 stating that the individual self, riding on the vital wind, ascends and descends along the *Suṣumnā* tube, cf. my translation.
Another interpretation is possible, taking *ūrdhvaṃ ca* together with *kurute*. In this case, it would mean: "The [vital] wind which has arisen from the place of the bulb, moving about the lower region (*adhaḥ*), makes the whole way of the sound (*sarvāṃ nāda-paddhatim*) in the upper region (*ūrdhvaṃ*), too (*ca*), rise up/upward (*uddhatāṃ*)."

279 My translation is based on SHARMA ibid., but slightly modified.

280 The term *guhā* "cave" usually means the hollow in the heart, but P.L. SHARMA 1992 considers it to be the navel. The minute sound, which is the most primordial one of the five vocal stages, must surely arise at the starting point of vocal manifestation, and SHARMA (ibid., p.154, note17) considers this point to be the navel.

the unmanifest (*avyakta*) in the palate (*tālu*) and the artificial one (*kṛtrima*) in the mouth. The five kinds of sound (*nāda*) correspond to the five stages of vocal manifestation. The higher it is, the more it develops.

After presenting his predecessors' opinions, the author Mataṅga (Bṛhaddeśī, anuccheda 1, following śl.24. SHARMA, P.L. 1992, p.8ff.) describes his own theory of vocal manifestation, explaining the production of the microtones (*śruti*).[281]

> [...] there in the beginning on account of the combination of the *dehāgni* (lit. bodily fire, battery of energy) and air, the sound propelled by the effort of the *puruṣa* (*ātman*), attacking the *ākāśa* (space) above the navel, ascending in many ways, in steps of a ladder like smoke, according to the will of the air, appears to be different by way of being composed of four *śruti*-s etc.[282] through being comprised of the inherent *pratyaya* (assured consciousness) of filling up (with air). (Translated by SHARMA, P.L. 1992, p.9)

In Bṛhaddeśī, verse 54, anuccheda 15, the author Mataṅga refers to the theory of Kohala, one of his predecessors. This theory seems to accord with Mataṅga's own theory mentioned above.

> By the will of the *ātman*, the *vāyu* [that is] moving upward from the base of the "earth" (= *nābhi*, navel) [and] is held on the 'wall' of the *nāḍī*-s and in the space, is known as *svara*, the delightful sound. (Translated by P.L. SHARMA 1992, p.29)[283]

Mataṅga quotes Kohala again in anuccheda 16:

281 P.L. SHARMA (ibid. pp.154–155) compares the Bṛhaddeśī's theory with that of the SR which is later than the Bṛhaddeśī and with that of the Pāṇinīyaśikṣā 6–9 (= 6,2ab–7cd of GHOSH 1991's edition of the Bṛhaddeśī) which precedes the Bṛhaddeśī.

282 The term *śruti* here means microtone. According to the Indian classical musical theory, the intervals between the tones (*svara*) of the octave contain either four, three or two microtones.

283 Bṛhaddeśī, verse 54, anuccheda 15: *ātmecchayā mahī-talād vāyur uddyan nidhāryate / nāḍī-bhittau tathākāśe dhvanī raktaḥ svaraḥ smṛtaḥ //*.

By the effort of the *ūrdhva-nāḍī*-s on account of rubbing and striking of all "walls", the sound that grows up to the cerebrum, is *svara*, it is *vyāpaka* (prevasive) [and] *para* (beyond). (Translated by P.L. SHARMA 1992, p.29)[284]

Besides that, Mataṅga quotes the opinion of his predecessor Tumburu whom he wants to refute (Bṛhaddeśī, 3, anuccheda 4; cf. P.L. SHARMA 1992, p.11). Tumburu insists that the microtone (*śruti*) is fourfold, according to the three morbific entities (*doṣa*) of the body, i.e. wind (*vāta*), bile (*pitta*) and phlegm (*kapha*), and the admixture of the three. Tumburu derives a high, hoarse (*rūkṣa*) voice from wind, a deep, full and clear voice from bile (*pitta*), and a creamy, soft and sweet voice from phlegm (*kapha*).

P.L. SHARMA 1992, pp.154–155, compares the musicological theory of vocal manifestation in SR 1,3, śl.3–4, and Bṛhaddeśī 20 with the linguistic theory of utterance in the Pāṇinīyaśikṣā (6,2ab–7cd[285]), a phonological text.

ātmā buddhyā sametyārthān mano yuṅkte vivakṣayā /2cd/
manaḥ kāyāgnim āhanti sa prerayati mārutam /
mārutas tūrasi caran mandra janayati svaram /3/
prātaḥsavanayogaṃ taṃ chando gāyatram āśritam /
kaṇṭhe mādhyandinayugaṃ madhyamaṃ traiṣṭhubhānugam /4/
tāraṃ tārtīyasavanaṃ śīrṣaṇyaṃ jāgatānugam /
sodīrṇo mūrdhny abhihato vaktram āpadya mārutaḥ /5/
varṇāñ janayate [...] /6ab/ [286]

In the SR, the process of vocal manifestation (SR 1,3, śl.3–4) and the positions of the three registers in the human body (SR 1,3, śl.7–8) are

284 Bṛhaddeśī anuccheda 16: *ūrdhva-nāḍī-prayatnena sarva-bhitti-nighaṭṭanāt / mūrchitā dhvanir āmūrdhnaḥ svaro 'sau vyāpakaḥ paraḥ //.*

285 Cf. the edition by GHOSH 1991 (pp.39–40). In the passages following this (6,2ab–7cd), the relation between the various kinds of articulation and parts of the body, i.e. the chest, throat, head, tongue, teeth, nose, lips and palate, is explained as "articulatory points".

286 P.L. SHARMA (ibid.) translates these verses as follows: "*Ātman*, having gathered or put together the content (*artha*) [of sound] with *buddhi* (intellect), activates the mind with the will to speak. The mind strikes the fire in the body. The fire propels the air. The air, moving in the chest-region, throat and cerebrum, manifests low, medium and high sounds respectively. Reaching the mouth cavity, the air manifests the *varṇa*s."

separately dealt with. The connection between the one topic and the other is not mentioned. In contrast to that, the connection of the two topics is clearly announced by the Pāṇinīyaśikṣā, which states that the three registers, i.e. low, middle and high, are produced by the process of vocal manifestation in which the vital wind gradually moves upwards in the body frame.

§4. Embryology, asceticism and music: Yājñavalkyasmṛti and SR

One of the well-known descriptions of embryology and anatomy contained in non-medical texts is that found in the Yājñavalkyasmṛti (= YS), in its section on the *āśrama* of the *sannyāsin* (YS 3,70–107).[287] The few verses following the description of embryology and anatomy have Haṭhayogic contents (YS 3,108–111).[288]

287 Cf. YAMASHITA 2001/2002, p.87. He precisely analyses the relationship between the YS's embryologico-anatomical verses and the CA's śārīrasthāna. The anatomical description in this part of the YS is the source of those in other Purāṇic texts, such as ViṣṇuP, ViṣṇudharmottaraP, AgniP etc., according to J.J. MEYER 1928 and HOERNLE 1907, p.42, p.44, p.59ff.
YAMASHITA ibid., p.88, explains the motive which led the author/authors of the YS to introduce embryology and anatomy in his/their account of the duties of ascetics (*yatidharma*) as follows: "the embryological passage (YS 3,75–83) is to be understood as an account of the development and delivery of the *ātman*, and the anatomical passage (3,84–107) as an explanation of the body parts of the *ātman*."

288 The opening passage of the embryologico-anatomical description, YS 3,63–64ab, instructs the ascetics to observe the prenatal stages (*garbha-vāsa*), diseases, old age, decaying of beauty, reincarnation, reversing of comfortable and uncomfortable etc. (*āvekṣyā garbha-vāsāś ca karmajā gatayas tathā / ādhayo vyādhayaḥ kleśā jarā rūpa-viparyayaḥ /63/ bhavo jāti-sahasreṣu priyāpriya-viparyayaḥ /*). According to this, the study of embryologico-anatomical science seems to be legitimate for the ascetics.
ZYSK 1990 postulates that medical science in South Asia was once developed in the ascetic milieu. But his methodology is sharply criticized by DAS 2003B as being imperfect.

The verses following those mentioned above treat the topic of music (YS 3,112–116). Both non-sacrificial vocal and instrumental music are mentioned as substitutes for the recitation of sacrificial songs, i.e. *sāman*-s.[289] The genres of songs mentioned as non-sacrificial vocal music in these verses[290] correspond to those mentioned in Bharata's Nāṭyaśāstra. The knowledge of playing the stringed instrument called *vīṇā*, i.e. the knowledge of the microtones (*śruti*), musical scales (*jāti*) and rhythms (*tāla*), is even considered to be a way to liberation (*mokṣa*) (YS 3,115).[291]

These three groups of verses, on embryologico-anatomical science[292], Yoga and music, seem to have been composed in different periods and inserted one after another into the original frame of the text.[293] The association of non-sacrificial music with asceticism or embryologico-anatomical science seems peculiar.

YS 3,58cd instructs the ascetic (*sannyāsin*), i.e. the one who lives in the fourth *āśrama*, to wander about (*pari√vraj*), i.e. to abandon both non-sacrificial and sacrificial activities.[294] But YS 3,112 considers the practice of the *sāman* songs to be an efficient way to attain Brahman. According to the comm. Vīramitrodaya on this verse, this

289 YS 3,112: *yathā-vidhānena paṭhan sāma-gānam avicyutam / sāvadhānas tad-abhyāsāt param brahmādhigacchati.*

290 YS 3,113–114, *aparāntakam ullopyam madrakam makarīṃ (prakarīṃ) tathā / auveṇakaṃ saro-bindum uttaraṃ gītakāni ca /113/ ṛg-gāthā pāṇikā dakṣa-vihitā brahma-gītikā / geyam etat tad abhyāsa-karaṇān mokṣa-sañjñitam /114/.* These two verses obviously contain misreadings. Compare with the names of the songs, *aparāntaka, ullopyaka, madraka, prakarī, oveṇaka, rovindaka* and *uttara*, mentioned in the Nāṭyaśāstra (32,200cd–201ab). These are the classes of *dhruvA* songs (cf. NIJENHUIS 1970, p.367. On the genre *dhruvā*, see BANSAT-BOUDON 1992, p.205). *ṛc, pāṇikā* and *gāthā* are also mentioned as classes of *dhruvā* songs in the Nāṭyaśāstra 32,1–2ab.

291 YS 3,116: *gīta-jño yadi yogena nāpnoti paramam padam / rudrasyānucaro bhūtvā tenaiva saha modate.*

292 I use the term "embryologico-anatomical science" to mean the matter of embryology and anatomy treated in the śārīrasthāna-s of the medical texts and in corresponding parts of non-medical texts (cf. §1.5. *Comparison with further medical and non-medical tetxts*).

293 Cf. KANE 1990, p.447.

294 YS 3,58cd: *ekārāmaḥ parivrajya bhikṣārthī grāmam āśrayet.* The commentary Vīramitrodaya explains *parivrajya* as meaning *laukika-vaidika-karmāṇi viśiṣyānuktāni santyajya.*

way is for those who are not capable of *samādhi*.[295] Another comm., Mitākṣara, calls it "the worship of Brahman [in the form] of word/sound (*śabda-brahmopāsana*)".[296] Further, the practice of the non-sacrificial music which is recommended by YS 3,113–116, as we have seen above, is the way for someone who is ignorant of the sacrificial music (*sāman*), according to the commentary Vīramitrodaya.[297]

Of course it is not difficult to imagine that some ascetics, who previously were educated musicians, did not cease their musical activity even after their renunciation. Indeed, for the ascetics in modern South Asia like the Bauls in Bengal, music is an essential part of their life.[298] So, YS 3,112–116 might perhaps imply the existence of such ascetics as active musicians, but these statements are so sparse that it hardly allows more than mere speculation. This problem is all the more difficult because the text of the YS is heterogeneous, as remarked above. The passages YS 3,112–116 are most likely a secondary insertion.

The structure of the text of the YS shows a striking similarity to that of the Piṇḍotpattiprakaraṇa of the SR. It first of all discusses embryologico-anatomical science (SR 1,2, śl.18–119), then Haṭhayoga (SR 1,2, śl.120–163ab). Finally it recommends the practice of non-sacrificial music as a substitute for the *nirguṇa* and *saguṇa* meditation (SR 1,2, śl.163cd–168ab), which is similar in its reasoning to that of the YS.

Śārṅgadeva, the author of the SR, might have had the beforementioned passages of the YS in mind, utilising them as his model. This theory is supported by the following two facts.

The first fact is that his predecessor[299] Abhinavagupta quotes YS 3,116 in his commentary (Abhinavabhāratī) on Bharata's Nāṭyaśāstra

295 Vīramitrodaya: *āśaya-śuddhy-abhāvena samādhy-aśaktam prati mokṣopāyam āha.*

296 Mitākṣara comm.: *yasya punaś citta-vṛtti-nirākarālambanatayā samādhau nābhiramate tena śabda-brahmopāsanaṃ kāryam ity āha.*

297 Vīramitrodaya (on YS 3,113): *sāmānabhijñasya mokṣopāyam āha.*

298 Cf. TROTTIER 2000, pp.62–93.

299 Śārṅgadeva mentions Abhinavagupta among his predecessors (SR 1,1, śl.19). Since his grandfather Bhāskara came from Kashmir (SR 1,1,2; SHRINGY 1999,

(on 28,11–12). The second is that the two commentators of SR, i.e. Siṃhabhūpāla (on SR 1,3, śl.2) and Kallinātha (on SR 1,1, śl.30), quote YS 3,115.[300] This is evidence of a tradition in which the YS passage had certain importance. It is, therefore, not at all improvable that the author of the SR was also aware of the statement of the YS.

Besides the SR and the YS, I found two musicological texts which associate embryology with music. They are Śāradātanaya's Bhāvaprakāśana and the Saṅgītopaniṣatsāroddhāra.[301] Śāradātanaya, the author of the Bhāvaprakāśana, was a contemporary of Śārṅgadeva, while the author of the Saṅgītopaniṣatsāroddhāra is of a later period. The outline of the passages in question in the two texts is similar to that of the SR. Therefore it is clear that this topic existed as an old tradition before the time of the SR, and was further handed down to later musicologists.

As to why anatomy and music are associated, we can think of the very old image of the body-*vīṇā*, which is traced back to Aitareya-āraṇyaka 3,2,5,[302] where the structure of the human body is compared to that of the *vīṇā*. This image seems to have acquired a durable tradition in the later periods, in the milieus of music, Yoga and

---

p.xiii), it is easy to imagine that he was also familiar with the works by Abhinavagupta, specially the commentary on the Nāṭyaśāstra.

300 YS 3,115. *vīṇā-vādana-tattva-jñaḥ śruti-jāti-viśāradaḥ / tāla-jñaś cāprayāsena mokṣa-mārgaṃ niyacchati.* Siṃhabhūpāla quotes *vīṇāvādana* in the Adyar edition (SASTRI 1943, Vol.I), p.76, l.8. He quotes the whole verse in his commentary on SR 6,418cd–421 (Adyar edition Vol.III = SASTRI 1986, p.430). Kallinātha quotes the whole verse, cf. Adyar edition, p.20.
Śārṅgadeva himself might be hinting at the thought of the body-lute, including *deha-sauṣṭhava* and *su-śārīra* in the qualities of a lute player in SR 6,422–424ab (Adyar ed. Vol.III = SASTRI 1986, p.430).

301 Śāradātanaya's Bhāvaprakāśana (G.O.S. No.45), a dramaturgical text contemporary with the SR, treats embryology in its description of the theory of music (adhikāra 7). It associates the seven elements (*dhātu*) of the human body with the seven tones of the octave (p.186, ll.5–6). The Saṅgītopaniṣatsāroddhāra (G.O.S. No.133) ll.11–24 treats embryology.

302 Cf. LATH 1997, p.201.

Tantra.[303] In the next chapter (§5. *"Comparison of the Human Body with the Musical Instrument in Indian Literature"*), I shall study the historical development of this image in Indian literature.

As a matter of fact, Nāṭyaśāstra 28,12, for which Abhinavagupta quotes YS 3,116 in his commentary, is on the comparison of the structure of the body frame with that of the stringed instrument, *vīṇā*.[304] In the same paragraph of his commentary, Abhinavagupta himself critically discusses this topic.[305] That Abhinavagupta, in the same paragraph of his commentary, mentions both YS 3,116 and the comparison of the body with the *vīṇā*, one after the other, suggests that he was well conscious of the association of the two topics. He was surely aware of the close relation of YS 3,116 to the foregoing embryologico-anatomical passages of the YS. Therefore I surmise that the musicologists at the time of Abhinavagupta were conscious of the discussion on the association between embryologico-anatomical science and music theory.

303 An intimation of this image is given in SR 1,3,8–9, cf. LATH ibid. Abhinavagupta mentions it (on Nāṭyaśāstra 28,13–15), immediately after he quotes the above-mentioned verse, YS 3,115.

304 Nāṭyaśāstra 28,12: *dvyadhiṣṭhānā svarā vaiṇāḥ śārīrāś ca prakīrtitāḥ / eteṣāṃ sampravakṣyāmi vidhānaṃ lakṣaṇānvitam /12/.*

305 Cf. LATH 1997, p.202. The Bhāratī on Nāṭyaśāstra 28,11–12: *[...] gāna-vṛttāntatvaṃ yad vīṇā-śarīra-vaṃśānām ārohaṇam avarohaṇaṃ ca / prāṇābhihananenaiva hi tīvrātīvreṇa śarīra iva vaṃze 'pi svara-niṣpattiḥ / vīṇāyāṃ tu ādarśe vāma-dakṣiṇa-viparyāsavat tāra-mandra-viparyāsa ity āśayenāha dvy-adhiṣṭhānāḥ svarā vaiṇāḥ śārīrāś ceti /.*

§5. Comparison of the human body with the musical instrument in Indian literature[306]

In SR 1,3, śl.8cd–10ab, the comparison of the human body with the stringed instrument called *vīṇā*[307] is mentioned to explain the mechanism of vocalisation.[308] The verses in question run as follows.

> Twenty-two horizontal tubes (*nāḍī*) are deemed attached to the tube [stretching] upwards from the heart. In them, so many (= twenty-two) microtones (*śruti*) are produced through the stroke (*āhati*) of the vital wind, endowed successively (*uttarottaram*) with increasing height (i.e., the higher the microtone is situated, the higher is the pitch). In the same manner (*evaṃ*) twenty-two microtones are deemed [to be produced] [each] in the throat, [and] further in the head.[309]

Here, the vocal production is explained with the model of a harp.[310] The vital wind arises from the digestive fire in the belly and blows up through the main tube running vertically in the body. The body frame is spanned with twenty-two horizontal tubes like the frame of a harp with its strings.[311] The vital wind, running up vertically, strikes the twenty-two horizontal tubes, which by turns vibrate in corresponding microtones, like a finger plucks the strings of a harp, which are by turns tuned to the twenty-two microtones.

306 This topic has been already treated by LATH 1997, p.207ff. The verses and passages from the SR, Dattila, Nāṭyaśāstra and Aitareyāraṇyaka are already discussed by him. Here I have added further references, and tried to develop some aspects.

307 For the form of the stringed instrument called *vīṇā*, cf. NIJENHUIS 1970, p.73ff. The *vīṇā* might denote various kinds of instruments such as lute, harp etc.

308 Cf. LATH 1997, pp.198–199.

309 *hṛdy ūrdhva-nāḍī-saṃlagnā nāḍyo dvāviṃśatir matāḥ /8cd/ tiraścyas tāsu tāvatyaḥ śrutayo mārutāhateḥ / uccoccataratā-yuktāḥ prabhavanty uttarottaram /9/ evaṃ kaṇṭhe tathā śīrṣe śruti-dvāviṃśatir matā /10ab/.*

310 In this context, the *vīṇā* is obviously the harp. It is not the lute, in which the pitch is modified by fingering with the left hand like the *vīṇā* of South India today or the sitar. But this stringed instrument here is provided with strings which are already tuned to the twenty-two microtones, namely, the harp.

311 According to ancient Indian music theory, an octave consists of twenty-two microtones, cf. NIJENHUIS 1992, p.7.

The comparison of the human body with the stringed instrument is found in various texts.

In Bharata's Nāṭyaśāstra (28,14–15), the expression *śārīra-vīṇā* "body-harp" is found. Here, the human body is conceived as a harp, while the real harp, the stringed instrument made of wood, is called *dāravī vīṇā* "wooden harp".[312] The musicological work Dattila (verse 8–9) mentions it, too.[313]

The oldest example of this comparison is a topos in the Black Yajurveda (Taittirīyasaṃhitā 7,5,9,2; Kāṭhakasaṃhitā 34,5).[314] The stringed instrument called *vāṇa* has one hundred strings, like a human being has a lifespan of one hundred years, one hundred faculties (*indriya*) and one hundred powers (*vīrya*). Here, the number one hundred associates the strings of the instrument with the lifespan, faculties and powers of a human being.[315]

The Aitareyāraṇyaka (3,2,5), composed in a later period than these texts, compares the human body with the stringed instrument (*vīṇā*) in more detail.[316] This text calls the human body *daivī vīṇā* "celestial *vīṇā*", in contrast to the wooden stringed instrument which

312 Cf. LATH 1998, p.201; BANSAT-BOUDON 1992, p.197.

313 Cf. LATH p.197ff.

314 Taittirīyasaṃhitā 7,5,9,2: *vāṇaś śatatantur bhavati śatāyuḥ puruṣaś śatendriya āyuṣyevendriye prati tiṣṭhanty ājiṃ dhāvanty.* Kāṭhakasaṃhitā 34,5: *yā vanaspatiṣu vāk tāṃ tenāvarundhate vāṇaś śatatantur bhavati śatāyur vai puruṣaś śatavīrya āyur eva vīryam avarunddhe,* (to which the editor, Leopold von Schroeder gives the parallels in TS 7,5,8–10; TBr. 1,2,6,6–7).
This topos is found in Brāhmaṇa-s, too. For example, Jaiminīyabrāhmaṇa 2,404 (also cf.2,418): *vāṇaṃ śata-tantrīm āghnanti, śatāyur vai puruṣaś śatendriyaś śata-vīryas, tasyaivendriyasya vīryasyāvaruddhyai.* Tāṇḍyabrāhmaṇa 5,6,13: *śata-tantrīko bhavati śatāyur vai puruṣaḥ śata-vīryaḥ.* The following verse (Tāṇḍyabrāhmaṇa 5,6,14) is intriguing: *tam ullikhet prāṇāya tvāpānāya tvā vyānāya tveti prāṇāpāna-vyānān eva tad āptāv avarundhate.* The strings of the harp might be here compared to the many tubes in the body, through which the streams of the vital wind flow.

315 However, we should not forget that this statement deals with "magische Äquivalenz" in which two things are associated with each other simply because they share the same number. In this case, the strings and the lifespan etc. are merely identified due to the number 100. Therefore this statement does not necessarily prenecessitate a theory which identifies the instrument with the human body.

316 Cf. LATH 1998, p.201.

is called *mānuṣī vīṇā* "human *vīṇā*". Although the passages in question are already dealt with by LATH 1998, I discuss them once more in the relationship to the SR.

> Now, indeed, this is the celestial *vīṇā*. The imitation of it is that human *vīṇā*. Like the head of this (= celestial *vīṇā* = human body), so is the head of that (= human *vīṇā* = wooden stringed instrument).[317] Like the belly of this, so is the sound box (*ambhaṇa*) of that.[318] Like the tongue of this, so is the plucking (*vādana*) of that.[319] Like the strings of this, so are the fingers of that.[320] Like the sounds (*svara*) of this, so are the voices (*svara*) of that. Like the touches (*sparśa*) of this, so are the touches of that.[321] Like this has, indeed, sound and is tightly bound, so, indeed, has that sound and is tightly bound.[322] Like this is, indeed, covered with hairy skin, so is this covered with hairy skin. Indeed, formerly[323], they covered the *vīṇā* with hairy skin.[324]

317 According to the commentator Sāyaṇa, the resonator made of a gourd which is attached to the upper part of the neck looks like the human head (*yo bhāgo gāyakasya vāmāṃsam āśrityāvasthitas tatrālābu-khaṇḍopetatvena śirasa ākāro dṛśyate*). The *vīṇā* of South India today also has two resonators or sounding bodies, cf. BEYER 1999, p.51. The resonator in the upper part of the neck is made of a gourd (ibid.).

318 More correctly, it is the hollow inside the neck, according to Sāyaṇa's commentary (*ambhaṇaṃ vīṇā-daṇḍa-madhya-varti cchidram*). Such a hollow neck of the South Indian *vīṇā* today is observed in BEYER 1999, Farbtafel XVI and XVII.

319 The comm. by Sāyaṇa: *śarīra-vīṇāyām avasthitā jihvā yathā svarotpatti-hetuḥ, tathaivāmuṣyai kāṣṭha-vīṇāyā hastena vādanaṃ svarotpatti-hetuḥ.*

320 Sāyaṇa explains: the fingers of the human being and the strings of the instrument have both various lengths (*asyāḥ śarīra-vīṇāyāḥ aṅgulayo yathā bahu-vidhā dīrghā vartante, tathaivāmuṣyāḥ kāṣṭha-vīṇāyās tantrayo dīrghatantavaḥ*).

321 The contact of the vital wind originating in the middle of the body [with the ducts] and the contact of a finger with the string, according to the comm. by Sāyaṇa (*svarābhivyakty-arthaṃ yathā śarīra-madhye vāyoḥ sparśa-viśeṣāḥ prayatnāt sampadyante, tathā kāṣṭha-vīṇāyā aṅguli-sparśa-viśeṣāḥ prayatna-sampadyāḥ*).

322 Sāyaṇa explains that the human body is tightly bound with ducts, while the wooden instrument is tightly bound with strings (*tardmavatī tardanavatī, dhamanībhiḥ śarīrāvayavānāṃ dṛḍha-bandhanaṃ tardanam, tathaiva kāṣṭha-vīṇāpi [...] tardmavatī tardanena tantrīnāṃ dṛḍha-bandhanena yuktā*). Thus Sāyaṇa associates the ducts of the human body with the strings.

323 That means, at the time when this text was produced, the instrument was not covered with hairy skin or fur any more, cf. Sāyaṇa who states *nanv idānīntanāḥ kāṣṭha-vīṇāś carmaṇā na badhyanta ity āśaṅkyāha.*

The commentator Sāyaṇa explains these passages in accordance with the idea that the vital wind originating in the middle of the body strikes the ducts. This idea is comparable to that of the SR, although, of course, this might be the commentator Sāyaṇa's own innovative interpretation rather than the original meaning of the text.

The Aitareyabrāhmaṇa's paragraph in which these passages are contained treats the worship of speech. The foregoing passages state that the syllables (*varṇa*) of speech correspond to various deities, various levels of the universe, various elements etc.

Sāyaṇa's commentary on this statement explains that each of the syllables (*varṇa*) constituting a *mantra* is worth worshipping.[325] This thought develops into the well-known notion *nāda-brahman* or *śabda-brahman* in later period, i.e. the notion that Brahman is sound, or the universe is music.[326]

Sāyaṇa also states that a musician who knows the celestial *vīṇā*, i.e. the human body, achieves fame in music halls (*sabhā*) of scholars and kings.[327] He mentions the expression "meditation on the *vīṇā*" (*vīṇā-dhyāna*) or "the adoration of the *vīṇā*" (*vīṇopāsti*). One adores the singing body as a gift from heaven, or conversely, one treats a musical instrument with love, as the symbol of the human body. This easily reminds us of the Indian musicians' custom of worshipping their instruments, which is observed even today.

---

324 Aitareyāraṇyaka 3,2,5: *atha khalv iyaṃ daivī vīṇā bhavati tad-anukṛtir asau mānuṣī vīṇā bhavati, yathāsyāḥ śira evam amuṣyāḥ śiro, yathāsyā udaram evam amuṣyā ambhaṇam, yathāsyai jihvaivam amuṣyai vādanam yathāsyās tantraya evam amuṣyā aṅgulayo, yathāsyāḥ svarā evam amuṣyāḥ svarāḥ, yathāsyāḥ sparśā evam amuṣyā sparśāḥ yathā hy eveyaṃ śabdavatī tardmavaty evamevāsau śabdavatī tardmavatī yathā hy eveyaṃ lomaśena carmaṇāpihitā bhavaty evam asau lomaśena carmaṇāpihitā / lomaśena ha sma vai carmaṇā purā vīṇā apidadhati.* These passages are already translated by LATH 1978, p.201, but I differ from him in some points in interpretation.

325 Sāyaṇa on Aitareyāraṇyaka 3,2,5: *akārādijñakārāntarānām mātṛkā-mantra-gatānāṃ sarveṣāṃ varṇānām atropāsanīyatvāt.*

326 For the notion *śabda-* or *nāda-brahman*, cf. BECK 1993.

327 Sāyaṇa on Aitareyāraṇyaka 3,2,5: *yaḥ pumān mānuṣa-vīṇā-sādṛśyānusandhāna-puraḥsaram etāṃ śarīra-rūpāṃ daivīṃ vīṇām upāste so 'yaṃ [...] vidvat-sabhāyāṃ rāja-sabhāyāṃ ca sarveṣām priyatamair vacanair atyantaṃ rañjako bhavati.*

The comparison of the human body with the musical instrument spread not only among musical theorists but also among Yoga practicians, as shown by the verses SR 1,2, śl.120–163ab on Haṭhayoga. For Haṭhayogic practice, acoustic perception is one of its important factors. During various stages of meditation, a Yoga practician is said to perceive various kinds of supernatural sounds which are imperceivable to ordinary people.[328]

The Dhyānabindūpaniṣad, one of the Yogopaniṣad-s, mentions the comparison of the body with the instrument.

> Sound (*nāda*) which arises in the neck of the *vīṇā* exists unmanifest (Verse 102ab)[329]

The commentary explains that "the neck of the *vīṇā*" means the tube (*nāḍī*) *Suṣumnā*.[330] According to Haṭhayogic theory, the *Suṣumnā* tube stretches vertically along the backbone. The vital wind (*prāṇa*) flows upwards through the *Suṣumnā*. The commentary states that the middle point of the neck of the lute is the place of sound, where sound resembling that of a conch arises.[331] The middle point of the neck of the lute, namely the middle of the backbone, is the place where the voice first arises. As stated in the above-mentioned verse of the SR (1,3,6), the voice is produced through the union of the vital wind and the fire of the body. The potential sound arising there goes up through the *Suṣumnā* along the backbone, piercing the navel and heart. During this process, sound gradually develops from a potential state into a manifest one.

Badarīnātha's Cakrakaumudī 1,12 also mentions the comparison of the body with the *vīṇā*, comparing the *Iḍā* and *Piṅgalā*, i.e. the tubes on the both sides of the *Suṣumnā*, to the strings of a *vīṇā* (*vīṇā-tantuvad*).[332]

328 Cf. Haṭhayogapradīpikā 4,64f. on *nādopāsana*, i.e. the practice of *anāhata-nāda*.

329 Dhyānabindūpaniṣad 102: *amūrto vartate nādo vīṇā-daṇḍa-samutthitaḥ / (śaṅkha-nādādibhiś caiva madhyameva dhvanir yathā).*

330 Comm. on verse 102: *vīṇā-daṇḍa-samutthitaḥ suṣumnāśraya-vīṇā-daṇḍe samutthitaḥ.*

331 Comm.: *vīṇā-daṇḍa-madhyam eva nādotpatti-sthānaṃ yathā śaṅkha-nādādibhiś caiva.*

332 Cakrakaumudī 1,11cd–12ab: *ūrdhva-koṇe suṣumnākhyā vāma-dakṣiṇayoḥ kramāt /11cd/ iḍākhyā piṅgalākhyā hi vīṇā-tantuvad āsthitā /12ab/.*

The comparison of the human body with the stringed instrument was handed down through the medieval to the modern period.

The Caryāgīti-s, the collection of mystic songs in the old form of the eastern dialects of New Indo-Aryan, contains a song mentioning this comparison. One line of Caryāgīti No.17 runs: *suja lāu sasi lāgeli tāntī / aṇahā dāṇḍī eki kiata avadhūtī* (cf. DASGUPTA 1976, p.98). DASGUPTA 1976 explains:

> In another song of Vīṇāpāda he [= Kānha] says that he has made a *vīṇā* (*i.e.*, lyre) of which the sun is the gourd (*lāu*) and the moon is the string and Avadhūti is the stand. On hearing the tune of the *Āli* and the *Kāli*, he says, the mighty elephant has entered Samarasa. Here the sun which is said to be the gourd and the moon which is said to be the string, are but the two nerves in the two sides, and the stand (*daṇḍa*) is the middle nerve. When the two nerves in the left and the right are controlled and fitted to the middle one, an *anāhata* sound is produced and it leads the elephant (i.e., *citta*) to the state of Samarasa. (ibid. p.98).

Kabīr, the poet in old "Hindi"[333] from the 15th century, sang as follows:

> Kabīr cannot [any more] play the instrument, all the strings have broken.[334]

In this verse, the poet calls his own body "instrument" (*jantra*) which he plays to express his love for God, and on which even God himself plays music. This idea is based on the picture of the heart attached with uncountable fine tubes radiating from it, according to Haṭhayogic theory. The verse describes the conflicting feelings of the poet full of love, who cannot sing his agony anymore, because all his blood vessels are torn up into pieces through the extreme pangs of broken love.

333 HEDAYETULLAH 1989, p.133, states that Kabīr''s language (in his work Bījak) "is said to be the Hindi dialect which was spoken in the neighbourhood of Benares, Mirzapur and Gorakhpur (Bhoj'purī)". But he at the same time mentions Grierson's opposing argument and Keay's theory that it is old Avadhī (ibid., pp.133–134, footnote 6). Actually, "Hindi" is historically a blanket term for a whole range of North Indian languages, and this has led to severe problems with regard to the modern usages of the term.

334 *kabīr jantra na bājai, ṭūṭī gae sab tār / jantra bicārā kyā karai, cale bajāvanahār*. In: Kabīr Granthāvalī, ed. by Parasnath Tivari, p.198.

Behind this idea of Kabīr, not only Hindu but also Islamic influence might be assumed, as he had sympathy for both cultures, and tried to integrate the elements of both. According to Islamic music theory, which was very strongly influenced by Hellenistic music theory, the structure of the lute (*al-'ūd*) is compared to that of the human body. According to this theory, the four strings of *al-'ūd* correspond to the four cardinal humours, i.e. yellow bile, blood, phlegm and black bile, as well as to the four types of disposition resulting from them.[335]

The song of Rabindranath Thagore, "Make me thy *vīṇā*" (*āmāẏ kara tomār bīṇā*) is a further echo of this notion:

> Make me thy *vīṇā* ; lift me in thine arms. All the strings of my heart will break out at my finger touch. With thy tender hands touch my life, and my heart will murmur her secrets in thine ears. In happiness and in sorrow she will gaze on thy face, and cry; and shouldst thou neglect her she will remain silent at thy feet. None knows in what strains her songs will rise up to the heavens and send a message of joy to the shore of the infinite.[336]

In his drama Śrābaṇ'gāthā ("songs of the *Śrāvaṇa* month"), an expression *citta-bīṇā* "the lute of spirit/consciousness" occurs. The king of dance, one of the figures in this drama, sings about the rainy season of *Śrāvaṇa*, the fourth month of the Indian calendar.

> *se jhaṙa yena saï ānande cittabīṇār tāre*
> *sapta-sindhu dik-diganta jāgāo ye jhaṃkāre*[337]

> "[It is] as though I delightfully bear the [press of the] storm on the strings of the lute of [my] spirit.
> Wake up the seven oceans, directions and horizons through that sound (*jhaṃkāre*) [of the storm]."

The lute of the spirit (*citta*) is seemingly the lute of the heart which is said to be the place of consciousness (*citta*).[338] The second line

335 See, e.g. the Arabian philosopher al-Kindī (about 801–870 AD), who was also intensively engaged in music, cf. PIETRUSCHKA 2001.

336 Translated by STRANGWAYS 1914, p.95. The original Bengali text is also quoted in his book.

337 ṬHĀKUR 1755, p.122.

reminds us of SR 1,3, śl.55cd–56ab, which states that the seven divisions of the earth (*sapta-dvīpa*) correspond to the seven musical tones. The seven oceans, directions and horizons are both external and inner. The noise of the thunder resounding from far away is at the same time the throbs or presentiments felt in the innermost of the heart. Through its rumbling, the seven oceans, directions and horizons, namely the whole world, are evoked. The notion that the macrocosm is contained in the microcosm is also one of the favoured topics of the Bauls of Bengal.[339]

Intriguingly, a similar notion of the lute invoking emotions can be found in the theory of affects (*Affektenlehre*) in the European renaissance. Thus the German Jesuit poet Jacob Balde (1603–1668) states that the poet plays the artistic, eleven-stringed lyre.

> Sie (= Lyra), gewichtig und wohlklingend, besteht, falls du es nicht weißt, aus elf Saiten. Diese muss man zum klingen bringen nach dem Gesetz der Natur, wenn nicht von Kunst belehrte Tüchtigkeit sie spielt. Wir alle werden von Hoffnung, Furcht, Sehnen, Haß, Schmerz, Freude und Zorn ständig bewegt. Der Leib erbebt von diesen Strängen und auch die Seele von ihrem Anschlag. Unter den hell klingenden Saiten ist diese erste die Liebe [...].[340]

MICHEL 1987, (pp.233–234) explains the background as follows:

> Wenn wir das Gedicht richtig verstehen, will der Autor folgendes zum Ausdruck bringen: Die Dichtkunst besteht im wesentlichen darin, Affekte zu erregen, wie es auch der Musiker tut, indem er die Saiten seines Instrumentes in Schwingungen versetzt. Dabei setzt der Dichter das System der elf Affekte voraus, das als erster Thomas von Aquin (1226–1274) zunächst im Sentenzenkommentar (3 Sent. 26) und ausführlicher in der *Sumna Theologica* (I–II, quaestio 22 sqq.) aufgestellt hat [...]. Ihm hat sich später die genaue aristotelisch-scholastische Tradition angeschlossen.

---

338 E.g. the heart is considered the place of consciousness (*cetanasthāna*) in SR 1,2, śl.83cd.

339 DAS 1992, p.389, note 7 and note 208 (on the seven division of the earth).

340 Translated by MICHEL 1987, p.233. The original Latin text is:
Illa, si nescis, gravis ac sonora / constat undenis fidibus movendis / lege naturae, nisi docta virtus / temperet arte. / Spe, metu, votis, odio, dolore, / gaudiis, ira variamur omnes. / Corpus his nervis animusque certo / contremit ictu. / Inter argutas resonae chordas / est amor princeps [...] /.

The Indian aesthetic theory of *rasa* and the European theory of "Affekt" show striking similarities. KOCH 1995 tried to investigate whether there could have been some exchange or contact between India and Europe on this matter, but the result of his study negates this. According to him, the two theories developed independently from one another.

In German, there is the expression, "Es findet im Herzen Widerhall". Japanese, too, has a similar expression, "it touches the strings of the heart" (*kokoro no kinsen ni fureru*). The same kind of feeling, which the human beings have in common, seems to exist at the base of such expressions, also of the Indian one examined above. But the uniqueness of the Indian theory is its treatment in detail of the anatomical parallelism between the human body and the stringed instrument.

Situating the text: Appendix I

In *Situating the text, Appendix I* and *II*, I discuss two authors who compare the human body to a musical instrument in detail. I discuss them here because I discovered these pieces of information only after I completed most parts of this work. I prefer to treat them separately here rather than disturb the coherence of the work by attempting to incorporate them in its main body.

Comparison of the human body to a string instrument in Kabīr's poetry

The comparison of the human body to a string instrument is dealt with in some songs of Kabīr, the medieval Hindi poet. In his *Sākhī*, Kabīr sings:

> *saba raga tāṃti rabāba tana, biraha bajāvai nitta /*
> *aura na koī suni sakai, kai sāīṃ kai citta //* [341]

> "All the tubes/nerves/sinews (*rag*) are the strings; the body is the *rabāba* (a kind of string instrument). [It] always plays [the song of] separation.
> No one else can hear either Lord or the mind."

The human body is compared to the *rabāb*, a string instrument of Central Asian origin, which is still being used in the countries of South Asia. The tubes which, according to Haṭhayogic theory, radiate in thousands from the heart are its strings. Someone is always playing the melodies or songs of the pain of separation on this

341 KABĪR, Sākhī (Veṃkaṭeśvara, 1972), p.51 (Birahako aṃga, No.36). This song is also given by DVIVEDĪ 1990, p.252 (No.177, verse 3). In the transcription of classical texts in New Indo-Aryan languages, I retain the short vowel [a] even in the case in which it is dropped out in the pronunciation today.

instrument. The player might be God, or the poet's own soul.[342] This inner melody is so subtle, that it is very difficult for an ordinary person to perceive this subtle sound of his innermost heart, the "sound not struck" (*anāhata nāda*).

Another song explains the correspondence between the human body and a string instrument in more detail.

sādho, yaha tana ṭhāṭha taṃbūre kā /
aiṃcata tāra marorata khūṃṭī, nikasata rāga hajūre kā //
ṭūṭe tāra bikharage khūṃṭī, ho gayā dhūrama-dhūre kā /
kahaiṃ kabīra suno bhāī sādho, agama paṃtha kāī sūre kā //[343]

"Oh, fitting person (*sādhu*), this body is the frame (*ṭhāṭha*) of the *tambūrā* (a string instrument). One stretches the strings, and winds the pegs. [Then] the melody/love (*rāga*) for the Lord appears (*nikasata*). [But] the strings broke; the pegs got crushed (*bhikhar-* lit. "to be scattered"), [the instrument] was steeped in dust (?) (*dhūrama-dhūre kā*). Kabīr' says, listen, [oh,] Brother, fitting person, the unwalkable path is [only] for some brave people."

DVIVEDĪ ibid., p.201, in his footnote explains:

Through stretching the strings and winding the pegs, namely, tuning the instrument, beautiful sounds are produced. In the same manner, through controlling the sense-organs (*indriy-daman*) and mind (*man ke saṃyam*), the *rāga* of the Lord manifests itself. The term *rāga* here has a double meaning, namely, music and love. Only some brave men are able to take this path which is difficult to walk upon (*agam paṃtha*).

This instrument, the body, is, however, in a damaged condition. Its strings and pegs are broken, and it is covered with dust. It needs to be repaired and cleansed.[344] This is a very troublesome task, which only

342 VAUDEVILLE 1993, p.187, translates a song from the Sākhī, "Allow the musician to play". In the footnote 65, she comments, "*bājamtari* [sic.], the 'Musician' alludes either to God, or, more probably, the human soul".

343 DVIVEDĪ 1990, p.201 (No.39 = 1–59 of Kṣitimohan Sen).

344 DVIVEDĪ ibid. fails to grasp the correct meaning of the third line (*ṭūṭe [...]*). He interprets, "when the assemblage of the sense-organs (*indriya*) and mind-consciousness (*man-buddhi*) dissapears/is_eliminated, and this subtle and gross body is crushed into pieces, the individual self becomes steady in its nature", which does not fit the context.

expert musicians and instrument makers are able to succeed in. Tuning the human body, that means bringing the sense-organs and the mind in harmony, is a very difficult operation to execute, and is only achievable by skilled Yogins.

The human body is said to be filled with sound not struck (*anāhata nāda*).

> *yahi ghaṭa caṃdā yahi ghaṭa sūra / yahi ghaṭa gājai anahada tūra //*
> *yahi ghaṭa bājai tabala-niśāna / bahirā śabda sune nahi kāna //*[345]

> "[In] this body (*ghaṭa*) is the moon. [In] this body is the sun. [In] this body, the drum (*tūra*) of the [sound] not struck thunders/roars. This body plays drum-signal (*tabala-niśāna*). A deaf (*bahirā*) ear does not hear the sound."

DVIVEDĪ 1990 explains this poem as follows (p.182, footnote):

> Its straight (=blunt) meaning is that, in this very [human] body, all the lights/celestial_bodies and all the auspicious musical instruments, which are visible in the external world, exist. In this very [body], the unstruck sound prevailing in the cosmos, too, is heard. But someone who does not have the inner eyes is not able to see this light/celestial_body.[346]

The "moon" and "sun" which are said to be contained in the body denote the two tubes (*nāḍī*) in the human body, *Piṅgalā* and *Iḍā*, which are often called "moon and sun" by Haṭhayogic texts.[347] We have already seen in the Caryāpada-s and Cakrakaumudī that the sound not struck arises from these two tubes which are compared to the strings of the musical instrument.

In this song, however, the source of the sound not struck is compared to that of percussion (*tūra, tabala*), rather than to that of string-instruments. Here, the human body is not directly identified

345 Quoted from the edition of Kṣitimohan Sen (1–83) by DVIVEDĪ 1990, p.182 (No.6). A similar statement (*isa ghaṭa aṃtara anahada garajai*) is quoted in p.184, too.

346 *sīdhā mat'lab yah hai ki isī śarīr meṃ ve sabhī jyotiyāṃ aur sabhī maṃgal-vādy vart'mān haiṃ jo bāhy jagat meṃ dikh'te haiṃ. isī meṃ vah viśv'vyāpī anāhat dhvani bhī sunāī detī hai. paraṃtu jis'ke bhītar kī āṃkheṃ nahīṃ haiṃ vah is jyoti ko nahīṃ dekh pātā.*

347 By SR śl.154cd, too. Cf. DAS 1992, p.403.

with the drums, but the drums are metaphorically said to be contained inside the body. As a matter of fact, in the Haṭhayoga texts, the not-struck sounds perceived by a Yogin during his Haṭhayogic practice are said to resemble those of various instruments, like *vīṇā*, drums and bells.[348] Ordinary people who do not possess the superhuman ability of Yogins are not able to perceive such a subtle sound.

jhī jhī jantara bājai / kara carana bihūnā nācai /
kara binu bājai sunai śrabana binu / śrabana śrotā loī /[349]

"The instrument (*jantra*) is sounding *jhī jhī*.
[Someone] dances without hands and feet.
Without the hands it sounds, [one] listens without ears."[350]

Inside his body, the Yogin perceives the sound resembling that of a musical instrument and feels the rhythm of dance, though he himself does not play the instrument nor does he dance.

The intensity of love in separation is sometimes described in a paradoxical way.

kabīra jantra na bājai, ṭūṭī gae saba tāra /
jantra bicārā kyā karai, cale bajāvanahāra //[351]

"O Kabīr', the instrument does not sound. All its strings are broken.
What does (= can) the poor instrument [do]. The player has gone [away]."

The poet cannot play the instrument any longer, namely the body. He has lost all his power to sing his sorrow of separated love, because he says all the strings of his heart were broken, due to the extreme intensity of this pain. He gives a sigh of discouragement: "What can

348 Cf. Haṭhayogapradīpikā.
349 DVIVEDĪ 1990, p.216 (No.83 = 3–84 of Kṣitimohan Sen).
350 I have decided not to translate the last line, *śrabana śrotā loī*, because *loī* is unclear.
351 In: Kabīr Granthāvalī, ed. by Paras Nath Tivari, Prayāg : Hindi Parishad, 1961, p.198. This song is translated by VAUDEVILLE 1993. p.190. The same song is contained in the Gurū Granth, too (cf. VAUDEVILLE ibid., pp.307–308, No.103).

the instrument do? The player, namely, God, has departed from it." It is no use maintaining the body, if God does not dwell inside it.

Many ideas of the Caryāpada-s seem to have been handed down to Kabīr. DAS GUPTA 1976, pp.416–417, presents an amazing example, a song of Kabīr of which every line is parallel to a line in the Caryā song of Ḍheṇḍhana. Through this tradition, Kabīr inherited the comparison of the human body to the stringed instrument, too.

Situating the text: Appendix II

The comparison of the human body to the stringed instrument, mentioned by the theologian of Gauḍīya Vaiṣṇavism, Rūpa Kavirāja

Rūpa Kavirāja, the theologian of Gauḍīya Vaiṣṇavism in the 17th century, mentions the comparison of the human body to the stringed instrument, *vīṇā*, in his work on the method of *rāgānuga bhakti* ("the devotional worship following desires").[352] The followers of this method imitate the behaviour of *gopī*-s, i.e., the cowgirls who are Kṛṣṇa's lovers, with the purpose to realise the union with Lord Kṛṣṇa. Some male followers even practice transvestism. According to the theory of *rāgānuga bhakti*, the follower has two different bodies: *siddha-rūpa* and *sādhaka-rūpa*. That means, his manner of existence has two stages: the *siddha-rūpa*, "the form as someone who is accomplished", denotes the higher stage to be acquired through the method of *rāgānuga bhakti*. In this higher stage, the follower identifies himself with the *gopī*, the lover of the Lord. In contrast, the *sādhaka-rūpa*, "the form as someone who is striving for accomplishment", is the physical body of the follower. The follower imitates the behaviours of the *gopī* (*siddha-rūpa*) with his own physical body (*sādhaka-rūpa*), like the actor changes into his role to play on stage.[353] Concretely, the follower clothes himself like a woman, and

352 For *rāgānuga bhakti*, cf. HABERMANN 2001, pp.100–102.

353 Intriguingly, HABERMANN 2001 compares this phenomenon with the dramaturgical method of the Russian Actor, Stanislavsky.
According to BHATTACARYA 1998 (p.231), Giriś Candra Ghoṣ, the founder of the modern theater of Bengal at the end of the 19th century, adopted, for the education of Vinodinī Dāsī, the star-actress of his theatrical group, a method which was strongly influenced by the *rāgānuga bhakti*. The modern theater of Bengal laid the groundwork for the later development of the South Asian film (including "Bollywood").

behaves like a woman. Rūpa Kavirāja tries to justify this practice in the following way:

> The *siddha-rūpa* and *sādhaka-rūpa* are similar to a *vīṇā* and a *vīṇā* player. Even though the two [*vīṇā* and *vīṇā* player] are distinct there is an oneness of their songs, because their essence is similar; just so, even though the two bodies are distinct their performances (*sevā*) are similar and even simultaneous. As the song produced on the *vīṇā* is situated in the mind of the *vīṇā* player; so the performance which occurs in the *siddha-rūpa* is situated in the *sādhaka-rūpa*. When separated there is no *rasa* in the music of the *vīṇā* and *vīṇā* player; likewise, when separated there is no *Vraja-bhāva*[354] born in the performance [of the *siddha-rūpa* or the *sādhaka-rūpa*].[355]

Like the musician and his instrument in playing melt into one, and like the actor and his role during performance are united with one another, the *sādhaka-rūpa* and the *siddha-rūpa* similarly melt into one during the ritual.

We find a comparable statement by SCHNEPEL 2005, a German ethnologist, who writes on the Odissi, the traditional dance form from Orissa. She mentions the theory of the body as "a body-of-ideas" by FOSTER 1997. She summarises FOSTER's theory:

> Die Ausbildung eines Tänzers, die ein langjähriges hartes Training voraussetzt, bringe zwei Körper hervor: den vom Tänzer selbst wahrgenommenen und fühlbaren, den er trainiert, und einen ideal-ästhetischen (imaginierten oder bei anderen Tänzern gesehenen und in Filmen dargestellten), der erreicht werden will. Beide Körper bedingten und beeinflussten einander und würden immer wieder eine Veränderung erfahren.

Actually, FOSTER 1997 deals with the western forms of dance and dancers like Isadora Duncan, Martha Graham, Merce Cunningham, contact improvisation technique etc.[356] FOSTER does not deal with a local dance form, but tries to formulate his theory based on the universal experiences of the western dancers. However, the connection between FOSTER's and Kavirāja's theory is obvious.

354 I.e., the state of the *gopī* who is united with her lover Kṛṣṇa.

355 Translated by HABERMANN 2001, p.102.

356 Cf. SCHNEPEL 2005, p.127. Unfortunately, I was not able to consult FOSTER 1997 myself.

English translation

On my translation method

The Piṇḍotpattiprakaraṇa has been already translated into English in SHRINGY & SHARMA 1999 (pp.21–107), which is so far considered as the representative study of the SR. This is a relative free and fluent translation which even a reader not specialised in Indology could read with ease.

A new translation therefore must be different from it in quality and serve a different purpose. It does not need to be free or fluent, but it must shed light on a new aspect of the original text, of which the proceding studies have not yet been conscious.

For this purpose, my translation tries to be as faithful as possible to the grammatical structure of the original language, so that readers are given the possibility to follow the way of grammatical interpretation exactly, word for word. Therefore my English translation may sometimes be very far from a literary or aesthetical one. However, the translation of this kind of genre, a scientific text, does not necessarily need to be beautiful. The original text studied here is not a literary but a scientific text, and the focus here is on something else than the literary or aesthetic value of the text. It is far more important to indicate to the reader the English equivalent of the meaning and grammatical function of each word. This is due to the fact that the original language, Sanskrit, has quite a complicated grammatical and syntactical structure, and because the cultural background of the original text is very different from that of many modern readers. Literally translating, one is able to pinpoint and precisely discuss subtle problems which would otherwise have escaped consideration in a free translation.[357] In this point, I follow, so

357 Cf. the discussion on the problems caused by the unreflected usage of the term "caste" in describing social categories in South Asian society, in DAS 2004 (p.94ff.).

to say, the method of translation which was once adopted by Tibetan Buddhist scholars who translated Sanskrit texts into their own language, or by Japanese monks who translated Chinese texts. They tried to reproduce every single subtle grammatical element of the original language in their mother tongue. In consequence they modified their mother tongue and invented an artificial language which enabled them to do mechanical one-to-one translation from the original language.

In the same manner, I adopt the same English word for the same Sanskrit term, but different English words for different Sanskrit terms, as far as possible. For example, in the translation of the commentary, in which each term of the *mūla* text (the original text to be commented on) is glossed with a synonym, I adopt different English words for the term quoted from the *mūla* text (*pratīka*) and that from the gloss. In many cases in which SHRINGY & SHARMA 1999 adopt the technical terms of modern Western medical science for the translation of the Sanskrit technical terms of Indian medicine, I translate them literally, so that I avoid the danger of tinging these terms with some bias.[358]

Remarks on the English translation

1. The particle *iti* is translated in various manners according to the context, often simply through quotation marks ("[...]").
2. Brahman refers to the neuter. Brahmā refers to the god (masculine).
3. The Sanskrit optative is usually translated with the English indicative.[359]

358 E.g., I translate the term *tvac* (SR 1,2,79) as 'skin' in contrast to 'serum' by SHRINGY & SHARMA 1999.

359 I follow DAS 1988, pp.508–509: "Ich mache hier darauf aufmerksam, dass der Optativ zur Kennzeichnung einer Möglichkeit oder eines Ereignisses, das erfolgt, wenn etwas in bestimmter Weise getan wird, oft auch in anscheinend elliptischen Sätzen steht; in solchen Fällen habe ich mit dem Indikativ wiedergegeben."

4. When the *mūla* text is reproduced *exactly* in the commentary, it is printed in bold letters. The *pratīka*, i.e. the term from the *mūla* text which is explained with its synonym by the commentary, is also printed in bold letters.
5. The variants are taken into consideration in the footnotes, if they seem to me to make sense. I do not note those, for which I was not able to work out any interpretation.
6. In translation of some Sanskrit terms, more than one meaning are given. In such a case these meanings are connected with a slash (/). I.g. *vacas* is translated as "sentence/speech", that is "sentence" or "speech". When a meaning consists of more than one English words, these words are connected with an underline (_) to avoid ambiguity. I.e. *doṣa* is translated as "fault/injurious_consequence"; that does not mean "fault" or "injurious", but "fault" or "injurious consequence".

Section: Arising/Origination of the [human] body (*piṇḍa*)[360]

SR śl.1–3

Song is that whose essence is sound (*nāda*). Instrumental music (*vādya*, lit. "that which is to be played") is esteemed because of the manifestation of sound. Dance[361] is accompanied [by] those (*tad*) two[362]. Therefore this triad rests on sound[363]. (śl.1)

360 The term *piṇḍa* might mean "an embryonic ball", but I provisionally translate it as "the human body".
One of the basic meanings of *piṇḍa* is "roundish mass", "ball", "piece", cf. SUNESON 1941–1942, p.115. For the term *piṇḍa* in embryology, cf. DOSSI 1998, p.93 and pp.154–156. For the term *māṃsa-piṇḍa* meaning "embryo, foetus", cf. SUNESON 1991. It means something coagulated, and is associated with a ball of flesh or rice, as the selection of examples below shows:
In ṚV 1,162,19, the term *piṇḍa* means [horse-flesh] dumpling (Opferklöße): *yā́ te gā́trāṇām ṛtuthā́ kṛṇómi tā́tā píṇḍānām prá juhomy agnáu* /19/. "So viele deiner Körperteile ich nach der Reihenfolge herrichte, so viele der Klöße opfere ich ins Feuer." (tr. by Geldner). In this verse, *piṇḍa*-s are compared to the limbs (*gātrāṇi*).
Piṇḍa also refers to rice balls eaten by the husband and wife who desire a child, cf. SHIVARAM 2001, p.10. "In one particular *śrāddha* rite, the *sapiṇḍīkaraṇa, piṇḍas* are used to reconstruct the bodies of dead ancestors in a way that images conception, foetal development, and birth" (ibid., p.10, note 24).
In the MārkaṇḍeyaP (10,5–6), an embryo is compared to a rice ball (*piṇḍa*). The text discusses the question why the embryo is not digested in the mother's belly, like a rice ball.
The term *piṇḍa* occurs later again in SR 1,2,75a. There it must be understood as "a human body" or "the body of an embryo". See my discussion in the footnote 796.

361 *Nṛtta*. There are three terms denoting "dance", i.e. *nāṭya, nṛtya* and *nṛtta*. On the semantic distinction between them, cf. VARMA 1957. But the term *nṛtta* in SR 1,2, śl.1 seems to simply denote "dance" in general, in contrast to vocal music (*gīta*) and instrumental music (*vādya*), cf. SR 1,1, śl.21: *gītaṃ vādyaṃ tathā nṛttaṃ saṅgītam ucyate*.
The manuscripts *ka* and *gha* read *nṛtya* instead of *nṛtta*.

362 The expression *taddvayānugataṃ nṛttaṃ* is translated by SHRINGY & SHARMA 1999, p.21, as "*Nṛtta* (dance) follows both", although *dvaya-anugata* usually means "[is] followed by the two". The comm. S, saying *aṅgenālambayed gītam*, seems to interpret in the same manner as SHRINGY & SHARMA. Besides, it

A phonemic unit (*varṇa*) is manifested through sound. A word from the phonemic unit. A sentence/speech (*vacas*) from the word. This (*ayam*) usage[364] from (=through) the sentence/speech[365]. Hereby the world rests on sound[366]. (śl.2)

Sound is announced [to be] of two kinds: struck and unstruck[367]

---

better suits the context, as the main topic of the SR is not dance but vocal and instrumental music. The reason of this awkwardness may be that the expression *taddvayānugataṃ nṛttam* originally belonged to a dramaturgical text which dealt with dance, and was taken in by the SR.

363 The manuscript D reads *idaṃ trayam* instead of *atas trayam.* "This triad rests on sound" or "The triad [in this world] (*idam*) rests on sound".

364 *Vyavahāra* is a grammatical term meaning "everyday verbal usage", cf. HOUBEN 1995, p.21, p.65, p.254 etc. The comm. S on SR 1,3,7 explains *vyavahāra*, as follows: *vyavahāre gāna-vyavahāre, ghaṭapaṭādy-abhidhāna-vyavahāre tu mukhotpannasyāpi dhvaner upayogitvāt.*
So, in the text here, SR 1,2, śl.2c, the demonstrative pronoun *ayam* "this" means "the usage [in this world]".

365 The manuscript D reads *vacasā* "through the sentence/speech" instead of *vacaso.*

366 Adopting the variant of the manuscript D *idaṃ*, śl.2cd would mean "This (*ayam*) usage from sentences, this (*idaṃ*) world rests on sound".
A similar statement is made by the Māṇḍūkya-upaniṣad 1,1, "Hari is *oṃ*. This syllable is this whole. The past, the present, the future – everything is just the phoneme *oṃ*" (tr. by PADOUX 1990, p.18).
In the Sāṅkhya school, too, the whole universe is considered an aggregation of sounds, according to HOUBEN 1995, p.60 and footnote 107.

367 The "unstruck sound" or "sound not struck" (*anāhata nāda*) in the Yogic context denotes the supernatural sound perceived by a Yogin in the various stages of Yogic practice, cf. Haṭhayogapradīpikā 4,64 ff. (SINH 1980, p.56).
The term *āhati* is contained in SR 1,3, śl.9, on the process of producing the microtones in the octave: *hṛdy-ūrdhva-nāḍī-saṃlagnā nāḍyo dvāviṃśatir matāḥ /8cd/ tiraścyas tāsu tāvatyaḥ śrutayo mārutā<u>hateḥ</u> / uccoccataratā-yuktāḥ prabhavanty uttarottaram /9/.* The vital wind strikes (*āhati*) the twenty-two oblique tubes in the body, and produces microtones. This explanation is based on the comparison of the human body to a *vīṇā* cf. *Situating the text §5, "Comparison of the Human body with the Musical Intrument in Indian literature".* So, *āhati* means striking the strings.
As a matter of fact, the term *āhati* is explained as meaning the striking of a string by fingers (*madhyamākrānta-tarjanyā tantrikāhatiḥ*) in the SR's sixth *adhyāya* (on instrumental music), śl.69. *āhata* (śl.732) and *āhatya* (śl.733) also seem to mean making sound through striking a string. In contrast to the case of the *vīṇā,* the term is not used for the flute, cf. SR 6, śl.424cd and the following

(manifested and not manifested).

This sound manifests in the [human] body (*piṇḍa*)[368], therefore, the *piṇḍa* is explained [or: from this, it is called *piṇḍa*[369]]. (śl.3)[370]

Comm. K on SR śl.1–3

Anticipating [the objection]: "Should not the explanation (*prati-pādyatva* lit. "the state of being to be explained") of sound be first in [the chapter] Svaragata[371] because of [sound's] being the cause of song?[372] [Of] what [use] with the investigation (*nirūpaṇa*) of the body?", he (= the author), doing away [with this objection] with [the argument (*iti*) of] the absence of the manifestation of sound alone[373] without the body, says in order to investigate the body: "**Song is that whose essence is sound**" (śl.1a, *gītaṃ nādātmakam*). This is the

---

(Adyar ed., Vol.III, p.430). The blowing of the flute is called *phūtkāra* (cf. SR 6, śl.431: *phūtkāra-prabhavo vāyuḥ pūryate mukha-randhrataḥ*).

On the other hand, the term *āhanti* in SR 1,3, śl.3–4 on the production of *nāda* seems to mean something else. It is stated that the self (*ātmā*), desiring to utter (*vivakṣamāno*), puts the mind into motion (*prerayati*); the mind strikes (*āhanti*) the fire of the body (*dehasthaṃ vahnim*); the fire puts the wind into motion; the wind, moving upward, manifests resonance (*āvirbhāvayati dhvanim*). In this case, *āhanti* seems to mean something like "to stimulate" (*prerayati*).

According to this statement, the *anāhata* sound seems to be the stage, in which the mind is already put into motion by the self, but does not yet strike the body fire.

368 Cf. SR 1,3, śl.82. The commentary K, explaining the term *śārīra* (lit. "the [talent] belonging to the body", i.e. in-born talent of singing), states that not only voice quality but also the ability to present a *rāga* depends on the body (*yathā dhvaniḥ śarīreṇa sahodbhavati, tathā tasya rāgābhivyakti-śaktatvam api śarīreṇa sahodbhavati. nahy abhyāsenāgantukam ity arthaḥ*).

369 For the various synonyms meaning the body, *śarīra, deha, piṇḍa, tanu* etc., cf. VATSYAYAN 1988, vol. I, p.85 (under the term *śarīra*).

370 The manuscript D reads *nigadyate* instead of *'bhidhīyate*. It would mean, "therefore, the *piṇḍa* is announced".

371 The Svaragata is the name of the first chapter of the SR, in which Piṇḍotpatti-prakaraṇa is contained.

372 I.e., should not sound – instead of the human body – be explained first in chapter Svaragata?

373 The manuscript C reads *nādasya* "of sound".

The Ānandāśrama edition (1896) reads *nāda-mātra-sparśa-śarīrasya* instead of *nāda-mātrasya*. But this does not make sense, and the editor corrects it into *nāda-mātrasya* in brackets.

meaning: Not only do the three, [namely] song etc., depend on sound, [but] because the usage (*vyavahāra*) of the whole world, whose base is speech, also depends on sound, [therefore] its (= sound's) base[374], [namely] the body, is indeed all the more to be observed first.

Comm. S on SR śl.1–3

Thus, having mentioned the list of topics[375] [of this work in the foregoing section], he (= the author) praises sound, to explain the human body, [namely] the place of manifestation of sound: "**Song**" (śl.1a, *gītam*). **That whose essence is sound**: whose **essence** (*ātman*)[376], [namely] nature (*svarūpa*), is **sound**. **Instrumental music**[377], the lute etc., **is esteemed**, attains pleasantness, **through the** very **manifestation of sound**[378]. **Dance is accompanied [by] those two**, because of [it] having been said that (*iti*): "One should support the song with the limbs."[379] So the meaning is: **the triad** in the form of song, dance and instrumental music **rests on sound**, [namely] depends on sound. [Explaining] that (*iti*) not only do song, dance and instrumental music depend on sound, but also the whole world, he relates: "**Through sound**" (śl.2a, *nādena*). **Through sound** (*nāda*), [namely] through resonance (*dhvani*), **the phonemic units**, [namely] *ka* etc., **are manifested**. What is this resonance (*dhvani*)? That which arrives from afar at the range of hearing (*karṇapatha*) of one unable to perceive a particular phonemic unit, and which causes the difference between weakness and sharpness[380] to attach to the phonemic units, that is called "resonance" (*dhvani*). **From the phonemic unit, a word** [is manifested] [like] "pot (*ghaṭa*)" etc. **From**

374 The variant D reads *tadādhāraṃ* instead of *tadādhāraḥ*. "The body whose base it (= sound) is [...]."

375 The list of topics to be dealt with by the SR is given in SR 1,1, śl.31–49ab.

376 I use boldface for the *pratīka*, i.e. direct quotation from the *mūla* text. The commentary explains *ātman* with its synonym *svarūpa*.

377 The term *vādya* might also denote "musical instrument[s]".

378 The term *nāda-vyakti* in the *mūla* text is explained as *nāda-abhivyakti*. I use boldface for it, too, as I consider to be a kind of quotation from the *mūla* text.

379 The first *prakaraṇa* of the first *adhyāya* does not contain this sentence.

380 In the musical theory, *manda* and *tīvra* mean that a tone is low or high, cf. NIJENHUIS 1970, p.72, on *mandra* "low", *madhya*, "middle" and *tāra* "high". These terms denote the three registers.

the word, **speech** (*vacas*[381]), [namely] a sentence (*vākya*), [namely] a group of words [is manifested]. **From speech**, [namely] from the sentence, **this** verbal **usage** [is manifested]. **Therefore** even the whole **world rests on sound**. Thus (*iti*) is the meaning.

It has also been said by Mataṅga in the Bṛhaddeśī:

> No song [can be] without sound, no [musical] tones (*svara*) without sound (*nāda*).
> No dance without sound. Therefore the world is that whose essence is sound.
> Brahmā[382] is remembered as having the form of sound, Janārdana (= Viṣṇu) as having the form of sound,
> the Supreme Power[383] as having the form of sound, Maheśvara (= Śiva) as having the form of sound.
> [In the human body,] between that which is called the place of Brahmā and that which is remembered as the knot of Brahman (*brahma-granthi*), the vital wind (*prāṇa*) is situated. The arising of fire is from the vital wind.
> Sound (*nāda*) is created from the union/contact of fire and the [vital] wind [in the body]. The point/drop (*bindu*)[384] arises from sound (*nāda*). Therefore all is made up of speech.[385]

He (= the author) explains the twofoldness of sound: "**Struck**" (śl.3a, *āhata*,). It is said that (*iti*) sound, of both (*api*) kinds, **manifests**, becomes clear, **in the body**.

SR śl.4–5

There is Brahman[386], [namely] intellect-joy (*cid-ānanda*), [namely] that which shines by itself, pure,

381 The reading *vāco* is obviously a mistake. Instead I read *vaco*.
382 The male god, i.e. one of the trinity.
383 *Parā śakti* "ability, energy". According to SHARMA 1992, p.152 (note 7), this term shows that the author of the Bṛhaddeśī was influenced by Śākta Tantra.
384 PADOUX 1992, p.110, explains that the *bindu* "is a concentrated sound vibration, a drop not only of energy but also of light. Sound and the Word eminently partake, in effect, of the luminous nature of consciousness, according to various Tantric traditions". Also see his footnote 72, "The luminous nature of *bindu* is clearly apparent in connection with the arousal of the *kuṇḍalinī*. […] *bindu* indicates a luminous dot and a one-pointedness of thought appearing during certain meditations […]". For the relation of the *bindu* to the arousal of the *kuṇḍalinī*, see op.cit. p.134ff.
385 Bṛhaddeśī 18–20 (in *adhyāya* 1), P.L. SHARMA 1992, p.6.
386 The *brahman* here is neuter, i.e. the Supreme Principle.

able/[that_which]_rules (*īśvara*)[387], [namely] that which is called "mark" (*liṅga*[388]), without a second, having no birth, omnipresent, (śl.4)
without change, shapeless, [that] ruling all (*sarveśvara*), indestructible[389], which has all power, and omniscient. Its parts are called individual selves (*jīva*[390]). (śl.5)

Comm. K on SR śl.4–5

And the body is dependent upon the deeds (*karman*) of the individual self (*jīva*). And the individual self, [which is] being enquired about [here], depends on (*apekṣate*) Brahman (the Supreme Principle), which is characterised by [having] it itself (i.e. the individual self[391])

I have translated śl.4a according to the interpretation of Comm. S. However, Śaṅkara's school of the later period (12–3th centuries) defines Brahman as *sad, cit* and *ānanda*. In accord with this, śl.4 could be translated as "Brahman is existence (*asti* = *sad*), [Brahman is] intellect (*cid*) and [Brahman is] joy (*ānanda*)." FUNATSU 1991, pp.86–87 also interprets it in the same manner.

387 The Ānandāśrama ed. (1896) reads *īśvaro 'liṅgaṃ* instead of *īśvaraṃ liṅgam*. In contrast to *īśvaraṃ* which is an adjective qualifying *brahman*, *īśvaro* "Lord" would be an apposition to *brahman*.

388 ROŞU 1978, p.130, explains according to CA śārīra. 1,62, as follows: "Liṅga <signes> qui laissent inférer l'éxistence de l'ātman. Celui-ci étant non-manifesté n'est pas perçu par les facultés sensorielles, mais il est saisissable à partir de certaines <signes>".
Referring to this, the text here, SR śl.4c, *īśvaraṃ liṅgam ity uktam*, might be interpreted, "it is said that *īśvara* is the mark". That could mean that the supreme self is perceived only through its marks, and people identify those marks with the supreme self itself.
For the term, *liṅga-deha* "sign body", cf. CA śārīra. 1,70–72; AS śārīra. 5,20. For its use in the Sāṅkhya texts, cf. Sāṅkhyakārikā 40 and Sāṅkhyasūtravṛtti 3, 9 (ROŞU 1978, p.131).
For the term *liṅga*, cf. SU śārīra. 1,17. ROŞU 1978 informs us of the explanation of this term in the Praśastapādabhāṣya 77–78 and 80.

389 The variants *ka., kha., ga.* and *gha.* read *atīśvara* "excessively able" instead of *anaśvara*.
"The lord of all" might be better for the translation of *sarveśvara*, for Brahman as the cause of the world is called "*īśvara*" by Śaṅkara.

390 The term *jīva(-ātman)* lit. "living self" means the individual self, in contrast with *paramātman* "the supreme self".

391 The reflexive pronoun *sva* refers to the subject of a sentence. Thus the meaning is: Brahman, a part of which is the individual self.

as a part [of Brahman].[392] When it (= Brahman) is enquired about because of [this] (*iti*), he (= the author) investigates its nature: "**There is Brahman**" (śl.4a, *asti brahma*). **Intellect-joy** (*cid-ānandam*) [means] that whose nature is the pleasure of knowledge. **That which shines by itself**, [means] [that which is] manifesting [itself] by itself. **Pure** (*nirañjana*), [namely] without adhering impurity. **Able** (*īśvaram*) [means] independent. **The mark** (*liṅgam*), because of the dissolution/ repose of the illusory [manifested] world.[393] **Without a second** [means] without another thing like itself. **Having no birth** [means] without birth. **Omnipresent** (*vibhu*) [means] all-pervading. **Without change** [means:] One is born, one exists, one grows, one matures/ ripens, one decays, one disappears—these are the six changes of being: free from them. **Shapeless** [means] free of shape. **[That] ruling of all** (*sarveśvara*) [means] the maker of the whole world. **Indestructible** [means] free of destruction.[394] **Which has all power (*sarvaśakti*)** [means] endowed with the powers (*śakti*) of desire, knowledge, action and experiencing (*bhoga*). **Omniscient** [means] knower of everything of past, future and present. Such (*iti*) is Brahman[395], as described in the Upaniṣad-s.

Having mentioned the nature of Brahman, he (= the author) investigates the individual selves and their body-relation through (= with regard to) [their] sameness of nature (= identity) with it (= Brahman): "**Its parts**" (śl.5d, *tad-aṃśāḥ*). **The parts (*aṃśa*)**, [namely] portions/measures (*mātrā*), **of it**, [namely] of Brahman.

392 The Ānandāśrama ed. (1896) reads *svāṃśe* instead of *svāṃśi*, but this reading does not seem proper. It is not noted by the Adyar edition.

393 The Ānandāśrama ed. (1896) reads it differently as *īśvaraṃ svatantraṃ prapañcayann aliṅgam*. This is not noted by the Adyar ed. It would mean, "Able, [i.e.] independent. [It is] without a mark (*aliṅgam*), [though] creating the illusory world". This reading accords with the variant of the *mūla* text SR śl.4c, *īśvaro 'liṅgaṃ*. The problem of this reading is, however, that it changes *īśvaro* of the *mūla* text into *īśvaraṃ*.

394 The Ānandāśrama ed. (1896) reads *vināśa* instead of *nāśa*.

395 The expression *brahmāsti* is a kind of quotation of *asti brahma* in SR śl.4. This *iti* concludes the explanation which has begun with *asti brahmeti* (p.28, l.1).

Declaring the explanation/name (*abhidāna*) of the body, he relates its root cause: "**There is**" (śl.4a, *asti*). The meaning is that (*iti*) being all the time and everywhere, [Brahman is] permanent and [all-]pervading. **Brahman** because of greatness (*bṛhattva*), or else, because of being increasing (*bṛṃhaṇatva*)[396]. **Intellect** (*cit*) is that whose form is knowledge; **joy** is that whose nature is pleasure, because of the *śruti* text (Bṛhadāraṇyaka Up., 3,9,34c) saying (*iti*) "[Brahman is] knowledge, joy". **That which shines by itself**, manifesting itself. "All shines/appears following him [who is] shining/appearing. All [in] this [world] shines/appears because of his shining/appearing."[397] **Pure** [means] void of adhering impurity of ignorance, because of ignorance's having its base in the individual self[398]. [Or else] even on the side[399] of the ignorance whose base is Brahman, [ignorance's] state of not causing illusion in respect to its own base (= Brahman) is mentioned by the word ["]pure["]. **Able** (*īśvaram*) [means] able to do, not to do or to do in another manner. **Mark** (*liṅga*) [means] the cause. **Without a second** [means] void of differentiation with respect to (*gata*) the own self [as regards] being of the same kind or being of another kind. **Having no birth** [means] having no cause. **Omnipresent** (*vibhu*) [means] [all-]pervading or mighty/competent (*samartha*). **Without change** [means] void of all changes[400] mentioned by Yāska (Nirukta 1,2 : 32,15–16): "It is born, it exists, it grows, it matures/ripens, it decays, it disappears/is_destroyed". **Shapeless** [means] deprived of[401] forms. **Ruling of all** [means] ruler of all, even of Brahmā[402] etc. **Indestructible** [means] having no destruction. **Which has all power** [means] that (*iti*) all the power called illusion (*māyā*), which is the cause of all the [illusory,

396 Here I translated *bṛṃhaṇa* in intransitive meaning. But it might be transitive "fostering, nourishing", too.

397 *Tam eva bhāntam anu […]* is not found in BLOOMFIELD 1906 (A Vedic Concordance).

398 I.e. ignorance is a property of the individual self (*jīva*).

399 I.e. even according to the position.

400 The manuscripts A and B read *Sad-bhāva-vikāra* "the six changes of being".

401 *Anavacchinnam* is obviously a mistake. I read *avacchinna*.

402 That is to say the male god Brahmā, one of the trinity, but not the Supreme Principle, the neutral Brahman.

manifested] world, is found in this (= Brahman)[403]. **Omniscient** [means] characterised by witnessing every object. **Its part**[404], [namely] not differentiated from it (= Brahman), **like the sparks of fire from fire**[405], because of [their] state of being fire. Conversely, it (i.e. *tad-aṃśaḥ* etc.) is mentioned of the individual self and the supreme self, admitting, indeed, the side [of the argument] pertaining to the state of being [both] a part and that containing the parts, because of it having been mentioned in the *sūtra* (Brahmasūtra 2,3,41): "[The individual self is] a part, because of the indication of [Brahman] in manifold ways[406]."

SR śl.6–10

They, covered with[407] beginningless ignorance, governed/restrained (*niyantrita*) by beginningless deeds (*karma*) which give pleasure and pain [and] whose forms are merit and sin[408], just as sparks of fire[409] are

403 The manuscript B reads *sadā* instead of *sarvā*. "[...] that (*iti*) the power called illusion, which is always (*sadā*) the cause of all [...]".
The manuscripts A and B read *māyā-śaktir vidyate yasminn iti* instead of *śaktir māyākhyā vidyate 'sminn iti*. It would mean "that (*iti*) [*sarvaśakti* is that] in which (*yasmin*) the power of illusion (*māyā-śakti*) [...] is found".

404 This commentary reads it as sg. *tad-aṃśaḥ* instead of the *mūla*-text pl. *tad-aṃśā(ḥ)*.

405 This passage of the commentary explains SR śl.6b.

406 I.e. in manifold ways, even as a slave or as a gambler, according to the Brahmasūtra 2,3,41.

407 The manuscripts *ka., kha., ga., gha.* and *ṅa.* read *upahata* "afflicted by" or "polluted with" instead of *upahita* "covered with".
The expression *anādya-vidyopahita* is also found in Rāmānuja's Śrībhāṣya on Brahmasūtra 1,1,13.

408 The Ānandāśrama ed. (1896) reads *pāpa-puṇya* instead of *puṇya-pāpa*.

409 This simile of sparks is found in Bṛhadāraṇyakopaniṣad 2,1,20 (RADHAKRISHNAN 1953, p.190), *yathāgneḥ kṣudrā visphuliṅgā vyuccharanti*. Also compare with YS 3,67: *niḥsaranti yathā loha-piṇḍāt taptāt sphuliṅgakāḥ / sakāśād ātmanas tadvad ātmānaḥ prabhavanti hi*. (On YS 3,67, cf. YAMASHITA 2001/2002, p.89.) KANE 1990 (vol.I, part 1), p.448, states that this verse is an imitation of Muṇḍakopaniṣad 2,1,1.
This simile is also mentioned in Śaṅkara's commentary on the *Brahmasūtra* 2,3,43.

united with the imposed properties (*upādhi*)[410] of wood etc., (śl.6–7b) attain in each birth a body connected with this and that (= particular) birth/genus (*jāti*)[411], a life span and experience born from the deeds (*karma*). (śl.7cd–8a)

They, in their turn (*punaḥ*)[412], have (*teṣām asti*) another [body], [namely] the minute subtle-body (*liṅga-śarīra*)[413]. It is deemed indestructible until liberation[414]. (śl.8bcd)

[They (= the wise ones)] know this (= subtle body) as that whose nature is the state of the minute[415] element[s], facultie[s]/organ[s] (*indriya*) and vital winds.[416] (śl.9ab)

410 For the term *upādhi*, cf. GARBE 1917, p.231, "Hier wird alles *upādhi* genannt, was zu einem Dinge in Beziehung steht, ohne ihm wesentlich anzugehören oder eine innere Verbindung mit ihm einzugehen". PREISENDANZ 1994 (Teil 1), p.69, translates it as "eine zusätzliche Bestimmung". INGALLS 1951, p.40, translates it as "imposed property".

411 *Tattajjātiyuta* and *karmaja* qualify all the three nouns, *deha, āyus* and *bhoga*.

412 The manuscript *kha.* reads *vapuḥ* instead of *punaḥ*. "There is another body, [...]".

413 Lit. "mark-body". The Ānandāśrama ed. (1896) reads *liṅgaṃ śarīraṃ*. This is not noted by the Adyar ed. But the Ānandāśrama ed. (1896) reads *liṅga-śarīraṃ* in the commentary K, like the Adyar ed. (p.30, l.4 from the bottom).
For the notion *liṅga-śarīra*, cf. Sāṅkhyakārikā 40 (cf. KAPANI 1992, vol.I, pp.132–133).
HALBFASS 2000, p.151, explains this term as follows. "Das klassische Sāṅkhya lehrt, daß das, was im Saṃsāra ‚wandert', ein ‚Feinkörper' (*liṅgaśarīra, sūkṣmaśarīra*) ist, der aus allen Produkten der Urnatur mit Ausnahme der fünf grobstofflichen Elemente besteht (SK [= Sāṅkhyakārikā], Vers 40)."
"Auf seiner Wanderung durch den Saṃsāra, so wird uns gesagt (SK, Vers 40), ist der Feinkörper mit gewissen Zuständen (*bhāva*) durchtränkt (*adhivāsita*, »parfümiert«). Diese speziell der Kenntnis (*buddhi*) zugehörigen Zustände oder Dispositionen werden in der Folge als eine achtfache Gruppe spezifiziert" (HALBFASS 2000, p.152). These eight groups are: *dharma, adharma; jñāna, ajñāna; virāga, rāga; aiśvarya,* and *anaiśvarya*. "Der Anteil, den ein Lebewesen an diesen teils positiven, teils negativen Zuständen hat, bestimmt seinen Status im Saṃsāra" (ibid., p.152).
The *liṅga-śarīra* or *sūkṣma-śarīra* is conceived as the receptacle of psychic elements (ROŞU 1978, p.208). Also cf. PREISENDANZ 1994 (Teil 2), pp.239–240.

414 Read *tad ā mokṣād.*

415 "Minute" qualifies all of the "element[s], facultie[s]/organ[s] and vital winds".

416 Cf. WINDISCH 1908, p.80: "Der feine Leib der Seele wird also gebildet vom Lebensodem, den fünf Sinnesorganen, dem *manas* genannten geistigen

[The one] having no birth[417] creates this world for the experience of individual selves. (śl.9ab)

And he, the self, [namely] the supreme self, then (*atha*), destroys [the world] for repose.

So, this creation and destruction is agreed upon as [being] beginningless in a stream/continuity (*pravāha*). (śl.9cd–10)

Comm. K on SR śl.6–10

"**Covered with beginningless ignorance**" (śl.6a, *anādya-avidyopahitāḥ*). **Beginningless ignorance** [means] the root ignorance; it means just (*iti yāvat*) not knowing of the true nature of the base (Or: not knowing the base as it is).[418] **Covered (*upahita*) with by** that **ignorance**, [namely] determined/limited (*avacchinna*)[419] [by that ignorance], they become [those] **called "individual selves"**. The meaning is that (*iti*) the name "individual self", pertaining to ignorance[420], [occurs] of the thing whose nature is the joy of the true intellect (*sac-cid-ānanda*) and the omnipresent, in [the case of] false attributing/superimposing (*āropa*) of properties (*dharma*) of un-truth/nonexistence, non-omniscience, the state of having pain, separateness etc., by the power of the imposed property (*upādhi*) of ignorance.

Zentralorgan, dem Wissen, dem Karma, der Erinnerung, umschlungen von den fünf Elementen in ihrer gleichfalls nicht sinnlich wahrnehmbaren transzendenten Form."

417 *Ajaḥ* (masculine nominative sg.) denotes Brahman, but grammatically qualifies *ātman* (SR śl.10a, *sa ātmā*).

418 The Ānandāśrama ed. (1896) reads *jñāna* instead of *ajñāna*. This is not noted by the Adyar ed. This reading is obviously a mistake.

419 The expression *upādhy-avacchinna* is often mentioned in Vācaspati's Bhāmatī commentary on the Brahmasūtra 1,1,1; 1,3,7; 1,4,6. The term *upahita* is glossed as *avacchinna* in the Bhāmatī 1,1,5. According to this statement, the term *avacchinna* is usually translated as "limited" or "restricted".
The Ānandāśrama ed. (1896) reads *avicchinnā* instead of *avacchinnā*. It would mean "[…] not separated (*avicchinna*) from [ignorance]" and could also make sense as an explanation of *avidyopahitā*. However, *avicchinnā* seems to be wrong, for one finds *dāru-tṛṇa-parṇādy-avacchinnāḥ* only a couple of lines below (Ānandāśrama ed., p.12, l.9; Adyar ed., pp.29–30), and *sthūla-śarīrāvacchinnānāṃ* (Ānandāśrama ed., p.12, l.18; Adyar ed., p.30, l.15).

420 The Ānandāśrama ed. (1896) reads *āvidyaka* instead of *āvidyika*.

There he relates an example: "**Just as of fire**" (śl.6b, *yathāgneḥ*). The rest [to be supplemented] is "parts". The meaning is (*iti*): Just as (*yathā*)[421] the parts of fire whose shape is heat (*tejas*), limited by the imposed property (*upādhi*) like wood, grass, jewel[422] etc., are called "sparks"; in that manner (*tadvat*).

"**They, [governed/restrained] by deeds**" (śl.6c, *te karmabhiḥ*). **They**, the individual selves, **by deeds which give pleasure and pain, [and] whose forms are merit and sin**[423] respectively. On the one hand (*tāvat*), [there is] the state of giving pleasure, of meritorious deeds, because of the *śruti*, "Verily this, indeed, makes him (*tam*) do a rightful deed, whom [this] wants to raise/save from these worlds"[424]. And [on the other hand, there is] the state of giving pain, of bad deeds, because of the *śruti*, "Verily, this, indeed, makes him do a non-rightful deed, whom [this] wants to bring_down/let_collapse."

"**By beginningless [deeds]**" (śl.6d, *anādibhiḥ*). Because of the beginninglessness of ignorance [which is] the imposed property (*upādhi*) of the state of being the individual self of the self (*ātman*) having the fancy of being a maker/agent (*kartṛtva*) etc.[425], [there is] beginninglessness also of the deeds whose maker/agent is the individual self, in the sense (*vivakṣā*) of the form of stream/continuity, according to the rule (*nyāya*), saying (*iti*) "Through the one origin (*yoni*) of the individuals (*vyakti*)[426] tinged[427] by its/that form

421 The Ānandāśrama ed. (1896) does not contain *yathā*. This is not noted by the Adyar edition.

422 The Adyar ed. notes the variant of the Ānandāśrama ed. (1896) *tṛṇa-parṇādy-acchinnā* "not cut off from wood, grass, leaf etc." But this note is presumably false, as the Ānandāśrama ed. (1896) actually reads *tṛṇa-parṇādy-avacchinnā*.

423 The Ānandāśrama ed. (1896) reads here *puṇya-pāpa*, though it reads *pāpa-puṇya* in the *mūla*-text.

424 In the Ānandāśrama ed. (1896), this line is missing, i.e. the part after *sukha-pradatvaṃ* up to *duḥkhapradatvaṃ*. This seems to be a mistake caused by manuscript transcription.
This sentence is contained in KauṣīakiUp 3,8 (cf. RADHAKRISHNAN 1956, p.782): *eṣa hy eva sādhu karma kārayati.* But the Kauṣītaki-upaniṣad does not read *eṣa u*. But the next line runs: *eṣa u evāsādhu karma kārayati.*

425 The Ānandāśrama ed. (1896) reads *kartṛtvād* instead of *kartṛtvādy*. "[...] having the fancy because of being an agent".

426 The term *vyakti* means the individual manifestation of the world. Cf. GARBE 1917, p.277.

(*ākṛti*)[428]" etc.[429]

Governed/Restrained, being bound[430]. **Connected with this and that (= particular) birth/genus**: They (=individual selves) **attain**, reach, **a body** connected with the birth/genus of mankind etc., **life-span and experience, born from deeds, in each birth**, in birth after birth.[431] Through that very statement (*ity anenaiva*), "**governed/**

---

427 *Uparakta* "tinged". Cf. GARBE 1917, p.377 (*uparāga*). In this context, this term seems to suggest that *karman* is only "tinged" with the attribute "beginninglessness"; namely, the relation of *karman* to beginninglessness is not absolute, but merely illusory.

428 The concrete meaning of the term *ākṛti* is not clear to me. HOUBEN 1995 (p.38, p.88) states that this term is sometimes used as a synonym of *jāti* "universal, class". But I am not sure as to the applicability of his interpretation in our context.

429 The Ānandāśrama ed. (1896) reads differently: *vyaktānām ekayā vinā*. It is not noted by the Adyar edition and this does not make sense.
The commentary here explains the qualifier *anādi* "beginningless" of *karman* "deeds". What is intended is: The *jīva*-hood (*jīvatva*) is caused by ignorance, therefore ignorance is the imposed property of *jīva*-hood. Beginninglessness, which is the quality of ignorance, is automatically transferred to the deeds, because the deeds are caused by *jīva*-hood.
The reason of this explanation seems as follows: If the *karman* were essentially beginningless, it would be inevitably endless, therefore permanent. This would result in the undesirable conclusion that liberation (*mokṣa*) from the *karman*-s, which is permanent, therefore indestructible, is impossible. In order to avoid this undesirable conclusion, the commentary tries to weaken the meaning of the adjective "beginningless" (*anādi*). The commentary explains that beginninglessness is not an intrinsical attribute of the *karman*, but one which is transferred to the *karman* from ignorance. That means, the *karman*, which is impermanent, has some permanence in it, but this does not suggest anything about the ontological status of the *karman* itself. Therefore, if ignorance is eliminated, the *karman* also becomes extinct. The permanence, to which the *karman* is related, is merely a relative, streamlike one (*pravāha-nitya* "permanent like a stream"), and not the absolute, unchanging absolute permanence. On the notion of *pravāha-nitya*, cf. HOUBEN 1995 (p.248; p.372, note 727).
The term *uparakta* seems to mean the transfer of an attribute from a substance to another. Thus, this term in this context suggests that the *karman*-s are merely tinged by, i.e. secondarily provided with the attribute "beginninglessness".
Unfortunately, I was not able to identify this *nyāya*.

430 The Ānandāśrama ed. (1896) lacks *santaḥ* "being", and reads *tatas tattajjātir* instead of *tattajjātiyutaṃ*.

431 The comm. K interprets, in this verse, *tattajjātiyuta* as qualifying *deha* and *āyus*, but *karmaja* as qualifying *bhoga*. This commentary gives an explanation

restrained by deeds"[432], when the origination of the bodies etc. from deeds too is proven without [special] qualification in connection with attainment of the body etc.[433] by the individual selves, the qualification "born from deeds" of experience is, furthermore, to express that (*iti*) the individual selves' state of being the agent is very valid even for the deeds being done in this birth, whose results are seen[434], [namely] agriculture etc.

"**They have**" (śl.8b, *teṣām asti*). The meaning is that (*iti*): **they**, the individual selves limited by a gross body (*sthūla-śarīra*), have **another** [body] other than the gross body, [namely] the **minute**, unmanifest, **subtle-body** (*liṅga-śarīra*); the meaning is that (*iti*) it is proved because of the *śruti* saying (*iti*) "A man (*puruṣa*) [whose] measure is that of the thumb" (Taittirīyāraṇyaka 10,38,i(a), Kaṭha-upaniṣad 4,13) etc.

And **until liberation**, until seeing directly the ultimate nature, **it** (= the subtle body) **is deemed**, proved in the *śāstra* [to be], **indestructible**, not having destruction because of non-extirpation of the own ignorance and karmic imprint (*vāsanā*)[435]. And **this** subtle body is that **whose nature is the state of the minute element, organ and vital winds. The elements (*bhūta*)**, earth etc., **the facul-**

as follows: it is obvious that *deha* and *āyus* are *karmaja,* therefore these two do not need the qualifier *karmaja*. In contrast, *bhoga* needs the qualifier *karmaja*, because the matter is not obvious.

432 The Ānandāśrama ed. (1896) reads *karmabhir yantritā* instead of *karmabhir niyantritā*. This does not accord with the *mūla* text.

433 The Ānandāśrama ed. (1896) reads a genitive (*dehādi-prāpter*) instead of a locative (*dehādi-prāptau*).

434 The Ānandāśrama ed. (1896) reads *kriyamāṇānirdiṣṭa-phalakāni* instead of *kriyamāṇāni dṛṣṭa-phalakāni*, and *pratijīvāna-kartṛtvam* instead of *pratijīvānāṃ kartṛtvam*. Neither of them makes sense. The editor himself marks the latter with a question mark.

435 Ignorance (*avidyā*) is the cause of reincarnation. Cf. HALBFASS 2000, p.252, "Das karmische Fehlverhalten, das uns an den Saṃsāra bindet, hat seine Wurzeln in der Begierde; dies wiederum ist nach vorherrschender Ansicht durch das Fehlwissen (*avidyā*), das Mißverständnis der eigenen Identität, bedingt."
Vāsanā-s are the impressions which were obtained in previous lives and are stored in the psyche (*citta*). These impressions influence the psyche in this and coming lives, causing the corresponding *karma*-s to manifest themselves (cf. HALBFASS 2000, pp.158–159).

ties/organs (*indriya*), eyes etc., **the vital winds (*prāṇa*)**, the five winds[436], [namely] *Prāṇa* etc.; **the state (*avasthā*)**, [i.e.] the state (*avasthāna*), of these **minute** ones, whose form is unmanifest, is united[437] with the internal_ instrument/organ (*antarindriya*), [namely] with the mind. [That] whose **nature (*ātman*)**, nature (*svarūpa*), is [such] a state (*avasthā* = *avasthānam*)[438]; that is mentioned in that manner.

Some agree on the eternity of the world itself in the form of a stream/continuity (*pravāha*), saying (*iti*) "The world is never other than such". For the sake of refutation of that opinion, he relates the self's state of being the maker/agent[439] of the creation and destruction of the world, [being endowed] with a motive (*prayojana*): "**Of individual selves**" (śl.9c, *jīvānām*). **The world**, whose nature is the [gross] element[s] and what pertains to the [gross] elements (= is material), is **for the experience** (*upabhoga*), for the experience (*anubhava*) of pleasure and pain, **of the individual selves** having the conceit/conception (*abhimānin*) of the two mentioned bodies. [That] **having no birth** [means] the supreme self. "He, the self" is the

436 The Ānandāśrama ed. (1896) reads *prāṇādipañcavāyavaḥ* instead of *prāṇādayaḥ pañca vāyavaḥ*. This is not noted by the Adyar ed.

437 Lit. "with the state of being united" (*saṅghātatvena*).

438 The Ānandāśrama ed. (1896) reads *saṅghātatvenāvasthānam* instead of *saṅghātatvenāvasthā avasthānam*. Namely, the *pratīka*, *avasthā*, is missing. This is not noted by the Adyar ed.

439 Namely, the self's being an agent. That means that the self is the agent of the creation and destruction of the world.
The commentator's logic might be as follows: Since the self as an agent creates the world, the world can not be beginningless; therefore the world is not eternal. However, this seems to contradict the statements in the last part of this commentary (p.39, ll.9–13). According to these statements, creation and destruction are beginningless, and the state of being an individual self is also beginningless. I can not solve this problem. Or, is the commentary p.39, ll.1–3 to be interpreted as "some never agree (*na kadācid [...] aṅgīkurvanti*) on the etermity of the world itself in the form of a stream/continuity saying 'the world is not such (*anīdṛśaṃ jagat*)'"?
The Ānandāśrama ed. (1896) reads *ātmakatvam* instead of *ātmakartṛkatvam*. This is not noted by the Adyar ed. The editor of the Ānandāśrama ed. corrects it into *ātmakṛtatvam* "the state of being made by the self". This could also make sense, meaning: "the state of the creation and destruction of the world being made with a motive by the self (*ātmakṛta*)."

sequence [of words].[440] The meaning is that (*iti*), even the supremacy of the self, [namely] the state of being the Lord (*īśvaratva*), is obtained with reference to the lowly individual selves.[441]

By (*iti*) "**he, the self**", Brahman mentioned before is referred to. The word "and" (*ca*) is out of order (*bhinnakrama*). "**And (*ca*) destroys**" is the sequence [of words].[442]

Then (*atha*), immediately after the stabilisation of the created world. **For repose**: "Of individual selves" is to be supplied (*anuṣañjanīya*), because of the creation's and destruction's being for the sake of the individual selves in the mentioned manner.[443] He relates their (dual) beginninglessness in a stream/continuity for making the individual selves' beginninglessness clear[444]: "**So this**" (śl.10c, *tad etad*). "**Creation and destruction**" [as a compound, i.e.] creation as well as destruction[445]: the neuter gender (*napuṃsakatva* lit. "the state of being neuter") is with reference to the state (*bhāva*) of a pair being like one.[446] "**Beginningless in a stream/continuity**" (*pravāhānādi*): The meaning is that (*iti*) [there is] being with a

440 I.e., this commentator considers *ajaḥ* to qualify the following *sa ātmā*.

441 The Ānandāśrama ed. (1896) reads *āpanna ity arthaḥ* instead of *āpannam ity arthaḥ*. This is not noted by the Adyar ed. If we take this reading, the implicit subject of the sentence would be *ātman*. In this case, *āpanna* which is the perfect participle of an intransitive verb cannot be considered passive. So we would have to translate it as: "[The self] obtains/attains even supremacy, [namely] the state of the Lord, with reference to (despite) the lowly individual self (*jīva*)." The problem of this reading is that *ātmano* becomes superfluous.

442 In this commentator's opinion, the conjunction *ca* is set in an anomalous order; *saṃharati ca* would be a correct order.

443 The Ānandāśrama ed. (1896) reads *bījānām* instead of *jīvānām*. This is not noted by the Adyar edition. It would mean: "of the seeds [from which the world sprouts]". This edition puts a *daṇḍa* after *anuṣañjanīyam*, while the *daṇḍa* after *jīvārthatvāt* is missing. With this, the meaning of the commentary would be different from the Adyar ed. It would mean: " "Of the seeds" is to be supplied. Because of the creation's and destruction's being for the sake of the individual selves in the mentioned manner, he (= the author) relates […]" etc.

444 The Ānandāśrama ed. (1896) reads *jīvānāditva-dyotanīyapravāhānāditva* "beginninglessness in a stream/continuity, to be made clear by the individual selves' beginninglessness".

445 The *dvandva* compound is decomposed: *sṛṣṭiś ca saṃhāraś ca*.

446 I.e. the coordinative (*dvandva*) compound *sṛṣṭi-saṃhāra* is n. sg., because creation and destruction are considered one single pair.

beginning[447] in the sense of (*vivakṣayā*) the individual (*vyakti*) [and] beginninglessness through the form of a continuity (*santāna*).

Comm. S on SR śl.6–10

[With regard to the question,] "Is it not (*nanu*) an unequal example, because of the sparks' difference through the mixture of opposite attributes of grossness and minuteness etc.?", he therefore says [as the answer]: "**Wood etc.**" (śl.6, *dārv-ādi*). By the word ["]etc.["], [there is] mention (*graha*) of the imposed property (*upādhi*), water, of the fire of lightning of the submarine flame (*abindhana*).[448] And consequently, the meaning is that (*iti*), with regard to sparks etc., the difference pertaining to the imposed property (*upādhi*) is not essential.

He relates those individual selves' being tied to the deeds: "**They by deeds (*karman*)**" (śl.6d, *te karmabhiḥ*). With regard to [the question], "Why indeed (*nanu*) [is there] a tie to the deeds, of the individual selves not separated/differentiated from Brahman?", he therefore says: "**By the beginningless [deeds]**" (śl.6d, *anādibhiḥ*). **Those** individual selves **in each birth obtain a body connected with this and that [particular] birth/genus (*jāti*)**, joined with the genus (*jāti*) of humanness, cowhood etc., **a life span and enjoyment (*upa*bhoga)**[449] **born from** the affairs of **the deeds** (*karma-vyāpāra-ja*) in accordance with this and that (= particular) genus (*jāti*). With regard to this (*atra*)[450]: "Why indeed (*nanu*) [is there] the acquisition of many births (*janma*) with regard to the individual selves' being tied to the gross body[451], because of (i.e. despite) the gross body's attaining destruction?"[452], he says: "**They have another**" (śl.8b, *teṣām*

447 The Ānandāśrama ed. (1896) reads *anāditvaṃ* instead of *sāditvam*. This is not noted by the Adyar ed. This reading does not make sense.

448 Or: "of the fire of the submarine flame [and] of lightning".

449 The comm. K makes the meaning of *bhoga* of the *mūla* text precise through *upabhoga* "enjoyment".

450 B reads *ity ata āha* instead of *ity atrāha*. "Hence".

451 A and B read *dehātmatve*, *dehātmakatve* respectively, instead of *deha-nibaddhatve*. "With regard to the individual selves' state of having the nature of the gross body", i.e. the opponent asks whether the individual self consists of the gross body. This is refuted by the author.

452 I.e., why are the individual selves reborn, even though the gross body, to which they are attached, is mortal and perishable.

asty aparam). **Until liberation**, making liberation the limit. He says: With regard to "Of what manner is the subtle body?" "**Minute**" (śl.9a, *sūkṣma*).[453] The meaning is that (*iti*) [there are] five minute elements (*tanmātra*) [and] eleven faculties/organs (*indriya*)[454]; thus it (= the body) is that which has sixteen parts (*kalā*).

[With regard to the question,] "What purpose/business (*prayojana*) indeed (*nanu*) is there of (i.e. in) the Supreme Lord's creation and destruction of the world?", he therefore says: "**[For the enjoyment/experience] of individual selves**" (śl.9c, *jīvānām*). [Thinking (*iti*),] "And because of the beginninglessness of the two, this is not a [valid] accusation[455] (*paryanuyoga*)", he says: "**So, this**" (śl.10c, *tad etad*). The combination of creation and destruction, [namely] **creation and destruction** [as a copulative compound], is **agreed upon** as [being] beginningless in a **stream/continuity**.

SR śl.11–12ab

Those individual selves (*jīva*) are not different from the self (*ātman*). The world is not different from the self (*ātman*), either (*vā*). Creating through the power (*śakti*), that (= self) is not different, like gold from an ear-ring.[456] (śl.11)
Others [say] that (*iti*) [it (= the self)] creates[457] through ignorance, like a rope[458] [creates the illusion of] a snake.[459] (śl.12ab)

453 The reading is not *sūkṣmam iti* but *sūkṣmeti*. That means that it refers to *sūkṣma-* in śl.9a and not in śl.8b.

454 According to the Sāṅkhya school, the faculties/organs are eleven: five faculties/organs of perception (*jñānendriya*), five faculties/organs of action, and the mind. Cf. GARBE 1917, pp.319–320.

455 B reads *paryanuyogya* instead of *paryanuyoga*. "It (or: He) is not to be accused".

456 The Ānandāśrama ed. (1896) reads *kuṇḍalādivat* instead of *kuṇḍalād iva*. The comparison of *ātman* to a golden ornament is found in YS 3,147 and in Śaṅkara's Śārīrakabhāṣya, Vedāntasūtra 2,1,14 (as informed by KANE 1990, vol.I, part I, p.447). Also Bhartṛhari's Vākyapadīya 3,125b (=3,2,15b) mentions it.

457 *Sṛlati* is obviously a mistake for *sṛjati*.

458 D reads *rajvā* (obviously a mistake for *rajjvā*), "like [someone creates] a snake through a rope (= mistakes a rope for a snake)".

459 The simile of a rope (i.e. Brahman) which is mistaken for a snake (i.e. the world) occurs in the Vedānta school, cf. DEUSSEN 1883, p.269.

Explaining the non-differentiation of the [supreme] self (*ātman*) and individual selves (*jīva*) by the relationship (*bhāva*) of the parts and that containing the parts (= the whole), he (= the author) relates the non-differentiation of the self and the world in (lit. "through") the relationship of result[460] and cause, according to the doctrine of differentiation and non-differentiation, with an example[461]: "**Those individual selves**" (śl.11a, *te jīvāḥ*). **That (*asau*)**[462], [namely] the self (*ātman*), **creating the world through the power**, [namely] through the power of action belonging to itself, **is not different** from it (= the world). The connection of (= with) the word "the world" is to be made by a change of the case ending.[463] Like the identity of **gold** and **an ear ring**, just so [it is]. Thus (*iti*) is the meaning. As they relate "The state of diversity is in (lit. "through") the form of (= due to) the result, non-differentiation through the nature of the cause, like non-differentiation through the nature of [being] gold, [but] differentiation through the nature of [being] an ear-ring etc." Resorting to the principal doctrine of the Vedāntin-s, he says[464]: "**[It] creates**" (śl.12a,*sṛjaty*). **Others** [means] Vedāntin-s.[465] **Through ignorance**, [namely] through non-knowledge of the basis (*adhiṣṭhāna*). With regard to that (*tatra*) [there is] an example: "**Like a rope a snake**"

460 The term *kārya* literally means "what is to be done".

461 The Ānandāśrama ed. (1896) lacks *ca* in *jīvānāṃ ca abhedam*. It also reads *jagataś cābhedamatānusāreṇa* instead of *jagataś cābhedam bhedābheda-matānusāreṇa*. These are not noted by the Adyar ed.
With this reading, it would be translated as: "[...] he (= the author) relates, with an example, according to the doctrine of non-differentiation between the self and the world in the relationship of result and cause".

462 *Asauḥ* is obviously a mistake for *asau*.

463 This sentence explains the foregoing *tasmād abhinnaḥ*. The term *abhinnaḥ* "not different" is to be supplemented by the ablative case of *jagat* "the world". The reader should "change the case ending", that means, change an accusative into an ablative *jagataḥ*. Thus *abhinnaḥ* implies "not different from the world".

464 The Ānandāśrama ed. (1896) reads *iti mukhya-vedānti-mata-tattvaṃ darśitam*, omitting the following part up to *brahma-vivartatvaṃ darśitam*. This is probably a corruption. It is noted by the Adyar ed. (p.32, variant 5) in an insufficient manner.

465 SR śl.12ab deals with the theories of *kevalādvaita*, *māyāvāda* (cf. THIBAUT 1890, i, p.cxx) and *vivartavāda* (ibid., i. p.xcv).

(śl.12b, *rajjur bhujaṅgamam*). By this, the manifested world's (*prapañca*) state of being the evolution of Brahman (or: the evolution due to Brahman) (*brahma-vivartatva*) is shown, because of [the manifested world's] being otherwise (*anyathā-bhāva*), due to [the manifested world's] being untrue/unreal (*atattva*) (i.e. not being as it seems).[466]

Comm. S on SR śl.11–12ab

He relates, with regard to whether he, the Supreme Lord (*parameśvara*), is different or not different: "**Like fiery energy (*tejas*)**".[467] **Creating the world through the power of the self** (i.e. through his own power), he is **not different** [from the world]. **That** (*asau*), [namely] the Supreme Lord. Through what sort of power? **Through power, different and not different from the self, like fiery energy**. Like fiery energy, regarded as (*iti*) different from the sun, is called "the fiery energy of the sun", [and regarded as] not different, is also indicated (*vyapadiśyate*) by "the sun, [namely] the fiery energy". (Or: "the sun [is] fiery energy") He mentions another

466 It seems to mean the following: by this, it is shown that the illusory manifestation of the world is evolved by Brahman; the world is [in reality] different [than what it seems to be], because it is not true/real (*atattva*).
However, the compound *anyathābhāva* could be analysed as *anyathā-abhAva* "not being otherwise". Actually the editor of the Adyar ed. seems to interpret in this way, as he puts *bhāvābhāvād* in parentheses after *anyathābhāvād.* He understands this compound to mean *anyathābhāvābhāvād* (= *anyathā-bhāva-abhāvād*) "because of the lack of being otherwise". With this, we would have to interpret this sentence in a different way: the world is evolved by Brahman; the world is [in reality] nothing else [than Brahman], because it is not that what it seems to be (*atattva*). This interpretation seems to me to fit better to the Advaita-Vedānta theory. Thus, *mukhya-vedānti-mata* "the doctrine of the main [stream of the] Vedāntin-s", two lines above, seems to denote the Advaita-Vedānta.
The Ānandāśrama ed. (1896) omits the part between *mukhya-vedānti-mata* (cf. Adyar ed., p.32, l.6 from the bottom) and *tattvaṃ darśitam* (ibid, p.32, l.3 from the bottom).

467 The comm. S reads SR śl.11ab differently from the *mūla* text. It reads *tejovad ātmano bhinnābhinnayā* instead of *te jīvā nātmano bhinnā bhinnaṃ vā*. This *śloka* then means: "Like fiery energy, different and not different from the self, creating the world through the power of the self, that (= the self) is not different, like gold from an ear-ring."

example: "**Gold**" (śl.11d, *suvarṇaṃ*). **Gold** is **not different from an earring** because of [the earring's] being gold, and **different** because of being an earring. He relates the [illusory] manifested world's pertaining to ignorance according to the Vedāntin-s' opinion: "**[It] creates**" (śl.12a, *sṛjaty*). The meaning is that, **like a rope** creates an imagined **snake**, from (= through) one's own non-knowledge, [namely through the false opinion:] "this is a snake", so the Supreme Lord too, through non-knowledge of one's own being the basis/substratum (*adhiṣṭhāna*), [creates] the [illusory] manifested world of the space (*ākāśa*) etc.

SR śl.12cd–13

From the self (*ātman*) first space (*ākāśa*), from it wind, from it fire, (śl.12cd)
from fire[468], water, from this, earth was born.

Those are the gross elements (*mahābhūta*). This is the body of the supreme intellect (*virāj*), of Brahman. (Or: Those gross elements are this body of the supreme intellect, of Brahman.) (śl.13)

Comm. K on SR śl.12cd–13ab

Through the order of creation of the elements (*bhūta*) etc., he (= the author) observes the human body (*piṇḍa*) [which is] material (*bhautika* lit. "pertaining to the elements") with its differences, by means of "**From the self first**" (śl.12c, *ātmanaḥ pūrvam*) etc. **From the self**, [this] being the cause, **at first**, before wind etc., **space (*ākāśa*)** was produced[469]: thus (*iti*) is the meaning. "**From it**": by the word "it", space is pointed to. From it, from space, [this] being the cause, wind was produced: thus (*iti*) is the meaning. "**From it fire**" etc. Here also, the meaning is that (*iti*) fire was born from wind. And thus (= in the same manner) (*tathā*) [says] the *śruti*, "From the self spacee is arisen, from space, wind." (Taittirīya Up. 2,1) **Those**, [namely] space etc., **the gross elements**. The relation [between the words of the sentence] is thus (*iti*): **this is the body (*tanu*) of the supreme intellect (*virāj*)**, [namely] of Brahman [which is] limited

468 D reads *tataś ca jalam* "from it" instead of *analāj jalam*.
469 The Ānandāśrama ed. (1896) reads *utpadyata* "is produced" instead of *utpanna*.

(*upahita*) by the gross elements[470], **of Brahman.**[471]

"This" (*eṣā*) is an indication (*nirdeśa*) of the aggregate of the gross elements, because of [their already] being [mentioned] before, being evident for all people.

"Of the supreme intellect" (śl.13d, *virājaḥ*): The self, creating the gross elements, entering them, is called the supreme intellect (*virāj*), because of thinking itself to be them (*tad-abhimānin*)[472]. Of it. **Those (*amūni*, n. pl.) are this (*eṣā*, f. sg.) body**: "Pronouns set forth the oneness between that being pointed at (*uddiśyamāna*) and that being referred back to (*pratinirdiśyamāna*), [and] occasionally (*paryāyena*) become possessor[473] of each other's gender."[474] Thus (*iti*), [there is] the appearance of the state of different genders.

Comm. S on SR śl.12–13

He relates the order of creation: **"From the self"** (śl.12c, *ātmanaḥ*). He says that (*iti*) the body of Brahman (*brahmaṇas tanur*), the supreme intellect (*virāj*), is of the nature of the five gross elements[475]: **"The gross elements"** (śl.13c, *mahābhūtāni*).

470 The Ānandāśrama ed. (1896) reads *mahābhūtasyopahitasya* instead of *mahābhūtopahitasya*. But it seems unsuitable that *mahābhūta* is in the same case as *brahman*.

471 The terms in boldface are those quoted from the *mūla* text. The commentator arranges them in a word order which is easier to comprehend.

472 *Tad-abhimānin* "regarding them as referring to its own self".

473 *Bhāñji* : neuter pl. nominative of the radical noun of √*bhaj*.

474 I was not able to identify the source of this citation.

475 The noun *virāj* "supreme intellect" is m. sg. nominative. It can only be in apposition to *tanuḥ*. Otherwise, we could read a *karmadhāraya* compound *virāḍ-brahmaṇas*, meaning either "the excellent Brahman" (in this case, *virāj* would be an adjective) or a *dvandva* compound "the supreme intellect and Brahman". The latter case seems unsuitable, as this compound is sg. genitive. I prefer the former, and so I would translate: "He says that (*iti*) the body of the excellent (*virāj*) Brahman is of the nature of the five gross elements."
The term *virāj* could be a female noun meaning "excellence, dignity", in which case, we would have to translate: "the body of Brahman is excellence/dignity".

SR śl.14

Brahman created Brahmā[476], and, bestowing on him the Vedas, it (= Brahman) created that which is material from the words of the Veda[s] through him (Brahmā).[477]

Comm. K on SR śl.14

"**Brahman** (nominative), **Brahmā** (accusative)" (śl.14a, *brahma brahmāṇam*): **Brahman** whose form is the supreme intellect (*virāj*) **created Brahmā**, [i.e.] the Four-faced. Again, it, [i.e.] Brahman, **bestowing the Vedas**, [i.e.] Ṛgveda etc., **on him**, [i.e.] the Four-faced, **created that which is material**, [i.e.] the body and faculties/organs etc., whose nature is (= which is consisting of) the five gross elements made fivefold[478], **through him**, [i.e.] through the one to be employed, the [divine] maker himself [being] the employer, **from the words of the Vedas**, [these being called] Veda because (*iti*) justice (*dharma*) and injustice (*adharma*) are (literally: is) instructed (*vedyate*)[479] through/by it, the words in it (= the Veda) expressive of form etc.; from them (= the words of the Vedas), [these] being the instrumental cause[480] due to[481] the[ir] being recallers of substances passed away in

476 Brahmā. In order to make the distinction between the neuter Brahman, i.e. the Supreme Principle, and the male god Brahman clear, I adopt the nominative form Brahmā for the latter.

477 Bhartṛhari's Vākyapadīya 1,124ff (RAU 2002, p.28) states that the world was made from word. It mentions also the *prajāpati*-s.

478 *Pañcīkaraṇa*: Of a *bhautika padārtha* (material matter), a half portion consists of its own kind of *mahābhūta* (gross element) and the other half consist of the other four *mahābhūta*-s. Thus a material matter contains all five *mahābhūta*-s. For example, one half of the earth matter consists of the *mahābhūta* of earth, and the rest consists of the other four *mahābhūta*-s, water etc., cf. BHATTACHARYA 1986, p.37. Also cf. DEUSSEN 1883, p.241, note 101, on Vedāntasāra 124.

479 The term *veda* is etymologically explained. *Vedyate* is the passive of causative.

480 The *nimitta* is the instrumental cause ("Veranlassung") in contrast with the material cause (*upādāna*), cf. GARBE 1917, p.293.

481 The Ānandāśrama ed. reads *smārakakartuṣā* instead of *smārakatvena*. The editor corrects it into *smārakatvāt*. It does not make a great difference in the meaning: the words of Vedas are the cause, "because of their being recallers" (-*smārakatvāt*), or "through their being recallers" (-*smārakatvena*).

another era (*kalpa*). The meaning is: "[Brahman is] the one having made[482] to carry out the creation of that which is material."[483]

Comm. S on SR śl.14

It is said: "Is it not (*nanu*) the general knowledge of people that (*iti*) only Gold-Embryo (*hiraṇyagarbha*)[484] is the maker (*kartṛ*) of the world? How is the Lord (*īśvara*) the maker?" Hence he says: "**Brahman**" (śl.14a, *brahma*). **Brahman** (n. nom. sg.), [i.e.] the Lord (*īśvara*)[485], **created Brahmā** (m. acc. sg.), [i.e.] the Gold-Embryo, **and bestowing on him the Vedas, created that which is material**, [i.e.] the manifested world, directed/set_in_motion the creating Brahmā. "**From the words of the Veda**": the ablative is [used] in the case of the elision (*lopa*) of the gerund (*lyap*).[486] And then, making [him] remember the words of Vedas, [it (= Brahman)] made [him] create[487]. From the *śruti,* "[He], who first directs (*vidadhāti*) Brahmā, [he] who indeed sends forth the Vedas for him."[488] And thence the meaning is thus (*iti*): the state of being the maker (*kartṛtva*) is the Gold-Embryo's (= Brahmā's), the state of being the causing/employing maker is the Supreme Lord's (Brahman's).

SR śl.15–16

By its (= Brahman's) order, Brahmā (m. sg.) created the *prajāpati*-s just with [his] mind.[489] The seminal (*raitasa*) creation of bodies from them, however, is being investigated. (śl.15)

482 The commentary glosses the periphrastic perfect *sarjayām āsa* by *sṛṣṭiṃ kāritavat.*

483 The Ānandāśrama ed. (1896) omits the part beginning with *nimittabhūtebhyaḥ* up to *ity arthaḥ*. This is not noted by the Adyar ed.

484 Comm. S identifies *hiraṇyagarbha* with Brahmā.

485 Comm. S identifies *īśvara* with Brahman.

486 The commentary considers the ablative *vedaśabdebhyaḥ* "from the words of the Veda" to stand for *vedaśabdān smārayitvā* "having reminded [Brahmā] of the words of the Veda". The commentary explains it to be the elision of the gerund *smārayitvā.*

487 The commentator glosses *sarjayām āsa*, periphrastic perfect for causative, through *asarjayat*, causative imperfect.

488 Not found in the Vedic Concordance.

489 This myth is contained in Manusmṛti 1,34.

The body is fourfold/[of]_four_kinds[490], because of a difference of origins, [namely] sweat, sprouting (*udbheda*), the foetal envelope (*jarāyu*)[491] [and] the egg. [The body] of lice etc. is from sweat, but of creeping plants etc.[492] is from sprouting. (śl.16)

Comm. K on SR śl.15–16

He tells the order of creation of that which is material from the personified maker, [i.e.] from the Four-faced (= Brahmā)[493]: "**By its order**" (śl.15a, *tad-ājñayā*). **By its**, [i.e.] the supreme intellect's, **order. From them**, [i.e.] from the *prajāpati*-s, **the seminal** (*raitasa*), [i.e.] that which has the form of a transformation (*vikāra*) of semen (*retas*). The [procreatory-menstrual] blood, too[494], is implied by the

490 The four kinds of birth are mentioned in ŚG 8,3 (cf. COMBA 1984, p.189ff), though it is not a verse identical to SR śl.16ab. However, it adds a kind called *mānasa* to these four kinds (cf. COMBA 1984, p.226). ŚG 8,3: *jarāyu-jo 'ṇḍa-jaś caiva sveda-jaś codbhij-jas tathā / evaṃ catur-vidhaḥ prokto deho 'yam pāñcabhautikaḥ //.*
SR śl.16cd–17ab shows an example for each kind of birth. The same topic is contained in ŚG 8,12 which is not parallel to the SR.
For the origin of the classification of living beings according to the manner of birth which goes back to Aitareyopaniṣad 5,3, see ROŞU 1978, p.181, and the references given in his footnote 4.

491 The term *jarāyu* means in very precise terminology "chorion" in contrast to *ulba* "amnion", as evinced by Aitareya-brāhmaṇa 1,3, 16–18. But the two terms are often equated. Often *jarāyu* is used in the sense of "afterbirth" and "foetal envelope", too. The meaning "uterus, womb" is a very late meaning. Cf. DAS 2003A, p.546 (*jarāyu*), p.528 (*ulba*) and p.512 (*aṇḍa*). For the distinction between *jarāyu* and *ulba*, see also ŚB 6,6,1,24.
Following DAS (ibid.), it would not be suitable to translate *jarāyu* here as "womb", but rather to take it in the meaning "foetal envelope".

492 The Adyar edition notes the variant of C.E. *latādayaḥ* instead of *latādinaḥ*. "[…] but creeping plants are from sprouting". Similarly, for the comm. S, the manuscripts A and B read *yūkādaya iti* "Namely (*iti*) lice etc. are […]" instead of *yūkādina iti*.
These forms with *ādinaḥ*, which I translate as genitive sg., might perhaps be falsely used as nominative pl.

493 The Ānandāśrama ed. (1896) lacks *sākṣātkartuḥ caturmukhād*, and reads *bhautikasrṣṭi-kramam* instead of *bhautikasr̥ṣṭeḥ kramam*. This is not noted by the Adyar ed.

494 The Ānandāśrama ed. (1896) lacks *api*. This is not noted in the Adyar ed.

word "semen" (*retas*). "**Sweat, sprouting**": the meaning is clear.[495] In "**of lice etc.**" etc., [the meaning is] "that whose first/beginning is lice", a *bahuvrīhi* [compound], through which there is the perception/understanding of its (i.e. the qualified one's) qualities (*tad-guṇa-saṃvijñāna*)[496].

Comm. S on SR śl.15–16

With regard to [the question:] "What did Brahmā create?", he says: "**By its order**" (śl.15a, *tad-ājñayā*). **The *prajāpati*-s**, [namely] *dakṣa* etc. With regard to [the question:] "What is born from them?", he says: "**From them, however**" (śl.15c, *tebhyas tu*). Born from semen, [namely] **seminal**. He relates the fourfoldness **of bodies**: "**Sweat, sprouting**" (śl.16a, *svedodbheda-*). With regard to [the question:] "What is born from what?", he says: "**Of lice etc.**" (śl.16c, *yūkādinaḥ*).

SR śl.17

[The body] of humans etc. is from the foetal envelope (*jarāyu*), but [the body] of birds etc.[497] from the egg. With regard to that, the human body is mentioned, because of being suitable for sound (*nāda*).[498] (śl.17)

Comm. K on SR śl.17

Among the mentioned fourfold bodies, [there is no sound] in three kinds, [i.e.] the bodies of lice etc.[499], because of the unconsciousness

495 The Ānandāśrama ed. (1896) reads *spaṣṭārthaḥ* instead of *spaṣṭo 'rthaḥ*. This is not noted by the Adyar ed.

496 *Tad-guṇa-saṃvijñāna*. It means a type of *bahuvrīhi* compound in which qualities denoted by the name are perceived along with the thing itself. Monier-Williams' Dictionary refers to Śaṅkara's Bhāṣya 1,1,2.

497 The Adyar ed. notes the reading of C.E. (= Calcutta Edition?) *vihagādayaḥ* "birds etc. are […]" instead of *vihagādinaḥ*.

498 Thus the SR, which is a musicological text, explains the reason of dealing with the human body with priority, stating that the human body is suitable for producing sound. But its parallel text, the ŚG, also mentions the priority of the *jarāyuja* beings, without explaining the reason, cf. COMBA 1981, p.189.

499 The Ānandāśrama ed. (1896) reads *trividhayūkādideheṣu* instead of *trividheṣu yūkādideheṣu*. This is not noted by the Adyar ed.

of creeping plants etc., because of the extreme minuteness of lice etc., [and] because of the lack of complete element[s] (*dhātu*)[500], tube[s] (*nāḍī*)[501] etc. of birds etc. And (*ca*), [there is no sound] in those whose birth is [from] the foetal envelope (*jarāyu*) either (*api*), because of the incapability of pronunciation of the body of the animals, due to [their] being beasts (*tiryañc*), despite the existence of completeness of element[s] (*dhātu*), tube[s] (*nāḍī*) etc. By elimination (*pāriśeṣyāt*), the human body[502] alone is suitable for sound. Thinking (*iti*) [that] thence the production only of it (= the human body) is investigated, he says: "**With regard to that**" (*tatra*).[503]

Comm. S on SR śl.17

He tells the reason for the investigation with particularity only of the human body among those bodies: "**With regard to that**".

500 For the term *dhātu*, cf. DAS 2003A, p.553ff. This term is explained in SR śl.79–80a. Śāradātanaya's Bhāvaprakāśana, a dramaturgical work, contains the musicological theory that the seven tones of the octave originate in the seven elements (*dhātu*) of the human body, cf. Bhāvaprakāśana, adhikāra 7 (G.O.S. No.XLV, p.186, ll.5–8; cf. my bibliography). Keeping this in mind, the commentator mentions the elements here. Also cf. *Situating the text §1.6.* and *§3.1* footnote 265.

501 For the term *nāḍī*, cf. DAS 2003A, p.560. In the Piṇḍotpattiprakaraṇa of the SR, this term, meaning the respiratory tube, has a Haṭhayogic background. That means, the term is not mentioned in the Āyurvedic part of this section (*prakaraṇa*), but in its Haṭhayogic part (SR 1,2, śl.120–163). An exception is SR 1,2, śl.40a, in which *nābhistha-nāḍī* denotes the tube situated in the umbilical cord. The other exception is SR 1,2, śl.65b, which belongs to a quotation from a Haṭhayogic text, Yogayājñavalkya (YY), cf. *Situating the text §2.3.3.*

In the Āyurvedic part of the SR, the tubular vessels are referred to with other terms, *śirā* or *dhamanī* (SR śl.101 and following verses). The commentator Kallinātha is not conscious of this fact.

502 The Ānandāśrama ed. (1896) reads *mānuṣadeham* instead of *mānuṣaṃ deham.* This is not noted by the Adyar ed.

503 The term *nāda* cannot mean simply "sound" in general. After all, birds do make sound, too.

For this problem, see also my study of the notion *nāda* (in *Situating the text §3.1, "The Notion of nāda as Suggested by the Text and its Commentaries"*).

The knower of the field (*kṣetrajña*)[505] is situated in the space (*ākāśa*). From the space, [he] has come to the wind. From the wind, to the smoke. And from it (*tatas*), to the cloud (*abhra*)[506]. From the cloud

504 There are two ways, *pitṛyāna* and *devayāna*, through which the souls of the dead ones wander, cf. WINDISCH 1908, pp.68–69. A similar theory is contained in the embryologico-anatomical passages of the YS (see below). The ŚG's eighth chapter on embryology does not mention this early theory of transmigration at all.
YS, 3,70–72: *sargādau sa yathākāśaṃ vāyuṃ jyotir jalam mahīm / sṛjaty ekottara-guṇāṃs tathādatte bhavann api /70/ āhuty apyāyite sūryaḥ sūryād vṛṣṭir athauṣadhiḥ / tad annaṃ rasa-rūpeṇa śukratvam adhigacchati /71/ strī-puṃsayos tu saṃyoge viśuddhe śukra-śoṇite pañca-dhātūn svayaṃ ṣaṣṭha ādatte yugapat prabhuḥ /72/* (Expressions parallel to the SR are underlined). On the relationship between YS 3,72 and CA, cf. YAMASHITA 2001/2002, p.91.

505 The term *kṣetrajña* "knower of the field" i.e. "perceiver of objects" denotes *ātman, puṃs* or *puruṣa*, cf. GARBE 1917, p.355. Also cf. WINDISCH 1908, p.84. For the term *kṣetrajña* used in embryology, cf. DOSSI 1998, pp.43–44; p.115. On the notion of *kṣetrajña* in YS 3,177cd–180, cf. YAMASHITA 2001/2002, p.119.

506 Obviously *abhra* and *megha* are differentiated here as the ordinary cloud and the raincloud.
According to MAYRHOFER 1992–1996, *abhra* etymologically means "Gewitterwolke, Gewölk, Regenwolke" while *megha* means "Wolke, trübes Wetter".
For the term *megha*, the Vaidikapadānukramaṇikā (BANDHU 1976) gives only three references in the Chāndogyopaniṣad (2,3,1; 2,15,1; 5,10,6) among the principal Upaniṣad-s. In ChāndogyaUp 2,3,1 (RADHAKRISHNAN 1953, p.361), a *megha* is obviously a raincloud. In ChāndogyaUp 2,15,1 (ibid., p.369) and 5,10,6 (ibid., p.433), both *abhra* and *megha* occur. According to these two passages, a *megha* is obviously a raincloud, and the term *abhra* seems to mean the stage of a cloud before raining (ChāndogyaUp 5,10,6: *abhram* (sic) *bhūtvā megho bhavati, megho bhūtvā pravarṣati*). The passages, 5,10,5–6, deal with the same topic as SR śl.18–21, namely the transmigration of *ātman* from space (*ākāśa*) through the cloud and plants into semen. The statement of the SR accords with that of the ChāndogyaUp.
The term *abhra* also occurs in ChāndogyaUp 5,5,1 and 8,12,2. In 5,5,1, it is called the "smoke" of the fire of lightning. In 8,12,2, it is mentioned together with wind (*vāyu*), lightning (*vidyut*) and thunder (*stanayitnu*). But in this context, *abhra* has nothing to do with rain.
The same topic as in SR śl.18–21, namely the transmigration of *ātman*, is also dealt with in TaittirīyaUp 2,1, but cloud is not mentioned there. It is stated that

(*abhra*), [he] enters/reaches the raincloud (*megha*)[507]. (śl.18)

And, satiated/strengthened (*apyāyita*)[508] by the oblation, as one by whom juice/sap (*rasa*)[509] has been swallowed in the summer, the sun deposits the thick juice/sap (*ghana-rasa*; "decoction")[510] through the rays of light in the raincloud (*megha*). [Or: [...] has been swallowed in the summer through the rays of light, the sun deposits the thick juice/sap in the raincloud.][511] (śl.19abcd)

When the raincloud (*balāhaka*) rains it (= thick fluid) [down], then the individual self, together with rain, passes over unperceived (*saṅkrāmaty avilakṣitaḥ*)[512] into the trees and herbs[513] grown/born

---

ātman enters into the sky, wind, fire, water, earth, plants and food one after another, before becoming a human being.

507 The manuscript D reads *megho* instead of *meghe*. This could be translated as: "From the cloud, the raincloud evolves." But that would be a strained translation for *avatiṣṭhate*.

508 D reads *āpyāhr̥to* instead of *āpyāyito*. It would mean something like "having been offered/brought near".

509 The term *rasa* means "water", too.

510 The term *ghana* might mean "cloud". In that case, *ghana-rasa* would mean "the water of the cloud."

511 In both cases, the rays of light seem to act like straws.

512 Cf. WINDISCH 1908, p.28, "Für gewöhnlich tritt das Seelenwesen unbemerkt zum Zeugungsakt hinzu." WINDISCH also states that the verb *ava-√kram* is frequently used to describe the *jīva*'s entrance into the *garbha* (ex. *garbhāvakrānti*).
Although expressions similar to these (*saṅ-√kram* and "unperceived") are used in SR 1,2, śl.20, the SR does not state that the *jīva* directly enters the uterus, but that it first enters into herbs, then into semen. This theory is different from that of the SU (śārīra. 3,4) and CA (śārīra., 3,3). The two classical medical texts state that the *jīva* directly enters into the mixture of semen and blood.
For the invisibility of the *jīva*'s entrance into the uterus, cf. AH śārīra. 1,3, "Wie das Feuer der Sonnenstrahlen vom Kristall verborgen nicht flammend gesehen wird, so der in die Embryostätte gehende Sattva." (tr. by WINDISCH 1908, p.28). WINDISCH (ibid., note 4) explains, "Komm. *evaṃ sattvo garbhāśayaṃ gacchan, vrajan, na dr̥śyata eva.* Der Eintritt des Sattva in den Mutterleib geht unsichtbar vor sich."
A similar statement is made by Vasubandhu, cf. HALBFASS 2000, p.191, referring to Abhidharmakoṣabhāṣya 3,18ff, "Um die Prozeßnatur des Eingehens in einen neuen Körper und das Wesen der damit verbundenen Ortsveränderung zu illustrieren, verweist Vasubandhu auf die Ausbreitung des Lichts. Es gibt hier nur eine Abfolge von Momenten des Leuchtens, kein durch den Raum wanderndes Etwas. Ebenso kann der kausale Strom der Momente

(*jāta*) from the earth.[514] (śl.19d–20)

From them, food is produced. That[515], eaten (*anna*, p.p.) by men[516], becoming semen[517] (lit. "gone to the state of semen"), is

---

des Existierens sich durch den Raum fortpflanzen und die Wiedergeburt in einem anderen, räumlich entfernten Körper verursachen."
The *jīva* is extremely speedy, like thought, cf. WINDISCH 1908, p.50, referring to AH śārīra., adhyāya 2: *svakarmacoditena manojavena jīvena*, "von ihren Taten getriebene gedankenschnelle lebendige Seele".

513 As to the question whether plants have souls, there was disagreement among the orthodox schools of philosophy, cf. HALBFASS 2000, p.201. For Śaṅkara's attempt to harmonize the theory of transmigration with the Vedic theory of rain incarnation, cf. ibid., pp.170–172. For the theory of the food chain in India, cf. WEZLER 1992.

514 According to the KauṣītakiUp, if the soul of a dead fails in the entrance examination into the moon, which is the door to the heavens, it is obliged to come back to the earth in the form of rain, and to be born as an animal or human being (cf. WINDISCH 1908, p.71). ChāndogyaUp 5,10,5 states that such unsuccessful souls go from rain into plants, which become the nutrition of human beings and animals, and then turn into the semen of these beings (WINDISCH, ibid., p. 76). In these Upaniṣad-s, the moon is the stop-over station of the souls of the dead ones. According to some Upaniṣad-s, there are two ways for the souls; one is the *devayāna*, through which the souls go from the world of gods into the sun; the other is the *pitṛyāna*, through which they go from the world of the ancestors into the moon (ibid., p.65, referring to Praśna-Up 1,9–10; BṛhadāraṇyakaUp 6,2,15). In contrast to these statements, the sun is the souls' stop-over station according to the SR.
For the expression *jīva-karma-preritaṃ,* cf. WINDISCH, ibid., p.76. The *karman*-s of the previous life determine the rank and class of birth which a soul enters, e.g. *brāhmaṇa*, *kṣatriya* or *cāṇḍāla*, human being or various animals etc.
The theory of rain incarnation is contained in AV 11,4, v.3: *yát prāṇá stanayitnúnā 'bhikrándaty óṣadhīḥ / prá vīyante gárbhān dadhaté 'tho bahvī́r ví jāyante /3/* "When breath (*prāṇa*) with thunder roars at the herbs, they are impregnated, they receive embryos, they are born many." (tr. by WHITNEY 1987).

515 The manuscript D reads *tu puruṣaiḥ śukratāṃ*, "But (*tu*) food becoming semen through men [...]".
The Ānandāśrama ed. (1896) lacks *tat* and reads simply *puruṣe*. The editor corrects it into *tat puruṣaiḥ*. This is not noted by the Adyar edition. Although this does not agree with the metre it could make another sense, "Food (*annaṃ*) is produced from them. Eaten (*annaṃ*), having become semen in men [...]".

516 This reflects an early embryological theory that semen already contains the *jīva* (cf. DOSSI 1998, pp.39–42).

517 The term *śukla* here might simply mean "white, pure".

poured into the temple of love (= vagina) of a woman[518] whose procreatory-menstrual fluid is pure[519]. (Or: That food, becoming semen through men, is [...].) (śl.21)

If [it,] pure, together with the procreatory-menstrual fluid, has (lit. "becomes something") gone into the womb, then that[520], urged by the deed[s] (*karman*) of the individual self (*jīva-karma-preritaṃ*)[521], forms/begins the embryo. (śl.22)[522]

518 The manuscript D reads *śuddhārtavāyāṃ yoṣāyāṃ* "is poured into a woman whose ..., into the temple of love".

519 *Śuddhārtavā-* (SR śl.21c). The term *śuddha* refers to the good quality of the procreatory-menstrual fluid (*ārtava*), according to the definition by the medical texts, e.g. AS śārīra., 1,20; SU śārīra., 2,12. For the term "procreatory-menstrual fluid" (*ārtava*), cf. DAS 2003A, pp.14ff (also see his *Index verborum*, under *ārtava-*); SLAJE 1995.

520 The manuscripts *ka., kha.* and *ga.* read *-preritaṃ sad* "then being urged by [...], [it] forms the embryo".

521 The term *jīvakarman* implies the theory that the mixture of semen and the menstrual-procreatory fluid alone is not able to coagulate and transform itself into the embryo; in order to do so, it needs to unite with the individual self (*jīva*) (cf. DOSSI 1998, pp.125–126). A similar theory is found in the Yogācāra philosophy that if *vijñāna* did not enter the mother's womb, the father's semen and the mother's blood would not be able to coalesce or/and coagulate, so as to become the proto-embryo (*kalala*), cf. SCHMITHAUSEN 1987, Part I, pp.37; 120; Part II, p.305 (see Part II, p.652, Index, *śukra-śoṇita*).
For the texts which locate the individual self's entrance in various moments after the mixing of the two fluids, cf. DOSSI 1998, pp.134–135. The CA (śārīra., 2,33; 3,3; 4,8) seems to consider that the *jīva* or *sattva* enters the mixture of semen and blood after the two fluids are mixed. Also see DOSSI 1998, p.133.

522 ŚG 8,13 contains a wording parallel to that of SR śl.21d and 22c.
ŚG 8,13: *janma-karma-vaśād eva niṣiktaṃ smara-mandire / śukraṃ rajaḥ-samāyuktam prathame māsi tad dravam //.* The wording *niṣiktaṃ smara-mandire* is identical to SR śl.21d. The wording *janma-karma-vaśāt* is also parallel to *jīva-karma-preritaṃ* of SR śl.22c. The wording *prathame māsi tad dravam* is also parallel to SR śl.23a, *dravatvam prathamemāsi.*
The edition which COMBA 1981 used seems to read ŚG 8,3a differently, and she translates it as "per effetto del desiderio di procreazione" (COMBA ibid., p.197). But she (p.197, footnote 16) informs us of a variant (ed. A) of the ŚG, *janmakarmavaśāt*, which is the same as the reading of the Bombay edition of the ŚG (1987).

With regard to the seminal creation of it, [i.e.] the human body, he relates the individual self's entrance/transition into the uterus/embryo, by means of the succession of the space etc., through [the passage with] the beginning: "**the knower of the field**" (śl.18a, *kṣetrajñaḥ*), [and] with the end "**has gone into the womb**" (śl.22b, *garbhāśaya-gatam bhavet*): **The field (*kṣetra*)** [means] the body. **The knower of the field (*kṣetrajña*)** is the individual self (*jīva*), in the sense of (*iti*) either: ["]cognises it (= field/body) as 'I'["], or: ["]knows 'this is mine'["], as mentioned by the Lord (= Kṛṣṇa): "This body is called 'the field', son of Kuntī (= Arjuna). Knowers of that [fact] (*tadvid*) call him who knows this 'the knower of the field'." (Bhagavadgītā 13,1)

The combination of smoke, light and wind is **the cloud (*abhra*)**. Only that having water is **the raincloud (*megha*)**. "**By the oblation**": A substance [such as] fuel, food, ghee etc., being thrown accompanied by mantras into the fire intended for (*uddeśena*) the gods by sacrificers (*yajamāna*, the patron/host of a sacrifice), is an oblation (*āhuti*). **Satiated/strengthened** by it (= *āhuti*): made satisfied. **In the summer, through the rays of light (*bhānu*)**, [i.e.] through the rays (*kiraṇa*), **the sun (*bhānu*)**, [i.e.] the sun (*sūrya*), **as one by whom juice/sap has been swallowed**, [i.e.] one who has obtained sap/essence (*sāra*) from the earth. **Thick juice/sap** is water. The order [of words] (*anvaya*) is: **the trees and herbs grown/born from the earth**[523]. **Unperceived**: because of not being visible to the eyes. **From them**, from the herbs. **Food** (*annam*): because (*iti*) it is eaten (*adyate* from √*ad*), [therefore] food (*anna* from √*ad*), [namely milk-]porridge (*odana*) etc. **That**, [milk-]porridge etc., **is eaten** (*anna*), [i.e.] consumed, **by men. Whose_procreatory-menstrual fluid is pure**[524]: the procreatory-menstrual fluid (*ārtava*) is blood (*śoṇita*)[525], which has the characteristic (*lakṣaṇa*) described as: "Whose colour/likeness is [that of] the juice of lac and the blood of a hare, and which, being washed, grows discoloured (*virajyate*)" (AH śārīra., 1,18ab).

523 The relevant words do not appear in the *mūla* text in the order which the commentator gives.

524 The manuscript D reads *śuddhārtavāyāṃ* "Into [her] whose […]".

525 The manuscript D's variant, *śuddham iti* "[means] that (*iti*) the procreatory-menstrual fluid is pure", seems unsuitable.

Of her, whose procreatory-menstrual fluid is pure. "**If pure**": semen (*śukla*), too, is pure, because of being said (*uktatva*) that (*iti*) "the pure procreatory-menstrual fluid is pure"[526], beginning from "[semen,] being (*sat*) capable (*śakta*)[527], white (*śukla*), heavy, viscous (*snigdha*), sweet, thick, copious, like (*-ābha*) ghee, honey and oil, [is fit] for [producing] an embryo".[528] "**The deed[s] of the individual self**": And that semen does not, only through entering (*gatatva*) the receptacle (*āśaya*) [of the embryo] (= the uterus), form/begin the embryo, but being **urged (*preritaṃ sat*) by the deed[s] of the individual self (*jīva*)**, [i.e.] urged by the deeds, [i.e.] by the commenced deeds, of the individual self (*jīva*) having the fancy for the body to be produced, only being (*sad eva*) connected (lit. "made to unite") with the action favourable to the existence/condition (*bhāva*) of the body, **forms/begins the embryo**. Here by the word "embryo" (*garbha*), the material [cause] of the body, [i.e.] the combination of semen and blood (*śoṇita*) which has gone to (= entered) the receptacle (i.e. uterus), is mentioned.

Comm. S on SR śl.18–22

He relates the manner of [its] appearance/arising (*utpatti*): "**The knower of the field**" (śl.18a, *kṣetrajñaḥ*). **The knower of the field** is the self (*ātman*). **Cloud** (*abhra*): a non-raining piece of raincloud (*megha*) is mentioned. **Raincloud (*megha*)**: a raining raincloud.[529] Thinking (*iti*), ["]How, pray (*nanu*), is the acquisition of water in a raincloud?["], he then says: "**By the oblation**" (śl.19a, *āhutyā*).

526 Or: "Pure, pure procreatory-menstrual fluid" (*śuddhaṃ śuddhārtavam*). This is illogical. It seems to be a corruption of *śuddhaśuklārtavam* "pure semen and procreatory-menstrual fluid" which is attested in AH śārīra., 1,18c.

527 The Ānandāśrama ed. (1896) reads *śuklaṃ śuklaṃ* instead of *śaktaṃ śuklaṃ*.

528 Or: "Semen (*śukla*), being capable, heavy, [...]". Both cases are problematic. In the first case, there is no mention of semen. In the second, the white colour is not mentioned.
The Ānandāśrama ed. (1896) reads *śukla* instead of *śakta*. This makes sense: "Semen, white, heavy [...]". The edition of AH śārīra., 1,17 (Lālacandra Vaidya 1963) which I consulted contains *śukraṃ śuklaṃ guru [...]*.
A similar description of the quality of semen is found in SU śārīra., 2,11.

529 *Megha iti varṣako meghaḥ*. This commentator probably considers the term *megha* to denote an ordinary cloud; therefore, he specifies it separately: "**Cloud**: the raining cloud."

Satiated/strengthened (*āpyāyita*), [i.e.] made satisfied (*tarpita*)[530], that is to say (*iti yāvat*), contented (*santuṣṭa*), **by the oblation** bestowed in a sacrifice etc., **the sun** (*bhānu*), [i.e.] the *sūrya*, **as one by whom juice/sap has been swallowed**, [i.e.] who has taken water to himself **through the rays of light**, **deposits the thick juice/sap**, [i.e.] water, in the raincloud.

He mentions the entering of the individual self into the herbs etc. together with the water rained from the raincloud: "**The cloud (*balāhaka*), it (= the thick juice)**" (śl.19d, *tam balāhakaḥ*).

He mentions the entering of the individual self into the embryo (*garbha*) from the herbs etc. through food, semen etc: "**From them**" (śl.21a, *tābhyo*).

530 A and B read *tṛptaḥ* "[being] satisfied".

SR śl.23–27[531]

A liquid state (*dravatva*) called "*kalala*" [532] arises in the first month[533],

531 PARALLELS
SU śārīra., 3,18: *tatra prathame māsi kalalaṃ jāyate. dvitīye śītoṣmānilair abhipacyamānānām mahābhūtānāṃ saṅghāto ghanaḥ sañjāyate, yadi piṇḍaḥ pumān, strī cet peśī, napuṃsakaṃ ced arbudam iti.*
AS śārīra., 2,9: *tatra prathame māse kalalaṃ jāyate. dvitīye ghanaḥ peśy arbudaṃ vā tebhyaḥ kramāt puṃstrīnapuṃsakāni.*
AH śārīra., 1,37ab–50a: For instance, *avyaktaḥ prathame māsi saptāhāt kalalībhavet /37ab/.*
AH śārīra., 1,49cd–50a: *dvitīye māsi kalalād ghanaḥ peśy athavārbudam /49cd/ puṃ-strī-klībāḥ kramāt /50a/.*
CA śārīra., 4,9–10: *[...] prathame māsi sammūrchitaḥ sarva-dhātu-kaluṣīkṛtaḥ kheṭabhūto bhavaty avyakta-vigrahaḥ sad-asad-bhūtaṅgāvayavaḥ /9/ dvitīye māsi ghanaḥ sampadyate piṇḍaḥ peśy arbudaṃ vā. tatra ghanaḥ puruṣaḥ, peśī strī, arbudaṃ napuṃsakam /10/.* The Nirnaya ed. reads the same.
YS 3,75: *prathame māsi saṅkleda-bhūto dhātu-vimūrchitaḥ / māsy arbudaṃ dvitīye 'ṅgendriyair yutaḥ /75/.* On the relationship between YS 3,75 and CA, cf. YAMASHITA 2001/2002, pp.93–95.
AgniP 369,19: *jīvaḥ praviṣṭo garbhaṃ tu kalale 'py atra tiṣṭhati / ghanī-bhūtaṃ dvitīye tu tṛtīye 'vayavās tataḥ /19/ caturthe 'sthīni tvaṅ-māsam (= māṃsam) pañcame roma-sambhavaḥ /20ab/.*
ViṣṇudhP 2,114,1: *jīvaḥ praviṣṭho garbhe tu kalalam prati tiṣṭhati / mūḍhas tu kalale tasmin māsa-mātraṃ hi tiṣṭhati /1/ dvitīyaṃ tu tadā māsaṃ ghanī-bhūtaḥ sa tiṣṭhati / tasya avayava-nirmāṇaṃ tṛtīye māsi jāyate /2/ tvak-carma pañcame māsi ṣaṣṭhe romṇāṃ samudbhavaḥ /3ab/.* A similar statement is also contained in ViṣṇudhP 2,111,5ff.
GaruḍaP (KIRFEL 1954) v.22–23: *aho-rātreṇa kalalam budbudam pañcabhir dinaiḥ / catur-daśair bhaven māṃsam miśra dhātu-samanvitam /22/ ghana-māṃsaṃ ca viṃśāhe, garbha-stho vardhate kramāt / pañca-viṃśati-pūrṇāhe balam puṣṭiś ca jāyate /23/.*
BhāgP 3, (Kapileyopākhyāna) 31,1–3. MārkaṇḍeyaP 11,2–6. Tandulaveyāliya v.17 (sūtra 1). Mahābhārata 12,308,115cd and followings.
The Hastyāyurveda presents the both theories of embryonic development, i.e. that of the classical medical texts (sthāna 3, adhyāya7, p.403) and that of the Purāṇas (3,8,89ff, p.413). For these theories of embryonic development, cf. SUNESON 1991.

532 For the term *kalala*, cf. DAS 2003A, pp.535–536. In Pāli it means "the state of the embryo immediately after conception" and "mud", cf. WINDISCH 1908, p.21, p.89.
For the listing of the stages of embryonic development, *kalala, abbuda, pesī, ghana* etc., in the Buddhist Canons like the Sa.myuttanikāya, Kathāvatthu etc., cf. WINDISCH (ibid.), p.89.

but in the second, [there are] a dense (*ghana*) *piṇḍa* (m. "a lump of flesh"), a *peśī* (f. "a mass of flesh"), a slightly dense *arbuda* (n. "a swelling").[534] [Or: but in the second, there is a *ghana*: a *piṇḍa*, a *peśī* and a slightly dense *arbuda*.][535] (śl.23)[536]

The Tandulaveyāliya, a Jaina text, mentions the stages, *kalala* (7 days), *abbuya* (7 days), *pesī* and *ghana*.

The term *kalala* is mentioned in the Sāṅkhyakārikā 43.

It does not occur in the Vedic texts, according to the Vedic Word-Concordance (BANDHU 1976).

533 The Ānandāśrama ed. (1896) reads *māse* instead of *māsi*. This is not noted by the Adyar ed.

534 D reads *peśīsañjñakabud*. It should be read *peśī-sañjñakam budbudam* "that whose appellation is *peśī*, [and] *budbuda*". The term *budbuda* belongs to the embryological theory of the non-medical texts like the Purāṇas, cf. SUNESON 1991.

The manuscripts *kha. gha.* and *ṅa* read *peśībandhana[m arbudam]* "the tie of *peśī*, and *arbuda*"; *peśī ca ghanam arbudam* "*peśī* and dense *arbuda*".

The Ānandāśrama ed. (1896) notes the variant of *ka.* and *ga.*, *peśīvadghana* "dense like a *peśī*".

535 The theory of SR śl.23–24ab accords with that of the classical medical texts. According to this theory, the stage in the second month is classified into three, and the three conditions are associated with the three sexes of the embryo. But the SU, AS, AH and CA are inconsistent in dealing with *ghana* and *piṇḍa*.

SU śārīra. 3,18 mentions *kalala* in the first month and *ghana* in the second. The terms *piṇḍa*, *peśī* and *arbuda* are respectively associated with male, female and the third sex (*napuṃsaka*).

AS śārīra., 2,12 mentions *kalala* in the first month. In the second month, the three conditions, *ghanaḥ, peśī* and *arbudaṃ*, are mentioned, and are respectively associated with male, female and the third sex. But *piṇḍa* is not mentioned. That means, the term *ghana* is something else than the *ghana* in the SU. It refers to the condition of the male embryo in the second month. The statement of AH śārīra., 1,37ab–50a accords with that of the AS.

The parallel passage in the CA is problematic. In SHARMA 1996's edition, CA śārīra., 4,9–10 does not contain the term *kalala*, but states that the state in the first month is turbid (*kaluṣī-kṛtaḥ*). The second passage on the second month (*dvitīye māsi ghanaḥ sampadyate piṇḍaḥ peśy arubudaṃ vā*) might mean, "In the second month, a *ghana* is born. Either *piṇḍa, peśī* or *arbuda*." But the statement in the third passage is contradictory to my interpretation. It states that *ghana*, *peśī* and *arbuda* are associated with male, female and the third sex (*ghanaḥ puruṣaḥ, peśī strī, arbudaṃ napuṃsakam*). Here, the male form is called *ghana*. Taking this into consideration, we have to interpret the second passage as "in the second month, a dense ball (*ghanaḥ piṇḍaḥ*), a *peśī* or an *arbuda* is born". This statement is in accordance with that of the AS and AH.

[When they arise,] these are (opt. *syuḥ*) the preliminary states (*prāg-avasthā*) of male, female[537] and third sex respectively.[538] (śl.24 ab)

On the other hand, SUNESON 1991, p. 114, quotes the same passages, CA śārīra., 4,9–10, from another edition by ŚRĪJAYADEVA VIDYĀLAṄKĀRA. This edition contains *kalanī-kṛtaḥ* instead of *kaluṣī-kṛtaḥ* in the first month. (For *kalana*, i.e. the exchange of n/l, cf. DAS 2003A, p.536.) Astonishingly it reads, in the third passage, *tatra pindaḥ puruṣaḥ, strī cet peśī, arbudaṃ napuṃsakam*, namely, the male state is not called *ghana* but *piṇḍa*, unlike in SHARMA's edition. In this case, the CA's statement accords with the SU's.
Thus we have two ways of interpreting SR śl.23cd–24ab. According to the SU, it would mean, "In the second [month, there is a stage called] *ghana* ("dense"). [In that stage,] a *piṇḍa,* a *peśī,* and a slightly dense *arbuda*. They are respectively the primitive conditions of male, female and the third sex". In contrast, according to the AS, AH and SHARMA's edition of the CA, it would mean, "In the second month, [there are] *ghana piṇḍa* ("dense ball"), *peśī*, [and] a slightly dense *arbuda*. They (= *ghanaḥ piṇḍaḥ, peśī* and *arbuda*) are the primitive conditions of male, female and the third sex, respectively." The comm. K seems to interpret it in this way (*ghanaḥ piṇḍaḥ janiṣyamāṇasya puṃsaḥ prāg-avasthā*).
To the expression *īṣad-ghanaṃ* "slightly dense" qualifying *arbudam* in SR śl.23d, no parallel was found in the SU, AS, AH, CA or the non-medical texts which I consulted.

536 The theory in SR śl.23–24ab accords with that of the classical medical texts. SR śl.23a (*dravatvam prathame māsi*) is parallel to ŚG 8,13d (*prathame māsi tad dravam*). Though most of the embryological verses in the SR are identical to those in the ŚG, SR śl.23–24ab have no equivalents in the ŚG. The ŚG (8,13d–14) contains a different theory which is in accordance with the Purāṇa-s (Cf. *Situating the text §2.1.1.*).
The embryological texts which locate the stage of *kalala* immediately after the stage of *dravatva*, are: SU śārīra., 3,18; BhāgP 3,31,2; ViṣṇudhP 2,114,1; AgniP 369,19; Tandulaveyāliya (informed by COMBA 1981, p.197, footnote 17). COMBA states that all the texts she consulted contain the same theory.
ŚB 6,6,2,9 states that an embryo has no bones.

537 The Ānandāśrama ed. (1896) reads *strīpuṃnapuṃsakānāṃ* instead of *puṃstrīnapuṃsakānāṃ*. But it is not suitable, because this order does not accord with the statements of the classical medical texts.

538 The theory that the embryo's sex is determined by the portion of semen and the procreatory-menstrual fluid is contained in, for example, AH śārīra., 1,5 and GaruḍaP (KIRFEL 1954, śl.21). On this matter, also cf. DAS 2003A, p.4; p.23 (note 57). (Cf. ibid., *Index rerum* under "embryo, sex of"). But the SR does not contain this theory of the portion of the two fluids.
For *napuṃsaka*, cf. DAS 2003A, pp.558–560; ANGOT *1993–94*.
NAMOUCHI 1995, p.197ff., deals with various aspects of the role of the third sex (*tṛtīyā prakṛti*) in classical Indian sexual life. For *napuṃsaka* in Jainism, cf.

But in the third, [those] deemed the five sprouts[539] of the hands, the feet and the head [arise]. (śl.24cd) [540]

Then, [when they arise,] the minute parts (*bhāga*) for/of the limbs (*aṅga*) and the secondary appendages (*pratyaṅga*) are (opt. *syuḥ*) (= appear) all at once[541], except for the beard, the teeth[542] etc., whose appearance is immediately after[543] birth.[544] (śl.25)[545]

HALBFASS 2000, p.80. JAINI 1991 informs us of a Jaina text discussing the discrepancy between physical and mental sex, including the problem of the third sex.

539 For the terms *aṅkura* and *praśākhā* in embryology, cf. DOSSI 1998, p.94. For the "seeds" (*bīja*) of the (rudimentary?) limbs/organs of the embryo, cf. DAS 2003A, p.21; p.25ff.

540 Parallels to SR śl.24cd–25 are found in SU śārīra., 3,32; AS śārīra., 2,12; AH śārīra., 1,54cd–55; CA śārīra., 4,11. On the relationship between CA 4,11 and YS 3,76–79, cf. YAMASHITA 2001/2002, pp.95–96.
ŚG 8,15ab deals with the state in the third month, like SR śl.24ab, but this verse is not identical to the SR: *karāṅghri-śīrṣakādīni tṛtīye sambhavanti hi* "Nel terzo mese si formano invero le mani, i piedi, la testa, ecc." (tr. by COMBA 1981).

541 CA śārīra., 4,11 (*yaugapatyena*) and SU śārīra., 3,32 (*yugapat*) state that all the members of the embryo develop simultaneously, cf. COMBA 1981, pp.175–176, (footnote 4) on ŚG 8,23. COMBA, however, remarks that ŚG 8, śl.23 and śl.34 (identical to SR śl.33cd and śl.36cd–37ab), which state that the body parts grow separately in the sixth and eighth month, are contradictory to this theory in ŚG 8, śl.23 (identical to SR śl.25).

542 Cf. ŚB 7,4,1,5 which states that creatures are born toothless.

543 "Immediately after (*anantara*) birth" is clearly not logical here. Maybe *anantara* can be translated in a different way, e.g. "[in the period] following/after birth]".
The parallel in AS śārīra., 2,12, contains *janmottara-kālaja-* "appearing in the time after birth".

544 SR śl.25 is identical to ŚG 8,18 (cf. COMBA 1981, p.199).

545 SU śārīra., 3,18 contains the expression *aṅga-pratyaṅga-vibhāga-* and *sūkṣma-*. But AS śārīra., 2,12 makes a different expression *sarvāṅgāvayavendriya-* and lacks a counterpart for *sūkṣma-*. The AS's *sarvāṅgāvayavendriya* is parallel to the CA's (śārīra., 1,11) *sarvendriyāṇi sarvāṅgāvayavāś ca*. But AS śārīra., 2,15, which describes the fourth month, contains *aṅga-pratyaṅga-vibhāga-*, i.e. an expression similar to the SU and SR śl.25a.
On the other hand, the SU does not mention the secondary appendages (*pratyaṅga*) appearing after birth, while the AS does, i.e. *janmottara-kālajebhyo dantādibhyaḥ*. Besides, the SU does not mention *prakṛti* and *vikṛti*, while they are mentioned in the AS.

This is agreed on by the wise ones[546] as the natural condition (*prakṛti*), but the other [is agreed on] as an unnatural condition (*vikṛti*). (śl.26ab)[547]

In the fourth their manifest state is produced, [and] also of the characteristics[548] (*bhāva*) [of the mind]. (śl. 26cd) [549]

They declare that (*iti*) the characteristics of males[550] are valour (*śaurya*) etc., but of females, cowardice (*bhīrutva*) etc., of the third sex, mixed (*saṅkīrṇa*). (śl. 27) [551]

Thus the AS seems to have integrated both expressions originating from the SU and CA. (For Vāgbhaṭa's intention to integrate both tradition of the SU and CA, cf. the last verse of the AH.) But the AS added a new theory of the appendages appearing after birth. So the SR contains both wordings of the SU and AS.

546 *sammatā satām* (genitive pl.) Cf. Pāṇini 2,3,71: *kṛtyānāṃ kartari vā*. SPEIJER 1993, p.50 (§66): "*Both instrumental and genitive are available to denote the agent with a kṛtya.* As a rule the instrumental is required, if the verbal sense prevails, but the genitive, if the *kṛtya* has the value of a noun adjective or substantive."

547 The ŚG (chap. 8) does not contain a parallel to SR śl.26ab.

548 The term *bhāva* occurs in SR śl.55, too. There, it means "[mental] condition", cf. my footnote 717 on SR śl.55, *bhāva*.

549 SR śl.26cd is identical to ŚG 8,19 (cf. COMBA 1981, p.199).
The Adyar ed. notes the variant of C.E. *bhāvānām upajāyate*, "[...] their manifest state is produced, [and] of the characteristics." This variant of C.E. is the same as the reading of ŚG 8,19 adopted by COMBA 1981 (p.199, footnote 22). But Jvālāprasādamiśra's edition of the ŚG (Bombay 1987) which I consulted contains *bhāvānām api jāyate*, i.e. the same reading as the Adyar ed. of the SR.

550 SR śl.27a *pusāṃ* is obviously a printing mistake for *puṃsāṃ*.

551 SR śl.27 is identical to ŚG 8,19cd–20ab. But the ŚG contains variants. ŚG 8,19cd–20ab runs: *puṃsāṃ sthairyādayo bhāvā bhīrutvādyās tu yoṣitām /19cd/ napuṃsake ca te miśrā bhavanti raghu-nandana /20ab/* (cf. COMBA 1981, p.200). In the first half of the verse, SR śl.27ab andŚG 8,19cd accord with each other, except for *śauryādayo* and *sthairyādayo*, but, in the second half, SR śl.27cd and ŚG 8,20ab considerably deviate from each other.
The edition used by COMBA 1981 (p.200 note 23) contains *bhūtatvād yāś ca yoṣitām / napuṃsake tu*. But COMBA (ibid.) also notes variants, *bhūtatvād yās tu [...] ca* (ed. B.) and *bhūtatvād yāś ca [...] tu* (ed. C.). Though COMBA (ibid.) separates *bhīrutvād* from *yās*, I prefer reading *bhīrutva-ādyās*.
The respective *bhāva*-s of the three sexes are not mentioned in the SU, but in CA śārīra., 4,14 and AS śārīra., 2,13–14. The statements of the CA and AS resemble each other in expression. The two texts mention *bhīrutva* "cowardice" among the *bhāva*-s of the female. On the *bhāva*-s of the male, the two texts

Thus he shows the various stages (*avasthā-bheda*), every month, beginning from the first month up to the ninth month, of the embryo gone into (= situated in) the womb (*āśaya* lit. "receptacle") in the mentioned manner, through "**a liquid state** (*dravatva*)" etc.[552] **Called "*kalala*" (*kalalākhya*)**: that which has the appellation (*ākhyā*) "*kalala*" is called thus. By (*iti*) "*kalala*", a liquid condition is mentioned, of the mixture of semen and blood (= procreatory fluid) which has not [yet] caught (= attained) hardness.

"**But in the second**". The [word-]order (*krama*) is thus (*iti*): A **dense (*ghana*) *piṇḍa*** is the **prior state** of a **male** to be born, a ***peśī*** is the **prior state** of a **female** to be born, a **slightly thick *arbuda*** is the **prior state** of [a person of] **the third sex** to be born.[553] **Dense (*ghana*)**[554] [means] endowed with hardness. A ***peśī*** [is] quadrangular[555], consisting chiefly of flesh. **Slightly dense (*īṣad-ghana*)**: so called because of its similarity both to a dense (*ghana*) [*piṇḍa*] and to a *peśī*. ***Arbuda***: that whose shape is [that of] the bud of *śālmalī* (silk-cotton tree).[556] As they say: "The *peśī* is borne in mind [by the

---

only state that they are opposite to those of the female. In contrast, the SR mentions *śaurya* "valour" as belonging to the male. For the *bhāva*-s of the third sex, the AS mentions *saṅkīrṇa* "mixed" which is the same expression as in SR śl.27c. But the CA makes a different expression, *ubhaya-bhāvāvayava*. For the mixed nature of the third sex, also cf. CA śārīra., 2,25b (*vyāmiśra-liṅgā prakṛtiṃ tṛtīyām*).

552 The Anandāśrama ed. (1896) reads the part for *prathama-mAsata* up to *dravatvam ity AdinA* in a totally different way: it reads *tata ārabhya prathamādi-nava(sic.)māsa-paryantam pratimāsagata-saṅghātasya dravībhāva eva*. "[There is] only a liquid condition of the embryo gone into the womb in the mentioned manner, coagulated (*saṅghāta*) every month, starting from then/it (i.e. the formation), beginning from the first month up to the ninth month." This edition omits the whole passage following *kalalākhya* up to *dravībhāva ucyate*. This is not noted by the Adyar ed.

553 The Ānandāśrama ed. (1896) omits this whole sentence; this is not noted by the Adyar ed.

554 *Ghana*. The commentator interprets it as a qualifier of *piṇḍa* (cf. my discussion in the footnote 535 on the *mūla* text SR śl.23).

555 The Ānandāśrama ed. (1896) reads *caturasrā* instead of *caturaśrā*; this is not noted by the Adyar ed.

556 The embryo is compared to the bud of *śālmali* in the Kādambarī by Bāṇa, cf. DOSSI 1998, p.94, "in seinem Aussehen der *Śālmali*-Knospe des Wollbaums

tradition] as quadrangular[557]. The dense *piṇḍa* is borne in mind as round. / Physicians know the *arbuda* as that whose shape is [that of] the bud of *śālmalī*."[558]

"But in the third": Of the hands, the feet and the head of/from the previously mentioned dense *piṇḍa* etc.[559] Two hands and two feet and the head: [this gives the compound] the hands, the foot and the head (*karāṅghri-śiras*). State of being like a single [thing] of a couple [of things] (*dvandva-ekavad-bhāva*), because of being the limbs of a living creature (*prāṇin*).[560] (I.e. A *dvandva* compound denoting a couple of limbs of an animal is singular.) Of that, [i.e.] of the hands, the feet and the head (*karāṅghri-śirasas*).

"**Sprouts**" **(*aṅkurāḥ*)** etc.: This designation (*vyapadeśa*) is [used] with the signification of (= to signify) the future limbs through the maxim of thread and cloth (*sūtra-śāṭaka-nyāyena*)[561]: "The sprouts, even while being produced, become the hands, the feet and the

---

ähnlich, der eine rote Blüte hat und dessen Samen in der Wolle liegt." Cf. her footnote 468, "[...] in Bāṇa's Kādambarī (p.61) werden frischgeschlüpfte Vögel sowohl mit Śālmali-Blüten als auch mit Embryonen verglichen". But in the SR, the comparison is restricted to *arbuda*.

557 The Ānandāśrama ed. (1896) reads here again *caturasrā* instead of *caturaśrā*; this is not noted by the Adyar ed.

558 In his commentary on SU śārīra., 3,18, Ḍalhaṇa quotes this verse as attributed to the author Gayin: *gayī tu bhoja-darśanāt piṇḍādīnām anyathākāram (ākāram) paṭhati; yathā, caturasrā bhavet peśī, vṛttaḥ piṇḍo ghanaḥ smṛtaḥ / śālmalīmukulākāram arbudam paricakṣate // iti.*

559 The Ānandāśrama ed. (1896) reads *ghana-piṇḍāt* "from the [previously mentioned] dense *piṇḍa*" instead of *ghana-piṇḍādeḥ*; this is not noted by the Adyar ed.

560 The Adyar ed. notes the variant of C adding *tasya* before *prāṇyaṅgatvād*. It would mean "... because of its (*tasya*) being the limbs of [...]".
The Ānandāśrama ed. (1896) reads *karāṅghriśirasām* instead of *karāṅghriśirasaḥ*, setting a *daṇḍa* between *tasya* and *karāṅghriśirasām*. This is not noted by the Adyar ed. The translation would be: "It has (literally: of it, there is) the state of [being] like [...] of a living creature". The following, *karāṅghriśirasām aṅkurā iti*, would be: 'The sprouts of the hands, feet and head".

561 Thread and cloth are often mentioned as an example for inherence (*samavāya*) in Nyāya-Vaiśeṣika. The relation of thread and cloth is that of the material cause and product. Cf. STRAUS 2004, p.128: "Das Tuch z.B. ist aus Fäden gebildet, ohne Fäden wäre es nicht denkbar, so werde die Fäden als die inhärente Ursache des Tuches angesehen."

head.[562]" **All at once (*yugapad*)**, in the very state of the sprout.[563] **"The parts for/of the limbs and the secondary appendages" (*aṅga-pratyaṅga-bhāgāḥ*)**: In the sprout of the head [there are] the parts for/of the neck, the eyes, the nose, the ears etc.; in the two sprouts of the hands the parts for/of the shoulders, the elbows, the wrists, the fingers etc.; in the two sprouts of the legs the parts for/of the thighs, the knees, the shanks (*jaṅghā*), the ankles, the toes etc. **Minute**, those whose form is unmanifest.

"This is [agreed on [...] as] the natural condition" (śl.26a *eṣā prakṛtir*): **The natural condition** is the appearance of the parts for/of the limbs and the secondary appendages in minute form in the very state of the sprout etc., **except for the beard, the teeth etc., whose appearance is immediately after birth**. That is to say (*iti yāvad*), [this is] the common nature of the viviparous[564] (*jarāyuja*)[565]. **"But the other [is agreed on as] an unnatural condition"** (śl. 26ab *anyā tu vikṛtir*): **The other (*anyā*)**[566]: The appearance of the beard, the teeth etc., arising immediately after birth at the very same time[567] as the limbs etc., the non-appearance of some [parts] in/among the limbs etc. or appearance with inverted order of position (*sthāna-vyatyāsa*) and the condition of lack or surplus of the number of the fingers etc.

"In the fourth" (śl.26c *caturthe*): **Their**, [i.e.] the limbs' and the secondary appendages', **manifest state (*vyaktatā*)**, [i.e.] the state of being made clear (*abhivyaktatā*)[568], [i.e.] the state of having separate individualities (*pṛthag-ātmatā*). The manifest state **also of the characteristics** (*bhāva*) [of the mind], [i.e.] of valour (*śaurya*) etc., [i.e.] of the qualities of the self [which are] merged in (or: attached to) [the self] (*līna*).

562 The Ānandāśrama ed. (1896) adds *ye* before *'ṅkurā jāyamānā*.

563 The Ānandāśrama ed. (1896) reads *-eṣa* instead of *-eva*. "All at once, this (*eṣa*) is in the state of a sprout".

564 Literally: "born of/from the foetal envelope (*jarāyu*)".

565 The Ānandāśrama ed. (1896) reads *jarāyujādīnām* "the viviparous etc." instead of *jarāyujānām*. This is not noted by the Adyar ed. But obviously it does not fit the context.

566 The Ānandāśrama ed. (1896) omits *anyā*. This is not noted by the Adyar ed.

567 The Ānandāśrama ed. (1896) reads *samakāla* instead of *samānakāla*.

568 In the Ānandāśrama ed. (1896), *abhivyaktatā* is missing. This is not noted by the Adyar ed.

Comm. S on SR śl.23–27

He relates the particular characteristics (*viśeṣa*) of the embryo's state every month[569]: "**a liquid state**" (śl.23a, *dravatvam*). He relates the previous state **of male, female and third sex** respectively: "**in the second**" (śl.23c, *dvitīye*). The *piṇḍa*'s similarity to a *peśī* is the state of a male, [the *piṇḍa*'s] hardness [is] the state of a female, [its] *arbuda*-hood, [i.e.] not being hard, [is] the state of the third sex.[570] Through the words (*śabda*) "**the beard, the teeth etc.**", the breasts of women etc. [are] to be known, [too], because of the mention of the cause (*hetu*), "**due to the appearance immediately after birth**"[571]. It is mentioned that in the fourth [month] the characteristics [of the mind] become clear. Distinguishing individually (*vibhajya*) he explains those very/same characteristics: "**Of males**" (śl.27a, *puṃsāṃ*).

SR śl.28–32

And its (= the embryo's) heart, born from the mother, desires objects.[572]

569 B reads *pratimāsasya* instead of *pratimāsaṃ*. "[…] of the embryo's state of (= in) every month".

570 The interpretation of the comm. S is as follows: The state in the second month is called *piṇḍa*. If the *piṇḍa* resembles a *peśī* ("a piece of flesh"), it is a male. If it is hard, it is a female. If it has the state of *arbuda*, or not hard, it is the third sex. But this statement does not accord with the classical medical theory.

571 The comm. S reads SR śl.25d differently from the *mūla* text as printed here, namely *sambhavād* instead of *sambhavān*. But the reading as an ablative seems to be wrong.

572 ŚG 8,20cd–21ab is identical to SR śl.28 (cf. COMBA 1981, p.200). The ŚG, however, contains the variants, *tato* and *-vivṛddhaye*, instead of *ato* and *-samṛddhaye*.
The expression of CA śārīra., 4,15 and that of AS śārīra., 2,16 are almost identical to each other, and are close to SR śl.28 (also compare with SR śl.69cd–70ab, on the umbilical cord). CA śārīra., 4,15: *mātṛjaṃ cāsya hṛdayam mātṛ-hṛdayenābhisambaddham bhavati rasa-vāhinībhiḥ*; AS śārīra., 2,16: *mātṛjaṃ cāsya hṛdayaṃ tad-rasa-hāriṇībhir dhamanībhir mātṛ-hṛdayenābhisambaddham bhavati*. The wording *mātṛjaṃ cāsya hṛdayaṃ* is identical to that in SR śl.28a. On the relationship between CA śārīra., 4,15 and YS 3,78cd–79, cf. YAMASHITA 2001/2002, p.95.
The statement of AH śārīra., 1,53 (1,56 in DAS & EMMERIC's (1998) ed.) accords with these statements, but its wording is different: *garbhasya nābhau mātuś ca hṛdi nāḍī nibadhyate / yayā sa puṣṭim āpnoti kedāra iva kulyayā*.

Therefore (*atas*), one should do that wished by the mother's mind for the embryo's growth/prosperity (*samṛddhi*)[573]. (śl.28)

And the wise ones call that woman who has two hearts characterised by two-heartedness (*dauhṛdinī*)[574]. (śl.29ab)[575]

From not bestowing of *dohada*-s[576] (i.e. "longings in the pregnancy period"), deficiency of limbs etc. arise (opt. *syur*) for (lit. "of") the embryo. (śl. 29cd)[577]

Which non-acquisition[578] of an object by (lit. "of") the mother

---

SU śārīra., 3,18 contains different expressions, though it deals with *dauhṛda*.

The Adyar ed. notes the variant of *ka., kha.* and *ga.*, *mātṛkaṃ* instead of *mātṛjaṃ*. This is obviously unsuitable, in comparison with the parallels of the CA and AS, which contain *mātṛja*.

The ŚB relates that the embryo receives the mother's prāṇa: ŚB 2,2,1,10 (WEBER's ed., p.144, ll.15–16): yadā vai jā́yate 'tha prāṇó, 'tha yāvan na jā́yate mātúr vaiva tā́vat prāṇam ánu prā́ṇiti, yáthā vā táj jātá évāsminn etát prāṇáṃ dadhāti. This text states that the embryo takes the mother's breath.

573 "The embryo's prosperity (*samṛddhi*)" seems to mean the child's future prosperity after birth rather than the increase of the embryo's body in the uterus, according to the following verses.

574 The manuscripts *kha.* and *ga.* read *dohadinīṃ*. The manuscript D reads *dauhṛdikāṃ*.

575 According to this verse, the terms *dauhṛdinī* and *dohada* are derived from *dvihṛd*. The same etymology is mentioned in SU śārīra., 3,18, CA śārīra., 4,15, AS śārīra., 2,16, and YS 3,79. Of these three texts, the SU and AS make an expression similar to that of SR śl.28ab. SU śārīra., 3,18: *dvihṛdayāṃ ca nārīṃ dauhṛdinīm ācakṣate* (the AS's wording is almost the same).

On the relationship between CA śārīra., 4,15 and YS 3,79, cf. YAMASHITA 2001/2002, pp.96–97.

576 D reads *dauhṛdānāṃ*. For the term *dohada*, cf. DAS 1988, pp.247–251; DAS 2003A, pp.470–471.

577 SR śl.29 is identical to ŚG 8,21cd–22ab (cf. COMBA 1981, p.200). The ŚG contains *tataḥ* and *dohṛdānāṃ* instead of *budhāḥ* and *dohadānāṃ* in the SR. The edition A of the ŚG consulted by COMBA (1981, p.200, note 25) contains *dauhṛdānām*, but COMBA reports the variant (ed. B) *dohadānām*, which is identical to the reading of the SR.

The topic of *dohada* is mentioned e.g. in YS 3,79: *dohadasyāpradānena garbho doṣam avāpnuyāt / vairūpyam maraṇaṃ vāpi tasmāt kāryam priyaṃ striyāḥ*.

578 The Ānandāśrama ed. (1896) reads *viṣaye lobhas* instead of *viṣayālābhas*. "The object for which the mother's greed/strong_desire [occurs], afflicted/diseased by that [object][...]".

[happens], [being] afflicted/diseased by that [object], the son[579] is born.[580] (śl.30ab)

The embryo is[581] rich (*arthavat*) and an enjoyer (*bhogin*) [of goods and pleasures] from [the fulfilment of] a *dohada*[582] pertaining to seeing the king, (śl. 30cd)
sportive/graceful (*lalita*) [from a *dohada*] pertaining to ornaments, most pious (*dharmiṣṭha*) [from that] pertaining to an ascetic or a hermitage[583], devotional/a_devotee (*bhakta*), [from that] pertaining to seeing a divinity (i.e. the image etc.), cruel (*hiṃsra*), [from that]

579 The variant D reads *tataḥ* instead of *sutaḥ*: "[the embryo] is then afflicted/diseased by that [object]".

580 ŚG śl.22cd is identical to SR śl.30ab. The ŚG, however, contains a variant *yad viṣaye lobhas* instead of *yad viṣayālābhas* in the SR. It is identical to the variant of the Ānandāśrama ed. of the SR.
The ed. A of the ŚG (cf. COMBA 1981, p.200, note 25) contains *tathārtho* instead of *tadārto* of SR. But COMBA notes the variant of ed. B and C, *tathārto*, instead of A *tathārtho.*
SU śārīra., 3,20–21 and AS śārīra., 2,19–20 deal with a topic similar to that in SR śl.30ab. SU śārīra., 3,21cd contains *prajāyate sutasyārtis*, an expression similar to the SR.
COMBA 1981 (p.201, note 25) reports that ŚG śl.22cd (identical to SR śl.30ab) accords with SU śārīra., 3,18, in locating *dohada* or double-heartedness in the fourth month; the other texts consulted by her locate it in the third month (e.g. YS 3,79; Tandulaveyāliya).
CA śārīra., 4,15 deals with *dohada* in the third month.
The pregnant woman's perception, including the auditory one, influences the embryo, cf. KAPANI 1992, p.137, "Pendant toute la période de gestation, la future mère doit être entourée de soins affectueux et de conseils judicieux, car tout ce qui se passe en elle et dans son entourage est censé influencer l'étre qu' elle porte". Sounds (music, recitation etc.) which the pregnant woman hears are one of the many factors determining the character of the future child (ibid, p.138). Cf. CA śārīra., 2,16 (*sattva-viśeṣakarāṇi [...] antarvatnyāḥ śrutayaś [...]*). CA śārīra., 2,21 prohibits the pregnant woman from listening to unpleasant or excessive sounds (*apriyātimātra-śravaṇair*) because it might damage the embryo.

581 Optative *syāt.*

582 According to the comm. K, *dohada* is the longing as well as the object longed for.

583 The pregnant woman has belief in a hermitage or the abode of ascetics, according to the commentator Ḍalhaṇa on SU śārīra., 3,24 (*āśrame saṃyatātmānaṃ dharmaśīlam prasūyate*); he explains, *yasyā āśrame tapasvinām adhiṣṭhāne śraddhā bhavet* etc.

pertaining to seeing a snake, (śl. 31)
sleepy, however, [from that] pertaining to eating a monitor lizard[584], powerful [from that] pertaining to eating beef (*go-māṃsa*).

[From that] pertaining to [the meat] of a buffalo (*māhiṣa*), [the pregnant woman] brings forth (*sūyate*)[585] a hairy son whose eyes are red like those of a parrot[586] (lit. "parrot-red"). (śl. 32)[587]

584 In CA śārīra., 2,21, a pregnant woman's excessive or customary eating of the meat of a monitor lizard/iguana (*godhā*), wild boar (*varāha*), and fish is mentioned among the factors damaging the foetus (*godhā-māṃsa-prāyā śārkariṇam aśmariṇaṃ śanairmehiṇaṃ vā, varāha-māṃsa-prāyā raktākṣaṃ krathanam atiparuṣa-romāṇaṃ vā, matsya-māṃsa-nityā cira-nimeṣaṃ stabdhākṣaṃ vā* [...]). The results of eating the meat of a monitor_lizard/iguana are different from that mentioned in the SR (sleepiness). The results of eating wild boar (*varāha*) are the same as those of eating *māhiṣa* mentioned in the SR. But the CA does not prohibit eating these meats, but only eating them excessively. If we follow the statement of SR śl.29cd, a pregnant woman who has such a *dohada* should eat those meats; otherwise, the child would be born damaged.

Flesh eating is mentioned in the 46th chapter of SU sūtrasthāna, on the rules of foods and drinks (*anna-pāna-vidhi*). SU sūtra., 46,81 mentions the monitor_lizard/iguana (*godhā*): *godhā vipāke madhurā kaṣāya-kaṭukā smṛtā / vāta-pitta-praśamanī bṛṃhaṇī bala-vardhanī.* But it does not mention the property of the meat of a monitor_lizard which prompts sleep. The property *nidrālu* is mentioned in SU sūtra., 46,226, as belonging to a river-fish named *pāṭhīna.*

The flesh of a cattle is mentioned in SU sūtra., 46,85, in the list of the flesh of the domestic animals. SU sūtra., śl.46,89: *śvāsa-kāsa-pratiśyāya-viṣamajvara-nāśanam / śramātyagni-hitaṃ gavyam pavitram anilāpaham.*

The flesh of a *mahiṣa* is mentioned in SU sūtra., 46,18. The properties, prompting sleep and manliness (*puṃstva*), are mentioned, which does not accord with the statement of the SR.

585 The verb √*sū* "bring forth, produce" is either the second or fourth class. The comm. S. contains that of the second class *sūte.*

586 That the parrot has red eyes sounds odd. I do not know if some kinds of parrots really have red eyes. The manuscript D reads *krūra-* instead of *śuka-*: "whose eyes are cruel_and_red/cruelly_red". The Ānandāśrama ed. (1896) reads *śūra-*; this is not noted by the Adyar ed.

As a matter of fact, the parallel in SU śārīra., 3,26ab runs as follows: *māhiṣe dauhṛdāc <u>chūraṃ raktākṣaṃ</u> loma-saṃyutam.* In comparison with this, the correct reading of SR śl.32c might be *māhiṣe <u>śūra</u>-raktākṣaṃ lomaśaṃ sūyate sutam.* According to the SU, *śūra-raktāśaṃ* is a *dvandva* compound. In this case, SR śl.32c would mean, "a brave red eyed son".

"**And [its heart,] born from the mother**" (śl.28a, *mātṛjaṃ ca*). **Its,** [i.e.] the embryo's, **heart, born from the mother**, [i.e.] arisen from the mother. Through "and/also" (*cakāra*) the mother's heart **also** (*ca*) [is denoted][588]; the relation [of the words] is that (*iti*): since (*yataḥ*) both these **desire**[589] **objects, therefore** (*ataḥ*). This is the meaning: Because of the desirousness for objects of the embryo's very heart [which is] connected[590] to the mother's heart, that wished by the mother[591] at that time is necessarily to be done (= fulfilled) by one whose desire is the embryo's growth/prosperity[592]. "**Characterised by two-heartedness**" (śl.29b *dauhṛdinīm*): The aggregate of the two hearts is "*dvi-hṛdaya*" (lit. "two-heart"); its state (*bhāva*) is "two-heartedness" (*dauhṛda*), [which is] correct due to being [of the same class as] *pṛṣodara*[593] etc. Because of the connection with that, the

587 The ŚG does not have a parallel to SR śl.31–32. It does not mention this topic, the relation of the mother's *dohada* to the child's future, at all.
The contents of the SR accords with that of SU śārīra., 3,22ff.
SU śārīra., 3,22: *rāja-sandarśane yasyā dauhṛdaṃ jāyate striyāḥ / arthavantam mahābhāgaṃ kumāraṃ sā prasūyate /22/ […] alaṅkāraiṣiṇam […] / āśrame […] / devatāpratimāyāṃ […] / vyāla-jātīnāṃ […] godhāmāṃsāśane […] gavām māṃse […] māhiṣe / […] /27/.*
Though CA śārīra., 4,15 mentions *dohada*, it does not contain the theory of the relation between the mother's *dohada* and the child's future. The AS and AH do not mention it.
In summary, it is not impossible that the passages SR śl.31–32 were originally not included in the common source text of the SR and ŚG; they were a secondary insertion from SU śārīra., 3,22ff., or a text close to the SU.

588 That means: SR śl.28ab is translated as "The mother-born [heart] and its heart desire (lit. desires) objects".

589 3rd person, sg. "desires", as *ubhaya* "both" is singular.

590 The Ānandāśrama ed. (1896) reads *sambandha* instead of *sambaddha*; this is not noted by the Adyar ed. "[…] of the embryo's very heart [whose] connection (*sambandha*) is to the mother's heart".

591 Originally *mātur abhīṣṭam*, which is a kind of quotation from the *mūla* text. I do not adopt boldface for this kind of quotation, as it is not a *pratīka* in the strict sense.

592 Originally *garbha-samṛddhi*. This is also a quotation from the *mūla* text, but I do not use boldface, due to the same reason as above.

593 *Pṛṣodara* "wind/air". APTE 1992 (on *pṛṣodara*) states: "The word is supposed to be compounded of *pṛṣat* and *udara*, the *t* of *pṛṣat* being dropped as the

embryo, too, is called "characterised by two-heartedness" (*dauhr̥da*). That two-heartedness is hers (= she has that two-heartedness), so (*iti*) she is characterised by two-heartedness; that[594] [woman].[595] He mentions the fault/injurious_consequence (*doṣa*) from not doing its/her wish: **"From not bestowing"** (śl.29c, *adānāt*). **Of *dohada*-s**: *Dohada* is the longing of a pregnant woman; the objects of it (= *dohada*) are also *dohada*-s; of them. He mentions the merit and demerit which will be produced on account of (*vaśāt*) this and that (= particular) *dohada*, by **"the embryo is"** (śl.30c, *garbhaḥ syāt*) etc.[596] **From the *dohada***, [namely] from the longing of the pregnant woman, **pertaining to seeing the king, the embryo** to be born becomes **rich** and **an enjoyer**. It is to be seen in the same manner in the following, too.

Comm. S on SR śl.28–32

"Born from the mother" (śl.28a *mātr̥jam*): **The heart, born from the mother**, [i.e.] related/connected to the mother, **desires objects wished for** by itself (= the heart). Or else, the relation (*sambandha*) is to be made that (*iti*) **its**, the embryo's, **heart** is **born from the mother** (*mātr̥ja*), [i.e.] arisen (*niṣpanna*) in the mother's [body-]members. He etymologises the word "*dauhr̥dinī*" ("characterised by two-heartedness"): **"and that [woman]"** (śl.29a, *tāṃ ca*). He mentions the fault/injurious_consequence with regard to not bestowing the *dohada*: **"from not bestowing"** (śl.29c, *adānāt*). **Deficiency of**

irregular case. The word is thus taken as the type of a whole class of such irregular compounds". He refers to Pāṇini 4,3,109.

594 *Tām*. The Ānandāśrama ed. (1896) does not contain it; this is not noted by the Adyar ed.

595 The part from *dauhr̥dinīm iti* up to *tām* is the explanation of *dauhr̥dinīm* in the *mūla* text.

596 In the Ānandāśrama ed. (1896), the part from *vaśāutpatsyamānasyo* up to *rājadarśanedohadād* is missing. This is presumably caused by the resemblance of *dohadavaśād* and *dohadādabhilāṣād* in the line below, in transcription. The omitted part is falsely inserted into the following part, between *tadāspandonmukhaṃ* (which is read *tadāsya svedonmukha-* by the Ānandāśrama ed.) and *ṣaṣṭha iti*. As a result, it runs: *tadāsya svedonmukha-hr̥daya-vaśād utpatsyamānasyotkarṣāya karṣāvāha-garbhaḥ* (?) *syād iti garbhiṇyā rājadarśane dohado bhavati*.
This is not noted by the Adyar ed.

limbs [means] one-eyedness, blindness, lameness etc. Through the word ["]etc.["] (*ādi*), weakness, listlessness **etc. (*ādi*)** are [also] to be known. He relates the particular results in the embryo through particular *dohada*-s: "**the embryo is**" (śl.30, *garbhaḥ syād*). "**Pertaining to [the meat] of a buffalo**": a woman whose *dohada*-longing is for (lit. "characterised by two-heartedness with regard to") **buffalo**-meat **delivers** (*sūte*)[597] **a son**, [i.e.] an offspring, **whose eyes are red** like [that of] **a parrot**, [i.e.] whose eyes tend towards red (*ālohita*), and [who is] **hairy**.

SR śl.33–41

In the fifth [month], the consciousness (*citta*) is awoken, [and there is] a state of well-nourishedness (*puṣṭatā*) of flesh and blood. (śl.33ab) [598]

597 The *mūla* text actually contains *sūyate*.

598 ŚG 8,23ab is identical to SR śl.33ab (cf. COMBA 1981, p.201).
COMBA 1981 (p.201, note 26) reports that the contents of CA śārīra., 4,24 is the closest to the theory of ŚG 8,23ab (= SR 33ab) on flesh and blood in the fifth month. The YS also mentions the appearance of blood in the fifth month (YS 3,80ab). But the CA does not mention the condition of consciousness (*citta*) in the fifth month, in contrast to SU śārīra., 3,30 (*pañcame manaḥ pratibuddhataraṃ*) and the AH. The Purāṇas consulted by COMBA (ibid.) locate the development of consciousness in the sixth month.
Although COMBA (ibid.) does not consult the AS, the AS (śārīra., 2,22) is actually most closely related to the SR, in locating the awakeness of consciousness and the appearance of flesh and blood in the fifth month. AS śārīra., 2,22: *pañcame manaḥ prabuddhataram bhavati māṃsa-śoṇitopayaś ca*. The wording of the AS resembles that of the SR. The AS's expression, *māṃsa-śoṇitopacayaś*, is almost the same as the CA's (śārīra., 4,21) *māṃsa-śoṇitopacayo*.
The Tandulaveyāliya (p.5, 7) mentions the accumulation of *pitta* and blood in the sixth month (*pitta-soṇiaṃ uvaciṇei*).
PARALLELS:
YS 3,80–81: *sthairyaṃ caturthe tv aṅgānām pañcame śoṇitodbhavaḥ / ṣaṣṭhe balasya varṇasya nakha-romṇāṃ ca sambhavaḥ /80/ manaś caitanya-yukto 'sau nāḍī-snāyu-śirā-yutaḥ / saptame cāṣṭame caiva tvaṅ-māṃsa-smṛtimān api /81/*
Cf. YAMASHITA 2001/2002, pp.97–99.
AgniP 369,20cd: *ṣaṣṭhe ceto 'tha jīvasya duḥkhaṃ vindati saptame.*
ViṣṇudhP 2,114,3cd: *saptame ca tathā māsi prabodhaś cāsya jāyate /3cd/ sa jīvo 'pi māṇḍūkaḥ śīte śītāditobhyasuḥ / mūḍhas tiṣṭhati, dharmajña, ṣaṇ-māsān garbhagas tathā /4/.*

In the sixth [month], [there is] the discernability of the bones, the cords (*snāyu*)[599], the fingernails, the hairs of the head (*keśa*) and the body-hairs (*roman*),[600] (śl.33cd)
and power[601] and colour are accumulated. (śl.34a)

But in the seventh [month], [there is] wholeness of the limbs.[602]

GaruḍaP (KIRFEL 1954) v.24: *tathā māse tu sampūrṇe pañca tattvāni dhārayet / māsa-dvaye tu sañjāte tvacā medaś ca jāyate /24/ majjāsthīni tribhir māsaiḥ keśāṅgulyaiś caturthake / karṇau ca nāsikā vakṣo jāyeran māsi pañcame /25/ kaṇṭha-randhrodaraṃ ṣaṣṭhe guhyādir māsi saptame / aṅga-pratyaṅga-sampūrṇo garbho māsair athāṣṭabhiḥ /26/.*

Hastyāyurveda, sthāna 3, adhyāya 8 (p.414): *buddhiḥ sañjāyate cāsya saptame māsi, pārthiva / aṣṭame sthira-sarvāṅgaḥ sandhi-snāyu-samanvitaḥ / tvag-asthi-māṃsa-medobhiḥ saṃyuktaḥ saṃvivardhate /109/.*

599 *Snāyu* "cord". On this term, cf. MÜLLER 1961, p.149 (item n. 239); DAS 2003A, p.584ff. (on *snāyu-*).

600 The Tandulaveyāliya states that in the seventh month there are 700 vessels (*sirā*), 500 muscles (*pesī*), 9 ducts (*dhamaṇī*), 9,900,000 pores (*roma-kūva*) excluding those of beard-hairs and head-hairs, and 35,000,000 including them. In SR śl.94cd–115, the same numbers are mentioned for these components.

601 *Bala*. The various functions of *bala* are listed in SU sūtra., 15,20. The list includes the clearness of voice and colour (*svara-varṇa-prasāda*). For *bala-varṇa* "power and colour", also cf. DAS 2003A, p.468.

602 COMBA 1981 (p.202, note 27) reports that the wholeness of the limbs (*aṅgapūrṇatā*) is mentioned in the CA and AH. But in my observation, the parallel of the CA (Sharma's edition, cf. Bibliography) peculiarly does not contain this expression, but describes the same state in a different manner (see the parallel quoted below). COMBA (ibid.) reports that the SU and CA do not deal with the respective developments of the various components, because the two texts consider that all the components develop simultaneously. COMBA (ibid.) concludes that the AH (śārīra., 1,57cd–58ab) is most closely related to ŚG 8,23cd–24ab (= SR śl.33cd–34ab).

But when comparing the AS with the AH, the AS (śārīra., 2,23) which COMBA did not consult is more complete in wording than the AH. Therefore, the AS is the closest to the SR śl.33cd–34ab.

PARALLELS:

AS śārīra., 2,23: *ṣaṣṭhe keśa-roma-nakhāsthi-snāyvādīny-abhivyaktāni bala-varṇopacayaś ca / saptame sarvāṅga-sampūrṇatā.*

CA śārīra., 4,22–23: *ṣaṣṭhe māsi garbhasya bala-varṇopacayo bhavaty, adhikam anyebhyo māsebhyaḥ [...] /22/ saptame māsi garbhaḥ sarva-bhāvair āpyāyyate [...] /23/.*

Hastyāyurveda, sthāna 3, adhyāya 8 (p.413): *caturthe māsi jāyante snāyvasthīni sirās tathā /96cd/ jāyate tasya nirvṛttir aṅgānāṃ tu mahīpate / nirvṛttir anupūrvyeṇa yathāvac chrotum arhasi /97/ śiro grīvā ca pṛṣṭhaṃ ca karṇau cā*

(śl.34b)[603]

Covering the openings of the ears with both hands, placed (lit. made to be) within the laps/lobes_of_the_ears (*pāli*) [604], he [= the foetus] (śl.34cd)
sits, anxious through dwelling as_a_foetus/in_the_womb (*garbha-saṃvāsād*), connected with the uterus (*garbhāśayānvitaḥ*). (śl.35ab)[605]

'syaṃ karas tathā [...] /98/ 99cd: *chavī ca pañcame māsi yathavat pravibhajya ca /99cd/ cakṣuḥ śrotram atho jihvā prāṇo 'pānas tathaiva ca / vṛṣkau nakhāni romāṇi yakṛd antrāṇi puṣphasam /100/ vyajyate hṛdayaṃ caiva ṣaṣṭhe māsi narādhipa / sambhavanti mahīpāla garbhasthasyeha dantinaḥ /101/ jīvopajīvam āśritya garbho bhavati kālataḥ /102ab/.*

603 ŚG 8,23cd–24ab is identical to SR śl.33cd–34ab (cf. COMBA 1981, p.202). The ŚG (Bombay 1987) contains *-nakha-* and *-loma-* instead of *-nakhara-* and *-roma-* of the SR. The ŚG's reading *-nakha-* is problematic, causing metric shortage. On the other hand, the SR's manuscripts, *ka.* and *gha.*, contain the same reading *-loma-* as in the ŚG.

604 According to this translation, in agreement with comm. K., the hands of the foetus are within the circumference of the ears. Alternatively we could translate: "covering the openings of the ears with both hands which have the lobes of the ears within"; in this case, the hands would be over the ears.
As to the translation of the term *pāli* as "lap", this is supported by the fact that ŚG 8,24cd which is identical to SR 34cd contains *pādāntarita-* instead of *pālyantarita-* (see below). The parallels in the GaruḍaP (KIRFEL 1954, śl.73–74) and the MārkaṇḍeyaP (11,6ff. also seem to have a similar meaning, both containing the term *jānu*.
Further: *pālyantarita-* could be interpreted in the meaning of *antarita-pāli-*, [i.e.] "with the lobes of the ear inside" or "with the outer ear inside", because, according to Pāṇini, *-ta-* participles can change their position in a compound. Cf. DAS 1990, pp.49–53.
Actually a Yogin in the womb posture is presented in a miniature attributed to the artist B, *Bahr al-Hayat* (year 1600–5), covering the ears with the hands, and burying his head in his laps (LEACH 1995, p.551). LEACH (ibid., p.559) comments, "a posture emulating a return to the womb and called by the name (*garbha*)".

605 ŚG 8,24cd–25ab (8,25 consisting of only two *pāda*-s) is identical to SR śl.34cd–36ab. The ŚG (Bombay 1987) contains readings *pādāntarita-, saṃvāsād asti* and *garbhalayānvitaḥ* instead of *pālyantarita-*, *saṃvāsād āste* and *garbhāśayānvitaḥ* of the SR. But edition A of the ŚG consulted by COMBA contains *duḥkhabhayānvitaḥ* instead of *garbhāśayānvitaḥ* in SR śl.35b (cf. COMBA 1981, p.202). COMBA translates, "Egli (il feto), avendo coperto le aperture delle orecchie con le mani, nascoste dai piedi, se ne sta angosciato, timoroso del dorore causato dal ricettacolo uterino" (COMBA ibid. p.202). She also notes two variants, the first is: ed. A *garbhalayānvitaḥ*, which is same as in

Remembering [606] various births [607] and torments experienced

the Bombay edition (1987) of the ŚG. She interprets, "connected with prenatal condition (*garbha*) and/up_to destruction/death (*laya*)", or "connected with [the life which lasts from] birth [until] death". But *garbhalaya* could be a mistake for *garbhālaya* which means the same as *garbhāśaya* "uterus". She notes the variant of the ŚG in ed. C and B as *garbhabhayānvitaḥ*.

The following verses, ŚG 8,26–32, describing the embryo's agony in the uterus, do not have parallels in the SR. ŚG 8,33, which is the closing word to the description of the agony in the uterus, is again identical to SR śl.36ab.

This might mean, that these verses from the common source text describing the agony in the uterus were omitted in the SR, or that, on the other hand, these verses were enlarged in the ŚG.

I prefer the latter, because I have the feeling that ŚG 8,26–32 describing the embryo's agony *in the uterus* and ŚG 8,33 (= SR śl.36ab) which is the closing word do not quite seem to fit. ŚG 8,33 mentions the agonies which the foetus experienced *in its previous lives*, but not the agony which it suffers in the uterus at this very moment. So, the connection between ŚG 8,33 and its foregoing verses is somehow awkward.

The foetus' posture in the uterus is not mentioned in the classical medical texts (SU, AS, AH, CA) or YS, AgniP or ViṣṇudhP. It is only mentioned in other Purāṇic texts, e.g. GaruḍaP (KIRFEL 1954, śl.73–74) and MārkaṇḍeyaP (11,6ff). The topics of SR śl.34cd–36ab, i.e., the foetus' posture, remembrance of previous lives (*jāti-smaraṇa*) and practice (*abhyāsa*), are not dealt with by the classical medical texts, but by the Purāṇas.

GaruḍaP śl.73–74: "Nachdem er (= der Embryo) die Handflächen an die Seiten der Knie gelegt hat, wächst er heran; dann, sobald er die Finger der Hand besitzt, /73/ sind die Daumen auf die Knie aufgelegt, ferner die Augen auf die Rücken der Knie und die Nase inmitten der Knie; so kommt der Mensch, der sich im Schosse des Weibes befindet, allmählich zum Wachstum. /74/" (tr. by KIRFEL 1954).

MārkaṇḍeyaP. 11,6ff: *tadvat prayāty asau vṛddhiṃ sa kośo 'dhomukhaḥ sthitaḥ /6/ tale tu jānu-pārśvābhyāṃ karau nyasya sa vardhate / aṅguṣṭhau copari nyastau jānvor agre tathāṅgulī / jānupṛṣṭhe tathā netre jānumadhye ca nāsikā / sphicau pārṣṇi-dvayasthau ca bāhujaṅghe bahiḥsthite /8/ evaṃ vṛddhiṃ kramād yāti jantuḥ strī-garbha-saṃsthitaḥ / anya-sattvodare jantor yathā rūpaṃ tathā sthitiḥ /9/.*

The description of the foetus' posture in ŚB 3,2,1,6, that the foetus has the hands closed, does not accord with the descriptions in the later texts mentioned above.

606 For *jāti-smaraṇa* "remembrance of (a) previous life/lives" in the embryological passages of Purāṇic texts, cf. HARA (1977 and 1980).

The term *jāti-smaraṇa* is explained in CA śārīra., 3,13 as follows: "If the mind is united with the attribute 'pure' (*śuddha*), it also remembers the previous life

before, (śl.35cd)
he remains (*vartate*)[608] eagerly engaged in exercise[609], ruminating

---

(*jāter atikrāntāyā api smarati*)." Also see SU śārīra., 2,57 (*bhavanti sattva-bhūyiṣṭhāḥ pūrva-jāti-smarā narāḥ*).
The knowledge of previous lives (*pūrva-jāti-jñāna*) is said to be caused by the *saṃskāra* of the previous lives (Yogabhāṣya 2,18, referred to by HALBFASS 2000, p.200).
The Śāradātilaka's (1,48cd–50) description of the foetus' agony and birth contains this topic: *[...] garbhe pūrva-janma-śubhāśubham* /48d/ *smaraṃs tiṣṭhati duḥkhātmā cchanna-deho jarāyuṇā* /49ab/.

607 *Nānā-jāti* "various births". A similar expression *bhavo jāti-sahasreṣu* "existence in thousands of births" is found in YS 3,64a. Vīramitrodaya's comm. on the YS explains it as meaning *manuṣya-go-śūkarādi-nānā-janma* "various births like human, cow, hog etc." Therefore *jāti* might also mean "kind", "species".
This statement perhaps refers to the actual embryonic development which goes through various phases like fish, amphibian, reptilian and mammal. The multiplicity of the embryo's form is mentioned in ŚB 4,5,2,12 (WEBER's ed., p.395, l.9): *víṣurūpa íti víṣu-rūpā iva hi gárbhāḥ* "The embryos have multiple forms".

608 "Eagerly engaged in exercise" (*abhyāsa-tatpara*) perhaps refers to the foetus' restless movements in the uterus. As a matter of fact, CA śārīra., 4,15 on *dohada* states that, when the sense organs are manifested, the foetus begins to perceive feelings and urges in its consciousness (*cetasi vedanā nirbandham prāpnoti*); from then onward, it begins to move (*garbhaḥ spandate*) and to desire what it experienced in its previous life (*prārthayate janmāntarānubhūtaṃ yat kiñcit*).
The foetus' movement is mentioned in ŚB 3,1,3,28 (WEBER's ed., p.230, ll.8–9): *antareṇa vai yóniṃ gárbhaḥ sáñcarati sa (= the sacrificer) yat sa tatraíjati tvat pári tvad āvártate tásmād ime gárbhā éjanti tvat pári tvad ā́vartante tásmād asyaiṣá sacaró bhavaty ā́ sutyā́yai.*

609 *Abhyāsa* "exercise". According to Nirukta, pariśiṣṭa 2,3, "excersise" is that of Sāṅkhya and Yoga, or studying the twenty-five elements beginning with *puruṣa* (*sāṅkhyaṃ yogaṃ samabhyasyet puruṣaṃ vā pañca-viṃśakam*), cf. WINDISCH 1908, p.90.
However, the term *abhyāsa* "habit" or "custom" is explained by the commentator Cakrapāṇi on CA śārīra., 8,16: *janma janma yad abhyastaṃ dānam adhyayanaṃ tapaḥ / tenaivābhyāsa-yogena tac caivābhyasate naraḥ.*
That means, an action which the person habitually did (*abhyāsa*) in the previous life/lives is recorded in the mind (*antaḥkaraṇa*), in the form of an imprint (*vāsanā* or *saṃskāra*); later, this imprint causes actions in the present life, which is comparable to a sprouting seed (ROŞU 1978, p.188).
If this is the case in SR śl.36ab, then the foetus is again engaged in the actions which were habitually done in its previous life.

upon the means of liberation.[610] (śl.36ab)[611]

610 D reads *dhyānatatparaḥ* "eagerly engaged in meditation/pondering" instead of *abhyāsatatparaḥ*.

611 ŚG 8,33 is identical to SR śl.35cd–36ab. The ŚG (Bombay 1987) contains different readings, *evaṃ smaran purā prāptā* instead of *smaran pūrvānubhūtāḥ sa*; then *api dhyāyan* instead of *abhidhyāyan*.
According to COMBA 1981, ed. C of the ŚG contains *api dhyāyan*; the ed. A contains *abhidhyāyan* which is identical to the SR. COMBA (ibid., p.205) translates ŚG 8,33, "Così ricordando le numerose nascite e le sofferenze sperimentate in precedenza, [il feto], meditando sul mezzo per sarvarsi, se ne sta unicamente intento alla ripetizione".
As to the term *mokṣa* "liberation", the question is whether it means liberation (*mokṣa*) from the cycle of births (*saṃsāra*), or literally from the uterus. Principally liberation as intended here seems to be the liberation from the cycle of births. But the liberation from the uterus also seems possible, as the agonies which the foetus suffers in the uterus are sometimes compared to those in hell (e.g. ŚG 8,32: *garbha-śayyāṃ samāruhya duḥkhaṃ yādṛk mayāpi tat / nātiśete mahāduḥkhaṃ niḥśeṣaṃ narakeṣu tat*).
ViṣṇudhP 114,16 seems to mean the liberation from the uterus: *tataḥ karma kariṣyāmi yena mokṣo bhaven mama / nāsti mokṣaṃ vinā saukhyaṃ garbha-vāse kathaṃ cana*, which roughly means "without liberation, there is no pleasure in staying in the uterus". MārkaṇḍeyaP 11,4 also contains the expression *mukta-mātram ihodarāt* "as soon as [I am] liberated from this uterus".
In BhāgP 3,31,17 the foetus wishes to escape (*vi-√vas*) from the uterus, but neither *mokṣa* nor the derivatives of *√muc* occur in this passage.
In contrast, in BhāgP 3,31,20, the foetus' wish is quite the opposite, i.e. to remain within and not to exit the uterus, although remaining in the uterus would mean having to suffer the agonies further. The reason for this wish is the foetus' fear that, after birth, it might forget its determination to attain liberation (cf. G.M. SHIVARAM 2001, p.99).
This topic is dealt with by numerous other texts. E.g. Nirukta, pariśiṣṭa 2; Garbhopaniṣad (cf. WINDISCH 1908, p.90). The Nirukta and Garbhopaniṣad locate this state in the ninth month. Nirukta, pariśiṣṭa 2,3: *avāṅmukhaḥ pīḍyamāno jantuś caiva samanvitaḥ / sāṅkhyaṃ yogaṃ samabhyasyet puruṣaṃ vā pañcaviṃśakam*. Garbhopaniṣad section 3: *atha navame māsi sarva-lakṣaṇa jñāna-sampūrṇo bhavati, pūrvajātiṃ smarati śubhāśubhaṃ ca karma vindati.* Section 4: *atha yoni-dvāraṃ samprāpto yantreṇāpīḍyamāno mahatā duḥkhena jāta-mātras tu vaiṣṇavena vāyunā saṃspṛṣṭas tadā na smarati janma-maraṇāni na ca karma śubhāśubhaṃ vindati.*
The foetus remembers its previous lives and does exercises (*abhyāsa*) in order to attain liberation. But at the moment of delivery, it forgets its determination and loses its remembrance due to the impact of the procreatory wind (*vaiṣṇava vāta*) and the excessive pressure of the parturient canal.

In the eighth [month] the skin and [the power of] remembrance[612]

PARALLELS:
VisṇudhP 114,5cd–18ab: *aṣṭame navame māsi bhṛśam udvijate tadā /5cd/ jarāyu-veṣṭito deho mūrdhni baddhāñjalis tadā / madhye klības tu vāme strī dakṣiṇe puruṣas tathā /6/. 114,7: tiṣṭhaty uttara-bhāge tu pṛṣṭhasyābhimukhas tathā / yasyāṃ tiṣṭhati sā yonau tāṃ tu vetti na saṃśayaṃ /7/ sarvaṃ smarati vṛttāntaṃ tvārabhya janmatas tathā /8ab/ andhakāre ca mahati pīḍāṃ vindati, bhārgava /8/ kīṭa-gandhena mahatā kalmaṣaṃ vindate param /9/ mātrānīte jale pīte paraṃ śītam upāśnute /9/ uṣṇe bhukte tadā dāham param āpnoti, bhārgava / vyādhibhiḥ paramām pīḍāṃ tīvrām prāpnoti duḥsahām /10/ vyāyāme ca tathā mātuḥ klamam mahad upāśnute / vyādhitāyāṃ tathā tīvrāṃ vedanāṃ samupāśnute /11/ bhavanti vyādhayaś cāsya tatra ghorāḥ punaḥ punaḥ / na ca mātā pitā vetti tadā kaś cic cikitsakaḥ /12/ saukumāryād rujaṃ tīvrāṃ janayanti tu tasya tāḥ / ādhibhir vyādhibhiś caiva pīḍyamānasya dāruṇaiḥ /13/ svalpa-madhye 'tha tat-kālaṃ yāti varṣa-śatopamam / santapyate tathā garbhe karmabhiś ca purātanaiḥ /14/ manorathāni kurute sukṛtārtham punaḥ punaḥ / janma ced aham āpsyāmi mānuṣye daiva-yogataḥ /15/ tataḥ karma kariṣyāmi yena mokṣo bhaven mama /nāsti mokṣaṃ vinā saukhyaṃ garbha-vāse kathaṃ cana /16/ garbha-vāsaś ca sumahalloke duḥkhaika-kāraṇam / evaṃ vicintayānasya tasya varṣa-śatopamam /17/ māsa-trayaṃ tad bhavati garbha-sthasya prapīḍyataḥ /18ab/.*
GaruḍaP (KIRFEL1954) 77cd–80ab.
MārkaṇḍeyaP 11,13: *smṛtiṃ tatra prayānty asya bahvyaḥ saṃsārabhūmayaḥ / tato nirvedam āyāti pīḍyamāna itas tataḥ /13/ punar naivaṃ kariṣyāmi mukta-mātram ihodarāt / tathā tathā yatiṣyāmi garbhaṃ nāpsyāmy ahaṃ yathā /14/ iti cintayate smṛtvā janma-duḥkha-śatāni vai / yāni pūrvānubhūtāni daiva-bhūtāni yāni vai /15/ tataḥ kāla-kramāj jantuḥ parivartaty adhomukhaḥ / navame daśame māsi sañjāyate tataḥ /16/* This text considers "liberation" as that out of the belly.

612 D *tvakśrutī* "the skin and [the power of] hearing". The readings of the two commentaries are different: comm. K reads *smṛti* while comm. S reads *śruti*. The ŚG's editions (Bombay 1987; the ed. A of COMBA 1981) read *śruti* in ŚG 8,34 (= SR śl.36cd–37ab). The ed. B of ŚG contains another variant, *-sṛtī* "going" or "conduct" (cf. COMBA 1981, p.205, footnote 37). But this variant does not fit the context. I guess it to be a misprint of *-smṛti*.
YS 3,81 mentions *tvac* and *smṛti* together in the eighth month, cf. YAMASHITA 2001/2002, pp.98–99. Except for YS, there is no text on embryology which mentions skin and remembrance, or hearing in the eighth month. If the SR's variant *tvaksmṛti* is correct, it would be included in the SR's small amount of verses parallel to the YS, cf. my footnote 186 in *Situating the text*.
In contrast, the Sāṅkhya system often lists the senses of hearing and touching side by side, cf. SR śl.49ab (*jñānendriyāṇi śravaṇaṃ sparśanaṃ [...]*); Ḍalhaṇa on SU sūtra., 15,20 (*ābhyantarāṇaṃ karaṇānāṃ śrotra-tvak- [...]*). The variant *tvak-śruti* might refer to this.

are[613] [there], and the vital fluid (*ojas*)[614]. And this[615], arising in the heart, (śl.36cd)
is pure and yellowish (*ā-pīta*) red[616], deemed[617] the cause (*nimitta*) with regard to life. (śl.37ab)[618]

Thus we are not able to definitely decide between *śruti* and *smṛti*.
I consider "[the power of] remembrance" (*smṛti*) to be problematic, as it is not logical for the foetus to remember its previous lives (SR śl.35) before it acquires the power of remembrance (SR śl.36).
The expression *tvak-śrutī* in ŚG 8,34 (= SR śl.36cd–37ab) is interpreted by COMBA 1981 (p.205) as meaning "[the formation of] the ear from skin" (Nell' ottavo [mese] si formano lo orecchie (śrutī) dalla pelle). But I am afraid that her interpretation is not relevant.

613 Optative *syātām.*

614 For *ojas* "vital fluid/force", cf. DAS 2003A, pp.530–535.

615 D reads *ojas tejaś ca* "[and] the vital fluid and fiery energy (*tejas*)" instead of *ojaś caitac ca.*
Perhaps, the report in COMBA's (1981) footnote 37, quoted below, might have something to do with this variant: According to Ḍalhaṇa's commentary on the SU, the *ojas* is the oily, subtle part, or the *tejas* of semen; Aruṇadatta calls *ojas* as "*sarvadhātūnāṃ tejaḥ śārīrasambhavaṃ*". Also cf. DAS 2003A, p.530.
The Ānandāśrama ed. (1896) reads *ojaś cetaś ca*, which is obviously a mistake. The editor corrects it into *ojaś caitaś ca*. This is not noted by the Adyar ed.

616 A similar statement on the yellowish-redness of *ojas* is found in CA sūtra., 17,74 (*hṛdi tiṣṭhati yac chuddhaṃ raktam īṣat sapītakam / ojaḥ śarīre saṅkhyātaṃ* [...]).
COMBA 1981 (p. 205, footnote 37) reports that the comm. A on the ŚG 8,34 (= SR śl.36cd–37ab) explains the prefix *ā-* as referring to all the three adjectives, *śuddha, pīta,* and *rakta*. She translates, "[...] e nel cuore ha origine il succo vitale [o] calore, che è in parte bianco (*śuddha*), in parte (*ā*) giallo (*pīta*), in parte rosso (*rakta*), ed è considerato causa della vita (*jīvite*)".
This is in accordance with Cakrapāṇidatta's interpretation of *īṣat* in CA sūtra., 17,74; he considers *īṣat* as qualifying both, *raktam* and *sapītakam.* Also cf. YANO 1988, p.121, note 17.
For the verses in the medical texts, parallel to ŚG 8,34 (= SR śl.36cd–37ab), dealing with *ojas*, cf. COMBA 1981, p.206, footnote 37 and 38.

617 The manuscripts *ka.* and *ga.* read *gatam* instead of *matam.* "[...] arising in the heart, pure and yellowish red, has become (*gatam*, lit. gone to [be]) the cause (*nimittaṃ*) with regard to life (*jīvite*)".

618 SR śl.36cd–37ab is identical to ŚG 8,34. The ŚG (Bombay 1987) contains *tvak-chutī, ojas tejaś ca,* and *jīvitam matam* instead of the SR's readings *tvak-śrutī, ojaś caitac ca,* and *jīvite matam.* The ŚG's *tvak-chutī* is obviously a misprint for *tvak-śrutī*, because its Hindi translation reads *śruti* correctly. Also the

Unsteady, it (*tat*)[619] runs now (*punas*) to the mother[620] and now (*punas*) to the embryo.[621] (śl.37cd)

Therefore, [if the embryo is] born in the eighth month[622], [it] does not live[623], abandoned by the vital fluid (*ojas*). (śl.38ab) [624]

---

manuscript D of the SR, which is noted by the Adyar edition, reads *jīvitam*, which is identical to the ŚG (Bombay 1987).

The edition A of the ŚG, which is consulted by COMBA 1981, also reads *-śrutī*. This edition, however, contains *jīvite* which is also contained in the Adyar edition of the SR. COMBA (ibid., p.32, footnote 37) notes variants: the edition B reads *-sr̥tī* instead of *-śrutī*, and *śubhram* instead of *śuddham*; the edition C also reads *jīvitam* instead of *jīvite* like the ŚG's edition which I used (Bombay 1987). I have already discussed the reading *tvak-smr̥ti* in the footnote 612 on SR śl.36cd ("remembrance").

619 The Ānandāśrama ed. (1896) notes the variant of *ṅa.*, *vipradhāvati* instead of *tatpradhāvati*. But ŚG 8,35ab (= SR 37cd) contains *tatpradhāvati*.

620 D reads *nārīṃ* "the woman" instead of *ambāṃ* "the mother".

621 I.e. through the umbilical cord. CA śārīra., 4,24 states that the *ojas* oscillates between the mother and foetus through the umbilical cord carrying nourishment, because the foetus is not yet complete (*aṣṭame māsi garbhaś ca mātr̥to garbhataś ca mātā rasahāriṇībhiḥ saṃvāhinībhir muhur muhur ojaḥ parasparata ādadāte garbhasyāsampūrṇatvāt*), cf. COMBA 1981, p.206, note 38. For the plural number of tubular structures joined to the umbilical cord, cf. DAS 2003A, pp.483–484. The AS contains almost the same wording as the CA. A parallel is found e.g. in GaruḍaP (KIRFEL 1954) v.27: *aṣṭame calate jīvo dhātrī-garbhe punaḥ punaḥ / navame māsi samprāpte garbhasthaujo dr̥ḍham bhavet /27/* (KIRFEL 1954 considers the first half of this verse to be a secondary insertion).

622 The Ānandāśrama ed. (1896) reads *māse* instead of *māsi*.

623 Contrastingly the Śatapathabrāhmaṇa states that the child who is born after the sixth month *does* survive. ŚB 9,5,1,63 (WEBER's ed., p.747, l.11ff.): *ṣáṇ-māsyā vā́ 'antamā gárbhā jātā́ jīvantī́ti. Sa yady ásaṃvatsara-bhr̥te mahád ukthaṃ śáṃsed [...].*

624 SR śl.37cd–38ab is identical to ŚG 8,35 (cf. COMBA 1981, p.206). The ŚG (Bombay 1987) contains *mātaraṃ ca punar garbham* instead of the SR's reading *punar ambām punar garbham*, and *jāto 'ṣṭame garbho* instead of the SR's *jāto 'ṣṭame māsi*. The edition A of the ŚG used by COMBA reads *māsi*, like the SR. COMBA also notes the variant of the edition C *garbho*, like the Bombay edition (1987) of the ŚG (Cf. COMBA ibid., p.206, footnote 38).

COMBA (ibid.) remarks that YS 3,82 (YS 3,71 in the edition used by her) is in vocabulary the closest to ŚG 8,35 (= SR śl.37cd–38ab), in comparison with the other texts: YS 3,82 (*punar dhātrīm punar garbham ojas tasya pradhāvati / aṣṭame māsyato garbho jātaḥ prāṇair viyujyate*).

[It has] stability (*avasthānam*) for [only] some while[625], due to the [subliminal] imprint (*saṃskāra*) [of body functions], like a severed limb. (śl.38cd)

The time of birth is[626] in the months beginning with the ninth. (śl.39ab) [627]

Besides the parallels noted by COMBA (ibid.), AS śārīra., 2,24 also deals with this topic. Yet, the YS's wording is closer to that of the SR than the wording of the AS is.

YAMASHITA 2001/2002, pp.99–100, states that YS 3,82 is related to CA śārīra., 4,24 and CA sūtra., 17,74.

625 The Ānandāśrama ed. (1896) reads *kañcit kālam* instead of *kiñcit kālam.*

626 Optative *syāt.*

627 SR śl.38cd–39ab is identical to ŚG 8,36 (COMBA 1981, p.207). The edition of the ŚG (Bombay 1897) which I used contains *pīḍitāṅgavat* instead of *khaṇḍitāṅgavat.* The ed. A of the ŚG used by COMBA (cf. ibid., note 39) contains *prasavasyāsya* instead of *prasavasya syān,* but COMBA notes the variant of the ed. E and C, *prasavasya syān.*

This verse deals with a premature birth in the eighth month. A baby that is born prematurely survives only for a short time, like a severed limb still moves by itself; then the baby dies. A parallel is contained in AS śārīra., 2,25, whose vocabulary is very close to that of SR śl.38cd. There it is stated that, even if the baby, being born in the eighth month, breathes for a while, it is merely a reaction due to the imprint of *ojas* like the reaction of a limb cut off (*yady api ca kiñcit kālam asyocchvasanaṃ syāt tac chinnasyevāṅgasyaujaḥ-saṃskārānuvṛtti-kṛtam*). Except for the AS, none of the texts that I consulted contains a parallel to SR 38cd.

Because of the ŚG's reading *pnḍitāṅgavat* instead of *khaṇḍitāṅgavat,* COMBA (ibid.) interprets ŚG 8,36ab (= SR śl.38cd) differently; that the condition (*avasthāna*) [of the foetus in the uterus remains] for some time [weak] like a tormented limb (*pīḍitāṅgavat*). But because of the parallel in the AS (*chinnasyevāṅgasya*), I prefer *khaṇḍitāṅgavat.*

COMBA ibid. (note 39) states that the CA and AH locate the suitable moment of birth between the first day after the eighth month (i.e. the first day of the ninth month) and the tenth month; but the SU locates it between the ninth month and the twelfth month; a birth after this period of time is considered abnormal. The AS integrates the two theories, mentioning a period from the first day after the eighth month until the twelfth month. The YS (3,82cd–83) locates the suitable birth in the ninth and tenth months, cf. YAMASHITA 2001/2002, pp.100–101. The SR mentions the ninth month as the starting point of birth, but does not mention a time limit.

The Atharvaveda already states that the child is born in the tenth month, cf. AV 5,25,10: *dhā́taḥ śréṣṭhena rūpéṇāsyā́ nā́ryā gavī́nyóḥ / púmāṃsam putrám ā́ dhehi <u>daśamé māsí sū́tave</u>.* ŚB 4,5,2,4 (WEBER's ed., p.394, l.3f.) states that the

The embryo's tube (*nāḍī*) situated at the navel[628], whose appellation is *"parā"*[629], carrying the mother's *āhāra-rasa*[630], is joined with

child is born after ten months: *dáśa-māsya íti yadā vai gárbhaḥ sámṛddho bhávaty, átha dáśamāsyas*. Parallels are also found in: GaruḍaP (KIRFEL 1954) v.31ab; Hastyāyurveda sthāna 3, adhyāya 8 (p.414).

628 The Ānandāśrama ed. (1896) reads *nābhisthā nāḍī* instead of *nābhisthanāḍī*.

629 My translation given above follows SU śārīra., 3,31. The SU mentions two tubes, i.e. the mother's tube carrying nutrition (*rasavahā nāḍī*) and the tube attached to the foetus' navel (*garbha-nābhi-nāḍī*) (SU śārīra., 3,31, *mātus tu khalu rasa-vahāyāṃ nāḍyāṃ garbha-nābhi-nāḍī pratibaddhā, sā 'sya mātur āhāra-rasa-vīryam abhivahati [...]*). The two tubes of the mother and foetus are connected to each other.
This is a very old theory which is also contained in the Jaina canon, Viyāhapannatti 406,15–32, cf. CAILLAT 1974 (I), pp.52–53. The Jaina text on embryology, the Tandulaveyāliya (sūtra 4), also mentions the two tubes, one of which is called "the other tube" (*avarā*), cf. CAILLAT (ibid.). There, it is stated that "[there are the tube] carrying the *rasa* of the mother's life (*māu-jīva-rasa-haraṇī*) [and that] carrying the *rasa* of the son's life (*putta-jīva-rasa-haraṇī*). [The tube] united with the mother's life touches the son's life. Thus, [the mother] eats, then it is digested (*pariṇāmei*). The other (*avarā*), too, which is united with the son's life, touches the mother's life, thus [the foetus] is accumulated." Since *avarā* is the Prākṛt form for Skt. *aparā*, the tube called *parā* of the SR might be identical to "the other tube" connected with the foetus. If this is the case, *parā* "other" is an adjective of *nāḍī*.
Or, it might be better to analyse *nāḍīmanubaddhāparābhidhā* differently from the Adyar edition, i.e. read *aparā* instead of *parā*.
However, AS śārīra., 2,31 contains another theory. It mentions *aparā* but this term has another meaning, namely "placenta". AS śārīra., 2,31: *tato vyaktībhavad-aṅga-pratyaṅgasyāsya nābhyām pratibaddhā nāḍī, nāḍyām aparā tasyām mātṛ-hṛdayam. tato mātṛ-hṛdayād āhāra-raso dhamanībhiḥ syandamāno 'parām upaiti* , "Thereafter (*tato*), with the navel of it (= the foetus) whose limbs and secondary appendages are developing, a tube is connected. With the tube, the placenta (*aparā*) [is connected]. To that [placenta], the mother's heart [is connected]. From that heart of the mother, the juice of food reaches the placenta, flowing (*syandamāno* lit. dripping) through the vessels". The commentator Indu explains *garbhasya nābhyām pratisambaddhā nāḍī bhavati, tasyāṃ nāḍyām pratibaddhā pūrvoktāparā tasyām pratibaddham mātṛ-hṛdayam* "There is a tube connected to the foetus' navel. To that tube, the before-mentioned placenta (*aparā*) is connected. To that (= placenta), the mother's heart is connected."
Due to this statement, we had better read, in SR śl.39d, *aparābhidhā* instead of *parābhidhā*, and interpret SR śl.39cd separately from śl.40ab. In this case, it would mean, "[That whose] appellation is "placenta" (*aparā*) is joined to the mother's tube carrying *rasa*. (39cd) The tube situated in the navel of the foetus

is that carrying the mother's *āhāra-rasa*. (40ab)" For the term *aparā* meaning "afterbirth", cf. DAS 2003A, p.482; pp.517–518.

CA śārīra., 6,23 also contains almost the same statement, mentioning aparā "afterbirth/ placenta" CA śārīra., 6,23, *nābhyāṃ hy asya nāḍī prasaktā, nāḍyāṃ <u>cāparā</u>, <u>aparā</u> cāsya mātuḥ prasaktā hṛdaye, mātṛ-hṛdayaṃ hy asya tām <u>aparām</u> abhisamplavate sirābhiḥ syandamānābhiḥ*. The commentator Cakrapāṇidatta states *<u>aparā</u> garbhasya nābhi-nāḍī-pratibaddhā <u>aparā</u> iti loke khyātā*, "it is stated by people that the placenta is connected with the tube of the foetus' navel".

Strangely, PRIYAVRAT SHARMA 1996 translates CA śārīra., 6,23 "The [umbilical] cord is attached to the umbilicus, placenta is attached to the cord on one side, and to the mother's heart on the other. Mother's heart floods the placenta (with nutritive fluid)". Namely, PRIYAVRAT SHARMA (ibid.) translates the second *aparā* "the other [side]" differently from the first and third "placenta". This seems to be an inconsistent translation.

With regard to the Jaina theory of *avarā nāḍī* mentioned above and my interpretation of *aparā-nāḍī* as "the other tube", we are not sure if the term *avarā* or *aparā* is really an adjective meaning "the other". We could imagine that it might have originally denoted "afterbirth", which was misinterpreted by these authors. Or, it might be the result of some contamination.

Anyway, all these statements seem to support my reading *aparā* instead of *parā* in SR śl.39d. But, in this point, neither of the two commentaries on the SR is helpful, as they do not mention this term.

AH śārīra., 1,56 (DAS & EMMERICK's ed.) runs *garbhasya nābhau mātuś ca hṛdi nāḍī nibadhyate / yayā sa puṣṭim āpnoti kedāra iva kulyayā*, namely, the umbilical cord connects the foetus' navel with the mother's heart; through this cord, the foetus gets nutrition. But the term *aparā* "placenta" is not mentioned.

PARALLELS:

GaruḍaP (KIRFEL 1954) 75b: *[...] bhukta-pītena jīvati / nāḍī cāpyāyanī nāma nābhyāṃ tasya nibadhyate /75/ strīṇāṃ tathāntra-suṣire sa nibaddhaḥ prajāyate / krāmanti bhukta-pītāni strīṇā garbhodare tathā /76/ tair apyāyita-deho 'sau jantur vṛddhim upaiti ca /77ab/.*

MārkaṇḍeyaP 11,11: *nāḍī cāpyāyanī nāma nābhyāṃ tasya nibadhyate / strīṇāṃ tathāntra-śuṣire sā nibaddhopajāyate /11/.* It roughly means: "The tube called *āpyāyanī* is connected with the foetus' navel and the end of the intestines (*antra*) of women." 11,12: *krāmanti bhukta-pītāni strīṇāṃ garbhodare yathā / tair āpyāyita-deho 'sau jantur vṛddhim upaiti vai /12/.*

Hastyāyurveda sthāna 3, adhy.7 (p.403): *nābhinibaddhayā nāḍyā mātur āhārajaṃ rasam bhuñjānaḥ saṃvatsarād vā vai jāyate /* Further sthāna.3, adhy.8 (p.414): *nābhyām pratiṣṭhitā nāḍī antare hṛdayasya ca /105/ mātropayuktān dehastho garbho vahati vai rasān /106ab/ [...] nāḍī rasa-vahā jñeyā tayā garbhaḥ sa jīvati / [...] jñānaṃ ca rasa-vīryābhyāṃ sambhūyā 'śu vivardhate /108ab/.*

the mother's tube carrying *rasa*.[631] (śl.39cd–40ab)

With both hands joined together on the forehead, that [embryo] is situated [facing] towards the mother's back.[632] (śl.40cd–41a)

---

630 *Āhāra-rasa* "juice of food". A substance which according to common current opinion seems to occupy a sort of intermediate position between the actual chain of seven *dhātu*-s (beginning with *rasa*) and food, but is actually not differentiated thus in the medical texts (cf. DAS 2003A, p.180ff., §7.29; p.528 on *āhāra-rasa*; pp.578–9, on *rasa*).
For the tube leading from the heart of the mother, which is joined with the umbilical cord, cf. DAS 2003A, p.470ff. This tube is called *rasa*-carrying tube (ibid.). Also cf. ibid., p.102.

631 SR śl.39cd–40ab is identical to ŚG (Bombay 1987) 8,37abcd (ŚG 8,37 consists of six *pāda*-s), cf. COMBA 1981, p.207. The ŚG (Bombay 1987) 8,37ab runs as *mātur asra-vahāṃ nāḍīm āśrityānvavatāritā* instead of the reading of SR śl.39cd, *mātū rasa-vahāṃ nāḍīm anubaddhā parābhidhā*. The last two *pāda*-s, ŚG 8,37ef, are not parallel to the SR.
According to COMBA ibid. (note 40), the statement of ŚG 8,37–38a (= SR śl.39cd–40ab) accords with the chronological scheme of the SU.
The SU (śārīra., 3,31) mentions two periods in the embryo's nutrition. (1) In the period until the appearance of the limbs and subsidiary appendages (*aṅga, pratyaṅga*), the vessel (*dhamanī*) carrying *rasa* sends nourishment into the whole body. (2) In the period after that, the umbilical cord is joined to the mother's tube carrying nourishment.
SR śl.39cd–40ab (identical to ŚG 8,37–38a) does not mention the former period, but its theory, in other points, accords with SU śārīra., 3,31 (*mātus tu khalu rasavahāyāṃ nāḍyāṃ garbhanābhināḍīpratibaddhā, sā 'sya mātur āhārarasavīryam abhivahati*), (cf. COMBA 1981, note 40).
AS śārīra., 2,30 also contains a parallel: *[...] nābhyām pratibaddhā nāḍī nāḍyām aparā tasyām mātṛ-hṛdayām / tato mātṛ-hṛdayād āhāra-raso dhamanībhiḥ spandamāno 'parām upaiti.* As to the term aparā, there are two possibilities of interpretation, as remarked in my footnote 629 on SR śl.39d, *parā*. The tube carrying *rasa* (*rasa-hāriṇībhir vāhinībhir*) is also mentioned in AS śārīra., 2,24.

632 The ŚG contains no parallel to SR śl.40cd–41a. After this verse, there is no further parallel to the SR in the eighth chapter of the ŚG. Parallels are again found in the ninth chapter of the ŚG, namely to SR 1,2, śl.44 (= ŚG 9,10cdef) and the following verses.
The relation of the foetus' position in the mother's belly to its sex (cf. SR 41cd) is mentioned in SU śārīra., 3,34; AH śārīra., 70cd–71ab; CA śārīra., 2,24cd–25ab. But in these texts, the foetus' posture (SR śl.40cd–41ab) is not described. The AS (śārīra., 2,29) is the unique medical text that mentions the foetus' sex and describes its posture. The AS's wording also resembles that of SR śl.40cd–

Contracting [its] body/limbs[633], the [male] embryo[634] settles, situated_on/going_to (*ga-*) the right side [of the womb]. The female is situated on the left side, the third-sex (*klība*[635]) is held to be situated in the middle[636]. (śl.41bcd) [637]

---

41 strikingly (see below). For the relation of the foetus' sex with the side of the mother's belly, cf. DAS 2003A, p.446ff.; p.457.

Thus, the AS is the closest to the SR. Also, AgniP 369,21 and ViṣṇudhP, which are parallel to each other, mention the foetus' sex in relationship to its position.

PARALLELS:

AS śārīra., 2,29: *garbhas tu mātuḥ pṛṣṭham abhimukho lalāṭe kṛtāñjaliḥ saṅkucitāṅgo garbha-koṣṭhe dakṣiṇa-pārśvām āśrityāvatiṣṭhate pumān, vāmaṃ strī, madhyaṃ napuṃsakam.*

AgniP 369,21: *jarāyu-veṣṭite dehe mūrdhni baddhāñjalis tathā / madhye klībaṃ tu vāme strī dakṣiṇe puruṣa-sthitiḥ /21/ tiṣṭhaty udara-bhāge tu pṛṣṭhasyābhimukhas tathā / yasyāṃ tiṣṭhaty asau yonau tāṃ sa vetti na saṃśayaḥ /22/.*

BhāgP 3,31,8: *ulbena saṃvṛtas tasminn antraiś ca bahir āvṛtaḥ / āste kṛtvā śiraḥ kukṣau bhugna-pṛṣṭha-śirodharaḥ /8/ akalpaḥ svāṅga-ceṣṭāyāṃ śakunta iva pañjare / tatra labdha-smṛtir daivāt karma janma-śatodbhavam / smaran dīrgham anucchvāsaṃ śarma kiṃ nāma vindate /9/ ārabhya saptamān māsāl labdha-bodho 'pi vepitaḥ / naikatrāste sūti-vātair viṣṭhābhūr iva sodaraḥ /10/ nāthamāna ṛṣir bhītaḥ saptavadhriḥ kṛtāñjaliḥ / stuvīta taṃ viklavayā vācā yenodare 'rpitaḥ /11/ [...].*

633 SR śārīra., 41a, *saṅkucad-gātro*, which literally means "[that] whose body/limbs is/are contracting", is an expression parallel to SR śl.42c *rujad-gātro.* Also compare it with CA śārīra., 6, 22 and 24 (CA śārīra., 6,22, *garbhas tu khalu mātuḥ pṛṣṭhābhimukha ūrdhvaśirāḥ saṅkucyāṅgāny āste 'ntaḥ kukṣau*).

634 In my translation, I adopted the reading of the Ānandāśrama ed. (1896) *garbho* instead of the Adyar ed. *garbham.* I cannot understand the reason why the Adyar ed. reads *garbhaṃ* (accusative sg.), which is obviously grammatically incorrect.

If an interpretation is to be made, perhaps the accusative form *garbhaṃ* meaning "uterus" might be the object of the verb *adhyāste*. In that case, it would mean, "[the male] settles in the uterus (*garbhaṃ*), contracting [his] limbs".

The manuscript D reads instead *naro* "the male".

635 The three sexes are mentioned in SR śl.24, śl.27 and śl.41. In the first two verses, *napuṃsaka* is contained as meaning "the third sex", but in the last one, *klība*. The SR does not seem to differentiate the two terms.

The CA and SU, in their parallels to these statements of the SR, call the third sex *napuṃsaka*, while the AH calls it *klība*. This fact seems to confirm ANGOT's hypothesis that the term *napuṃsaka* is not tainted negatively, but *klība* is considered a person with a defect. In the CA, *napuṃsaka* is considered

"**In the fifth**" (śl.33a, *pañcame*). **Awoken**: the internal_instrument (*antaḥkaraṇa* = faculty of thought), [which was] concealed (*līna*) before, then is on the verge of palpitation (*spanda*).[638]

"**In the sixth**" (śl.33c, *ṣaṣṭe*). **Cords (*snāyu*)**, [namely] fine vessels (*sirā*); **hairs of the head (*keśa*)**, [namely] those which grow on the head; **body-hairs (*roman*)**, [namely] those which grow on the body/limbs; **power (*bala*) and colour (*varṇa*)**: **power** is essence/energy (*sattva*), **the colour** is whiteness/fairness (*gauratā*) etc.

"**But (*tu*) in the seventh**" (śl.34b, *saptame tu*). **With both hands placed (lit. "made to be") within the *pāli***. The *pāli* is the thigh, because of the dictionary [saying]: "*pāli*, female [gender], in [the meanings] a corner/edge, lap, row". **Both hands, made to be within**, [namely] covered, **with the two *pāli*-s**, [namely] the thighs; with them (= with the hands).[639] "**Covering the openings of the ears**": through this, the state of the contracting body through being face down (= head down) is mentioned. **Anxious**, [namely] frightened.

---

to be a *prakṛti* "natural disposition" but not a *vikṛti* "anomaly" (Cf. ANGOT 1993–94, p.22).

DAS 2003A, p.559 (footnote 1930) expresses his doubt about ANGOT's simplification, but the examples he quotes from the CA, SU and AH (ibid., pp.538–541) in themselves seem to confirm ANGOT.

636 *Madhyasthitaṃ*. The manuscript D reads *madhye sthitaṃ*. The manuscripts *gha.* and *ṅa.* read *madhyāśritaṃ* "attached to the middle".

637 For the association of right/left with male/female, cf. DAS 2003A, pp.445–447; pp.452–453 (note 1549); p.457. Also cf. W. KIRFEL 1951.

ŚG śl.16cd–17ab locates this topic in the fourth month, cf. COMBA 1981, p.198. But the wording is not strictly identical to that of SR śl.41bcd. (ŚG 8,16cd–17ab: *putraś ced dakṣiṇe pārśve kanyā vāme ca tiṣṭhati /16cd/ napuṃsakas tūdarasya bhāge tiṣṭhati madhyataḥ /17ab/*). So, the two texts do not share a same source in this case.

This topic is found e.g. in the texts mentioned below:

AH śārīra., 1,70–72.

AgniP 369,21cd: *madhye klībaṃ tu vāme strī dakṣiṇe puruṣa-sthitiḥ.*

Tandulaveyāliya 16: *dāhiṇa-kucchī purisassa hoi vāmā u itthīyāe ya / ubhyaṃtaraṃ napuṃse tirie aṭṭheva varisāiṃ.*

638 In these passages, the Ānandāśrama ed. (1896) is totally corrupted, inserting a false line, as I have already noted (cf. my footnote 596 on comm. K on SR śl.28–32, *"the embryo is" etc.*).

639 I use boldface for the terms functioning as *pratīka*-s.

Connected with the uterus, [namely] one whose body is enclosed by the placenta (*jarāyu*). **Eagerly engaged in exercise**: exercise here so_to_say/certainly (*nāman*) is pondering over the self (*ātman*) with constant repetition, with the desire to leave transmigration (*saṃsāra*) behind.

"**In the eighth**" (śl.36c, *aṣṭame*). "**The skin and the [power of] remembrance are [there]**". The meaning is that (*iti*), even though/if the skin and the [power of] remembrance exist already (*api*) in the seventh [month], nevertheless here the denseness of the skin and the state of having manifoldness as the object of [the power of] remembrance appear. "**And the vital fluid (*ojas*)**". "Is"[640] is the complement [of the ellipsis] (*ūha*). **The vital fluid** is the essence of semen. **And this** vital fluid, **arising in the heart**, [i.e.] which has the heart as its resting place, **pure**, [i.e.] uncorrupted by the corrupted wind (*duṣṭa-vāta*) etc.[641], **yellowish red**, [i.e.] whose colour is slightly yellow_and_red/yellow-red, **is deemed the cause** (*nimitta*), [i.e.] the reason (*kāraṇa*), **with regard to life**, [i.e.] with regard to the maintenance of the vital wind (*prāṇa*). Thus (*iti*) is the meaning of the connection of the sentence-members (*anvayārtha*). That vital fluid, **unsteady**, [i.e.] not being settled in one place, **continuously (*muhur*)**[642] runs (*dhāvati*) **to the mother, and to the embryo**. Thus/That is said by the teacher Vāgbhaṭa (AH śārīra., 1,62cd)[643]: "In the eighth [month] the vital fluid moves continuously to mother and son successively". The meaning is that (*iti*): Because the vital fluid is not settled in the eighth [month], therefore [it is so]. "**Therefore (*atas*), born in the eighth month, [it] does not live**" (SR śl.38ab):

640 Opt. *syāt.*

641 I.e., by the corrupted *doṣa*-s, i.e. wind etc.

642 The comm. reads *muhur* for *punar*; and *dhāvati* for *pradhāvati.* So, this commentator possibly reads the verse SR śl.37cd as *muhur ambāṃ muhur garbhaṃ cañcalaṃ tat pradhāvati.* Because of this presumption, I have not chosen boldface for "run" (*dhāvati*), which differs from *pradhAvati* in the original text. Contrastingly, the commentator could have read it as *punar ambām punar garbhaṃ cañcalām muhur dhāvati.* In that case, I would have to use boldface for "run", too.

643 The comm. K has also used citations from the AH while explaining SR śl.18–22 (Adyar ed., p.37, ll.4–5). The parallel contained in the AS differs considerably from it. AS śārīra., 2,18: *aṣṭame garbhaś ca mātṛto garbhataś ca mātā rasa-hāriṇībhir vāhinībhir muhur muhur ojaḥ parasparam ādadāte.*

The meaning is that (*iti*) [there is] no totality/fullness of the months.[644] The meaning is thus (*iti*): But (*api tu*) if the vital fluid, leaving the embryo, moves[645] to the mother, then the born [child] does not live because of being abandoned by the vital fluid. If, however, the vital fluid, leaving the mother, moves/should_move[646] to the embryo, then, on the other hand (*tu*) the mother does not live at [the time of] the delivery. If, however, the delivery [should occur] just at the time of rapid flow (*pradhāvana*) of the vital fluid, then *both* do not live, because of its not being settled/situated[647] in [either] *one* [of them][648]. Thus (*iti*) is the intention. Fearing [the objection]: "Is not, in the world, in some cases the stability of an embryo, even of one abandoned by the vital fluid, of one living for some while immediately after birth, observed? So how come (*katham*) there is non-living?" he (= the author) tells the reason (*upapatti*) with an example: "**For some while**" (śl.38c, *kiñcit kālam*). **Due to the imprint (*saṃskāra*)**, [i.e.] because of the imprint of the circulation of the vital fluid. **Like a severed limb**: [*khaṇḍitāṅga*] is a *karmadhāraya* compound: "that is severed and a limb"[649]. The meaning is that (*iti*), as a severed limb is observed to be active in the subsequent moment (*kṣaṇa*), because of the imprint of the circulation of the vital wind (*prāṇa*), thus.

"**In the [months] beginning with the ninth**" (śl.39b, *navamādiṣu*). The utilisation (*grahaṇa*) of the word "beginning" (*ādi*) is to show that (*iti*) in some cases the delivery occurs/may_occur (opt. *syāt*) in the ninth month, in some cases in the tenth, in some cases even in the eleventh.

Comm. S on SR śl.33–41

The meaning is that (*iti*) **in the fifth** month **the consciousness/intellect (*citta*)** becomes **awoken**, characterised by knowledge. **Dis-**

644 I corrected *na māsa-sākalyārthaḥ* into *na māsa-sākalyam ity arthaḥ*.
645 Opt. *saṅkrāmet*.
646 Opt. *saṅkrāmet*.
647 *Avasthāna* can also mean "stability", i.e. "not having stability".
648 *Ekatrāpi* : lit. "even in *one* [place]".
649 A *karmadhāraya* compound is usually analysed in a manner like *khaṇḍitaṃ ca aṅgaṃ ca*. Maybe *tad* in our text is used in order to emphasize the apposition: "[that which is] *khaṇḍitam*, that is *aṅgam*".

cernability (*viviktatā*) is being evident, when separateness (*pṛthaktva*) is present. **Accumulated**, [i.e.] having obtained growth. **"Placed within**[650] **the *pālī*"**: the *pālī* is the_lobe_of_the_ear/the_outer_ear (*karṇa-pālī*).[651] **Covering the openings of the ears with both hands made to be within (*antarita*)**, [i.e.] hidden, by it (= the *pālī*).[652] **Anxious**, having_attained (*āpanna*) aversion (*vairāgya*). ***Garbhāśaya***[653] is the pre-eminence (*atiśaya*)[654] of meditation/rumination (*dhyāna*) within the womb, [i.e.] inquiry about the self; **connected with** that.[655] **Torments**, [i.e.] oppressions/pains. **Ruminating upon**, [i.e.] thinking about, **the means of liberation**, [i.e.] the expedient (*sādhana*) for [accomplishing] liberation. **Eagerly engaged in exercise**, [i.e.] intent on inquiry about the self.

He describes the condition in the eighth month: **"In the eighth"** (śl.36, *aṣṭame*). **Hearing**[656] [means] the reception of the ear of an external sound, but not the faculty (*indriya*) of hearing, because of it, through [its] permanence[657], not being produced due to its having the nature of space (*ākāśa*). **The vital fluid (*ojas*)** is a certain element (*dhātu*).[658] **Pure**, [i.e.] not dirty. It is said that (*iti*) the vital fluid is **the**

650 *Antarita* lit. "made to be within".

651 I consider *pālī* to be feminine sg. But the form *pālī* could be the feminine dual of *pāli*. In that case, we could translate it as "the two *pāli*-s are the two lobes_of_the_ears/outer_ears". Then, however, *tayā* (feminine sg.) would be problematic.

652 It is unclear to me what kind of posture is meant. Does it mean that the lobe of the ear is big enough to cover the entire hand? But picturing this, it seems a bit strange to me. Another possible interpretation is to consider *antarita* and *vyavahita* to be intransitive: "Covering the openings of the ears with both hands, which_the_lobes_of_the_ears_are_made_to_be_within/through_which_the_lobe_of_the_ears_are_hidden." So, the ears are completely concealed by the hands. But I am rather skeptical as to whether these past participles could be really interpreted in this manner.

653 The Adyar ed. notes the variant of B *garbhaśatha*. But it makes no sense.

654 Explaining *āśaya* through *atiśaya*, this is obviously far-fetched.

655 Saying that *garbhāśaya* does not mean the uterus, but the foetus' ruminations, is clearly untenable.

656 Comm. S. contains *tvak-śrutī*, instead of *tvak-smṛtī*.

657 That means, the faculty of hearing is by contrast permanent.

658 Peculiarly, this commentary considers *ojas* to be a *dhātu*. Usually in the classical medical theory *ojas* is not a *dhātu*, but the essence of the *dhātu*-s (*dhātusāra*).

cause with regard to life. He makes that same [matter] clear: "**Now (*punar*) to the mother**" (śl.37c, *punar ambām*). Asking (*iti*) how the nourishing of the foetus [is brought about], he now says: "**The mother's**" (śl.39c, *mātur*). **Joined with the tube (*nāḍī*)**, [i.e.] mingled with it. **Towards the mother's back**, pointed at [it]. From the indication of the sign of a male (= the ending of a male noun), [namely] "**situated_on/going_to the right side**" (*pārśvagaḥ* m. nom. sg.), it is understood (lit.: is attained) that (*iti*) [the baby is] a male.

SR śl.42–46

Then [the child] is made [to have its] head down[659] by the powerful procreatory winds.[660] The boy, whose limbs are paining, is expelled

659 Cf. Śāradātilaka 1,50: *sa piṇḍita-śarīro 'tha jāyate 'yam avāṅ-mukhaḥ / kṣaṇaṃ tiṣṭhati niścesto bhītyā roditum icchati /50/.*

660 The procreatory wind is explained by DOSSI 1998, p.136: "Durch Kontakt mit dem *vaiṣṇava*- oder *prājāpatya*-Wind bei der Geburt verliert der Embryo sein Bewußtsein (*cetanā*), gewinnt es aber wieder, wenn er vom Wind der Außenwelt (*vāyu-sparśa-samanvita*) berührt wird. Doch dann raubt Viṣṇu's *Māyā* ihm sein Wissen". For this topic, also see HARA 1980, p.152ff. The *sūtimāruta* is also mentioned in YS 3,83 and CA, cf. YAMASHITA 2001/2002, p.100.
The procreatory wind is *apāna*, according to CA cikitsā., 28,5–12; SU nidāna., 12,20a (cf. ZYSK 1993, p.207).
The Śatapathabrāhmaṇa states that the vital wind dealing with discharge is in charge of delivery, too: ŚB 7,1,2,15 (WEBER's ed., p.576, ll.8–10): *tráyo vā́ 'íme vāñcaḥ prāṇā́ḥ, [...] ékáṃ hy évaitád rūpaṃ yónir eva, prájātir eva yád eté 'vāñcaḥ prāṇā yad dhi mū́traṃ karóti, yat púrīṣám praiva táj jāyate.* Paraphrase: There are three downward winds; [the altar is the womb]; the procreatory wind is the same one that brings forth faeces and urine.
The oldest evidence of the procreatory wind is probably R̥V 5,78., i.e. the hymn to the Aśvins, wishing for an easy delivery: *7a. yáthā vā́taḥ puṣkarínīṃ samiṅgáyati sarvátaḥ / 7c. evā́ te gárbha ejatu niraítu dáśamāsiyaḥ // 8a. yáthā vā́to yáthā vánaṃ yáthā samudrá éjati / 8c. evā́ tváṃ daśamāsiya sahā́vehi jarā́yuṇā // 9a. dáśa mā́sāñ chaśayānáḥ kumāró ádhi mātári / 9c. niraítu jīvó ákṣato jīvó jī́vantiyā ádhi //.*
GELDNER (1951)'s translation runs: "Als Atri in den Glutofen hinabstieg und euch anrief wie eine Frau in Kindesnöten, da kamet Aśvin mit des Falken frischer, glückbringendster Eile herbei (v.4). Tu dich auf, o Baum, wie der Schoß der Kreißenden ! [...] (v.5). Wie der Wind allerwärts den Lotusteich bewegt, so soll sich deine Leibesfrucht regen, sie soll zehn Monate alt herauskommen (v.7). Wie sich der Wind, wie der Wald, wie das Meer bewegt, so geh

du Zehnmonatskind samt der Nachgeburt ab (v.8). Nachdem der Knabe zehn Monate in der Mutter gelegen hat, soll er lebendig, unversehrt, lebendig aus der Lebenden herauskommen ! (v.9)" (Underlined by me.)
The Atharvaveda also mentions the procreatory wind. E.g., AV 1,11, i.e. the hymn wishing for an easy delivery, 2cd: *devā́ gárbhaṃ sámairayan táṃ vyū́rṇúvántu sū́tave*, "May the gods set the foetus in motion (*samairayan*) and make it open for birth". The verb *sam-√ īr* "to set something in motion" is associated with the movement of breath and wind, cf. AV 3,31,7ab: *prāṇéna viśvátovīryaṃ devā́ḥ sū́ryaṃ samaíraуan,* "By breath did the gods set in motion the sun [...]" (tr. by WHITNEY); AV 5, 30,14: *prāṇéna [...] sáṃ sṛjemáṃ [...] sámīraya bálena.* (Paraphrase: "unite it through breath, set it in motion by force"). The derivative noun *samīra* means "wind, breeze". For the term *samīraṇa* associated with wind in the medical texts, cf. MÜLLER 1961, p.145.
AV 1,11, v.6 is translated by ZYSK 1993, p.201, as follows: AV 11,4,14, "A human being breathes out (*apānati*) and breathes in (*prāṇati*) when inside the womb (*garbhe*). When you, O *Prāṇa*, urge him on he is born again."
PARALLELS
AS śārīra., 2,32: *evaṃ jaṭharastho garbho janmakāle tu prasūti-māruta-yogāt parivṛtyā 'vāk-śirā niṣkrāmaty anu cāparā cyutā mātṛ-hṛdayāt.*
CA śārīra., 6,22–24: *garbhas tu khalu mātuḥ pṛṣṭhābhimukha ūrdhvaśirāḥ saṅkucyāṅgāny āste 'ntaḥ kukṣau. /22/ [...] /23/ sa copasthita-kāle janmani prasūti-māruta-yogāt parivṛtyāvāk-śirāḥ niṣkrāmaty apatyapathena, [...] /24/.*
The AH mentions the wind, but is quite different from the SR in wording. AH śārīra., 1,66: *tasmiṃs tv ekāhayāte 'pi kālaḥ sūter ataḥ param / varṣād vikārakārī syāt kukṣau vātena dhāritaḥ /66/.*
The SU's śārīrasthāna does not contain a parallel, either. SU śārīra., 5,45, describing the moment of delivery, does not mention the procreatory wind. But SU cikitsā., 15,14, dealing with difficult delivery might be associated with this theory, states that an abnormal position of the coming child is caused by the excitement of the wind (*garbhasya gatayaś citrā jāyante 'nila-kopataḥ*).
GaruḍaP (KIRFEL 1954) *tataḥ kāla-kramāj jantuḥ parivartaty adhomukhaḥ /80/ navame daśame vāpi māsi sañjāyate tataḥ / niṣkramyamāṇo vātena prājāpatyena pīḍyate /81/ niṣkramate ca vilapan tadā duḥkha-nipīḍitaḥ / niṣkramāṃś codarān mūrcchām asahyām pratipadyate /82/ prāpnoti cetanāṃ cāsau vāyu-sparśa-sukhānvitaḥ / tatas taṃ vaiṣṇavī māyā samāskandati mohinī /83/ tayā vimohitātmāsau jñāna-bhraṃśam avāpnute /84ab/.*
ANALYSIS:
Parallels to SR śl.42 are found in the AS and CA. The wordings of the two medical texts are very close to each other. In contrast, the SU and AH do not mention the procreatory wind. Parallel expressions to the SR's *rujad-gātro yantracchidreṇa* are not found in the medical texts but in the non-medical texts (Purāṇa-s).

through the cleft of the organ (= vagina)[661]. (śl.42)[662]

661 The term *yantra*, translated here by the neutral term "organ", obviously denotes the vagina. Literally "fetter/fastening", it is also widely used in the sense of "apparatus". In Tantric texts, certain diagrams etc. representing the female genitals are also called *yantra* and seem to presuppose the meaning "apparatus". Here, however, it is probably not this meaning which is intended, but another one, namely "torturing instrument". In delivery the child is tormented, because the opening of the *yoni* is by one finger narrower than its own size, cf. DOSSI 1998, p.84, quoting PadmaP 2,8,12ff.: *yonir vikāśam āyāti caturviṃśāṅgulaṃ tadā / pañcāviṃśāṅgulo garbhas tena pīḍā vijāyate //* "Dann öffnet sich die *yoni* vierundzwanzig Finger breit. Der Embryo ist fünfundzwanzig Finger groß. Daraus entsteht Leid". Compare this passage with the comm. S's explanation of *yantracchidreṇa* in SR śl.42d.
DOSSI 1998 (p.136, note 692) translates *yantra* as "Einzwängung", quoting Garbhopaniṣad 4: *atha yoni-dvāraṃ samprāpto yantreṇāpīḍyamāno mahatā duḥkhena jāta-mātrāstu (-mātras tu ??) vaiṣṇavena vāyunā saṃspṛṣṭas tadā na smarati janma-maraṇāni na ca karma śubhāśubhaṃ vindati /*.
YS 3,83, however, uses the term *yantra* in the meaning of "apparatus/instrument". It compares the uterus from which the baby comes out to a bow-machine (*yantra*), maybe a crossbow slotting the arrow. On YS 3,83, cf. YAMASHITA 2001/2002, p.102. Also see its parallel, ViṣṇudhP 113,18cd.

662 ŚG 8,38b–40 contains the same topic, i.e. the process of birth, but its wording is different from SR śl.40ab–41a. The ŚG, like the Purāṇa-s, emphasizes the agony which the child suffers at the moment of birth. In contrast, the SR merely hints at the agony through the expression, *rujad-gātro yantracchidreṇa* "whose limbs are paining, through the cleft of the organ".
COMBA 1981, (note 42) comments as follows: The CA (śārīra., 6,24) and SU do not describe the moment of delivery very much; the two medical texts do not mention the foetus' agony at all. In contrast, the Purāṇa-s put emphasis on the foetus' agony in their description of delivery.
In this point, the SR, containing fewer descriptions of the foetus' agony, is closer to the medical texts than the ŚG is. The ŚG's attitude is like that of the Purāṇa-s, dealing with the foetus' agony in detail.
The SR mentions the position of the foetus shortly before the moment of delivery, facing the mother's back, cf. SR śl.40, *mātṛ-pṛṣṭham abhi sthitaḥ*. This statement is inconsistent with SU śārīra., 5,45, *'bhimukhaḥ* "facing forward". But, in other points, SR śl.40d–41a and SU śārīra., 5,45 are close to each other in wording:
SR śl.40d–41a: *mātṛ-pṛṣṭham abhi sthitaḥ // adhyāste saṅkucad gātro garbhaṃ* (Variant *garbho*) *[...].*
SU śārīra., 5,45: *ābhugno 'bhimukhaḥ śete garbho garbhāśaye striyāḥ.*
In describing the foetus' position in the uterus, the CA accords with the SR, cf. CA śārīra. 6,22, *garbhas tu khalu mātuḥ pṛṣṭhābhimukha ūrdhvaśirāḥ*

saṅkucyāṅgāny āste 'ntaḥ kukṣau (Underline marks the expressions parallel to the SR).
According to SR śl.42 on delivery, the procreatory wind pushes the baby out, turning it upside down. SU śārīra., 5,45 does not accord with the SR so well, except for stating "upside down" (*śirasā yāti [...] prasavam prati* "[the child] goes towards birth, with the head [first]"). The SU does not mention the procreatory wind. CA śārīra., 6,24 accords with the SR in mentioning the procreatory wind (*prasūti-māruta*) and the child's position with its head turned down (*parivṛtyāvāk-śirāḥ*). The description of the AH (śārīra. 1,77–82) does not have so much to do with that of the SR, focusing on the practical treatment of the parturient woman. YS 3,83 is the closest to the SR in wording: *navame daśame vāpi prabalaiḥ sūti-mārutaiḥ / niḥsāryate bāṇa iva yantra-cchidreṇa sa-jvaraḥ /83/* (parallel expressions are underlined). The YS lacks a counterpart for "having the head upside down", but the ViṣṇudhP (113,18cd) and AgniP (369,27), which are the two descendants of the YS, contain it.
PARALLELS:
SU śārīra., 5,45: *ābhugno 'bhimukhaḥ śete garbho garbhāśaye striyāḥ / sa yoniṃ śirasā yāti svabhāvāt prasavam prati /45/*, "The foetus lies in the uterus of a woman, [having the limbs] bent and facing forward; it moves, by nature, to the *yoni*, [having] its head [first], towards delivery." The commentator, Ḍalhaṇa, glosses *ābhugnaḥ* with *saṅkucitāṅgaḥ* which is a close expression to SR śl.41a, *saṅkucad-gātro*. Ḍalhaṇa might perhaps have referred to a text which is close to the tradition of the SR.
YS 3,83: *navame daśame vāpi prabalaiḥ sūti-mārutaiḥ / niḥsāryate bāṇa iva yantra-cchidreṇa sa-jvaraḥ /83/* Explaining the term *yantra*, the commentary states, *dhanur-yantreṇa sudhanva-prerito bāṇa ivātivegena nirgama-samanantaraṃ ca bāhya-pavana-spṛṣṭo naṣṭa-prācīna-smṛtir bhavati*. The term *yantra* is associated with the shooting machine (a cross bow?). This comparison is further handed down to the ViṣṇudhP. Besides, the commentary quotes Nirukta section 18: *jātaḥ sa vāyunā spṛṣṭo na smarati pūrvaṃ janma maraṇaṃ karma ca śubhāśubham*.
AgniP 369,27 (HOERNLE 1907 states that it borrowed its embryological passages from the YS): *sūti-vātair adho-bhūto niḥsared yoni-yantrataḥ / pīḍyamāno māsa-mātraṃ kara-sparśena duḥkhitaḥ*. The new-born child in the first month is so sensitive, that he feels tormented even by the soft touch of its mother's hand, cf. HARA 1980.
A parallel is found in a Japanese text titled, *Hou-motsu-shū*, an anthology of Buddhist stories, whose author is traditionally said to be *Taira no Yasuyori*: "They first describe that which is called the agony of life/birth (*shou-ku*), as follows. After staying in the mother's belly for three hundred days, or two hundred sixty days, [the baby] is at last pushed out by the wind of karma. It is torn off like the skin of a living cow. [...] Or, [the baby feels] as if it were torn with thousands or hundreds of swords, though it is caught in a [soft and fine] mattress [made] in the land of *Nagoya*. Therefore, the baby's first voice is a

Now[663] the exertion/proclivity of him who has just been born is relating to (*gocara*) milk (lit. "that whose scope/object (*gocara*) is that pertaining to the [mother's] breast"), because of the [subliminal] imprint (*saṃskāra*) of the perceptions (*bodha*) in the previous birth. Thus/Therefore (*iti*) is the permanence of the individual self (*jīva*).[664] (śl.43)

---

cry, "*ku kana*" ("What a sufferance!")." Cf. HARA 1977. HARA notes parallels in the Chinese Buddhist canon, too.

ViṣṇudhP enlarged the description which it borrowed from the YS. ViṣṇudhP 114,18cd: *tatas tu kāle sampūrṇe prabalaiḥ sūti-mārutaiḥ /18cd/ bhavati avāṅmukho jantuḥ pīḍām anubhavan parām / adhomukhaḥ saṅkaṭena yoni-dvāreṇa vāyunā /19/ niḥsāryate bāṇa iva yantra-cchidreṇa sajvaraḥ / yoni-niṣkramanāt pīḍāṃ carmotkartana-sannibhām /20cd/* (This comparison was handed down even to Japanese Buddhist literature, as shown above.) 21ab: *prāpnoti ca tato jātaḥ tīvraṃ śītam asaṃśayam / janma-jvarābhibhūtasya vijñānaṃ tasya naśyati /21/* (Due to the fever/pain of birth, the child loses its consciousness, memory of previous lives and decision to strive for liberation.) 22: *kara-saṃsparśān mātur na ca jānāty asau tadā / kara-patrasya saṃsparśān māsa-mātraṃ vimohitaḥ /22/* On the above-mentioned passages in the ViṣṇudhP and AgniP, cf. YAMASHITA 2001/2002, p.101.

BhāgP 3 (Kāpileya), 31,22: *evaṃ kṛta-matir garbhe daśa-māsyaḥ stuvann ṛṣiḥ / sadyaḥ kṣipaty avācīnam prasūtyai sūti-mārutaḥ /22/ tenāvasṛṣṭaḥ sahasā kṛtvā 'vāk śira āturaḥ / viniṣkrāmati kṛcchreṇa nirucchvāso hata-smṛtiḥ /23/ patito bhuvy asṛṅ-mūtre viṣṭhābhūr iva ceṣṭate / rorūyati gate jñāne viparītāṃ gatiṃ gataḥ /24/ para-cchandaṃ na-viduṣā puṣyamāṇo janena saḥ / anabhipretam āpannaḥ pratyākhyātum anīśvaraḥ /25/ śāyito 'śuci-paryaṅke jantu-svedaja-dūṣite / neśaḥ kaṇḍūyane 'ṅgānām āsanotthāna-ceṣṭane /26/ tudanty āma-tvacaṃ daṃśā maśakā matkuṇādayaḥ / rudantaṃ vigata-jñānaṃ kṛmayaḥ kṛmikaṃ yathā /27/ ity evaṃ śaiśavam bhuktvā duḥkham paugaṇḍam eva ca /28ab/.*

GaruḍaP (KIRFEL 1954) v.72cd: *adhomukham ūrdhva-pādaṃ garbhād vāyuḥ prakarṣati /72cd/.*

MārkaṇḍeyaP 11,17: *niṣkrāmyamāṇo vātena prājāpatyena pīḍyate / niṣkrāmyate ca vilapan hṛdi duḥkha-nipīḍitaḥ /17/ niṣkrāntaś codarān mūrcchām asahyām pratipadyate / prāpnoti cetanāṃ cāsau vāyu-sparśa-samanvitaḥ /18/ tatas taṃ vaiṣṇavī-māyā samāskandati mohanī / tayā vimohitātmāsau jñāna-bhraṃśam avāpnute /19/.*

663 The manuscripts *kha.* and *ga.* read *api* instead of *atha*: "The activity/proclivity [...] of milk, even though (*api*) he is just born (lit. even of him (*tasyāpi*) who is just born)".

664 Neither the CA nor the SU contains this topic in their śārīrasthāna. But Hastyāyurveda 3,9,33 (p.439), *evaṃ hi jāyate hastī jātaś cāpi pibet payaḥ.*

His[665] substances (*bhāva*) are[666] six-fold/of_six_kinds: those born from the mother, then those born from the father[667],
those born from the nutrient fluid (*rasa*)[668], those born from the self (*ātman*), those arisen from nature/disposition (*sattva*)[669], then those born from suitability (*sātmya*)[670]. (śl.44)[671]

The new-born baby's exertion to milk is often considered to be a proof of transmigration and rebirth. This is one of the topics which Hindus discussed to prove the immortality of the soul (*ātman/jīva*), cf. HALBFASS 2000, pp.196–198, referring to Nyāyavārttika 3,1,18f.

665 D reads *tatra* instead of *tasya*: "The substances are six-fold; with regard to that, [there are] those born from the mother [...]".

666 Optative.

667 AS śārīra., 5,18–25 lists *mātṛja, pitṛja, ātmaja, sātmyaja, rasaja* and *sattvaja.*

668 On the term *rasa* "nutrient fluid", cf. DAS 2003A, p.578.

669 For the term *sattva*, cf. WINDISCH 1908, pp.52–55. This term originally denotes the intermediary condition between the previous and new life, namely the state before entering into the uterus (ibid., p.193). The *sattva* delivers the characteristics derived from the previous *karma* to the new life (ibid, p.55). But, in CA śārīra., 4,44 (parallel to SR śl.54–55), the term *sattva* is used in a broader sense, namely it denotes a nature or character which a personality inherits from the previous life through the intermediary state. Here, the term *sattva* does not denote the intermediary state in itself. In these passages, the CA lists, as typical characters, seven of *śuddha-sattva,* six of *rājasa-sattva* and three of *tāmasa-sattva*, cf.WINDISCH 1908, pp.55–56. For the problem of *sattva*, also cf. ROŞU 1978 (see his index under *sattva*); R.F.G. MÜLLER 1955, p.41.

670 The manuscripts *gha.* and *ṅa.* read *sātmya-gocarāḥ* instead of *sātmyajās tathā.* "Those whose scope/object is *sātmya*".
The term *sātmya* is explained in CA vimāna., 1,20, *tad yad ātmany upaśete* "that which does good to/agrees with the self".
Ḍalhaṇa on SU sūtrasthāna 35,39 explains: *sātmyaṃ nāma sukhaṃ [= ārogyam] yat karoti tad ucyate sātmya*, "that which does/makes pleasure [= diseaselessness] is called *sātmya*". (Cf. ROŞU 1978, p.174). Ḍalhaṇa lists various kinds of *sātmya*, e.g. suitability of locality (*deśa*), species (*jāti*), season (*ṛtu*), exercise (*vyāyāma*) etc.
Thus, the term *sātmya* seems to mean "suitability" or "agreeableness to one's individual nature". SHARMA 1996, p.305, translates *sātmya* as "suitableness". The tolerable level of taste, climate, exercise etc. depends on individuals, cf. SU sūtra., 35,39. Each individual has a suitable level for himself, e.g. one person can consume more of salty taste than others. According to SU sūtra., 35,39 and Ḍalhaṇa's commentary, *sātmya* seems to mean the capacity with which the body can adjust itself to circumstances.
The sixth chapter of CA's vimānasthāna deals with the theory of the six tastes (*rasa*). There, the term *sātmya* is used in association with combinations of

The soft (*mṛdu*) substances, however (*tu*[672]), namely (*iti*) blood (*śoṇita*), fat (*medas*), marrow (*majjan*), the spleen (*plīhan*), the liver (*yakṛt*), the rectum/anus (*guda*), the heart and the navel, and so on (*evam ādyāḥ*), are considered [to be] arisen/derived from the mother. (śl.45)[673]

---

tastes. A person should not consume a taste excessively, but in the suitable portion for his own nature or habit. CA vimāna., 1,20 states that *sātmya* is of three types, superior, inferior and medium, and of seven types according to the six tastes individually, or collectively. For example, a collective use of all the tastes is considered superior.
SU śārīra., 3,33 lists vigour (*vīrya*), diseaselessness (*ārogya*), power (*bala*), colour (*varṇa*) and intelligence (*medhas*) as the substances derived from *sātmya*. SR śl.56ab, too, lists the substances derived from *sātmya*, but the statement of the SR is inconsistent with that of the SU. This matter is discussed further in my footnote 724 on SR śl.56ab.

671 ŚG 9,10cdfg is identical to SR śl.44. The ŚG contains *satya-sambhūtāḥ svātmyajās tathā* instead of *sattva-sambhavāḥ sātmyajās tathā* of the SR. But, ŚG 9,22cd, identical to SR śl.56ab, contains *sattvajāḥ* instead of *sātmyajāḥ* of the SR. If we compare it with the parallel statements in the other medical texts (see below), the reading of the ŚG is obviously unsuitable; it should be read *sātmyajāḥ*.
The six groups of substances (*bhāva*) are mentioned by the SU (śārīra., 3,33), CA (śārīra., 3,4; śārīra., 3,14; śārīra., 3,6–13) and AS (śārīra., 5,18–25). The SU does not call them *bhāva*, but *śarīra-lakṣaṇa*. To the six, the AH (śārīra., 3,4cd —8c) adds the seventh group called *caitana* "belonging to *cetas*". The YS does not mention them, except for the *ātmaja* group (YS 3,73–74). On YS 3,73–74, cf. YAMASHITA 2001/2002, p.92. The two descendant texts of the YS, i.e. the ViṣṇudhP (115,12cd and following) and AgniP (369,31cd–36), which are in these verses parallel to each other, mention only *mātṛja, pitṛja, ātmaja,* and the three sub-groups of *sattvaja* (i.e. *tāmasa, rājasa,* and *sāttvika*), though the two text do not mention the term *sattvaja* in itself.
The BhāgP (3,31) contains no parallel.

672 The manuscript *ṅa.* reads *-ādyās te* instead of *-adyās tu*. "The soft substance, namely, blood [...] the navel, and so on: they (*te*) are considered [to be]".

673 ŚG 9, śl.11 is identical to SR śl.45.
PARALLELS:
SU śārīra., 3,33: *tatra garbhasya pitṛja-mātṛja-rasajātmaja-sattvaja-sātmyajāni śarīra-lakṣaṇāni vyākhyāsyāmaḥ / garbhasya keśa-śmaśru-lomāsthi-nakha-danta-sirā-snāyu-dhamanī-retaḥ-prabhṛtīni sthirāṇi pitṛjāni, māṃsa-śoṇita-medo-majja-hṛn-nābhi-yakṛt-plīhāntra-guda-prabhṛtīni mṛdūni mātṛjāni, etc.*
AS śārīra., 5,18 and 5,19.
AH śārīra., 3,4cd–5ab: *mṛdv atra mātṛjaṃ rakta-māṃsa-majja-gudādikam /4cd/ paitṛkaṃ tu sthiraṃ śukra-dhamany-asthi-kacādikam /5ab/.*

The firm (*sthira*[674]) ones, namely (*iti*) the beard, the body-hairs (*loman*), the hairs [on the head] (*kaca*), the cords (*snāyu*), the vessels (*sirā*), the ducts (*dhamanī*), the nails, the teeth, the semen[675], etc., arise/come_into_existence from the father. (śl.46)[676]

CA śārīra., 3,3: *[...] jāyate samudāyād eṣām bhāvānām, mātṛjaś cāyaṃ garbhaḥ, pitṛjaś cātmajaś ca, sātmyajaś ca rasajaś ca, asti ca khalu sattvam aupapādukam iti [...] /3/.* The same line again occurs in CA śārīra., 3,14.

CA śārīra., 3,6–7: *[...] tvak ca lohitaṃ ca māṃsaṃ ca medaś ca nābhiś ca hṛdayaṃ ca kloma ca yakṛc ca plīhā ca vṛkkau ca bastiś ca purīṣādhānaṃ cāmāśayaś ca pakvāśayaś cottaragudaṃ cādharagudaṃ ca kṣudrāntraṃ ca sthūlāntraṃ ca vapā ca vapāvahanaṃ ceti mātṛjāni /6/ keśa-śmaśru-nakha-loma-dantāsthi-sirā-snāyu-dhamanyaḥ śukraṃ ceti pitṛjāni /7/.*

AgniP 369,31cd: *mātṛjāni mṛdūny atra tvaṅ-māṃsa-hṛdayāni ca /31cd/ nābhir majjā śakṛn-medaḥ-kledānyāmāśayāni ca /32ab/.* AgniP, 369,32cd: *pitṛjāni śirā snāyuḥ śukraṃ caivā, ['tmajāni tu] /32cd/.*

ViṣṇudhP 2,115,12cd: *mātṛjāni mṛdūny atra tvak ca māṃsaṃ ca, bhārgava /12cd/ hṛdayaṃ ca tathā nābhiḥ svedo majjā yakṛt tathā / klomāntaṃ ca gudaṃ rāma āmasyāśayam eva ca /13/.* ViṣṇudhP, 2, 114,14: *pitṛjāni sthirāṇy agra-bhūmijānīha yāni tu / snāyu-śukra-śirāś caiva ... /14cd/.*

Hastyāyurveda, 3,9, v.41: *jarāyu-raktam māṃsaṃ tu mātṛjān viddhi hastinaḥ / śukram majjāsthi-medāṃsi śirā-roma-nakhāḥ pituḥ.*

ANALYSIS:

Among the parallels of the consulted texts, that of the SU is the closest to SR śl.45 in wording (see the words underlined). The SR does not mention the bones (*asthi*) among the father-born group, nor the flesh (*māṃsa*) among the mother-born group, though the SU does. The SU does not call these substances *bhāva* but *śarīra-lakṣaṇa* (marks of the body), while the CA calls them *bhāva*. The AS also contains these groups of substances, but does not mention the terms like *bhāva* or *śarīra-lakṣaṇa*; this text simply adapts neuter pl., e.g., *mātṛjāni*. The parallel of the AH is too brief, but contains *kaca* (the same term as in SR śl.46a) which is a synonym of *keśa* mentioned by the SU and CA. The YS does not mention these substances at all. The AgniP and ViṣṇudhP, which are parallel to each other, mention them.

674 D reads *bhāvāḥ* instead of *sthirāḥ*: "The substances, namely the beard [...]". The Ānandāśrama ed. (1896) reads *śukram ityādi* instead of *śuklam ityādyāḥ* ;this is not noted by the Adyar ed.

675 D reads *śuddha* instead of *śukla*. It would mean "[that which is] purified", but does not fit the context.

676 ŚG 9, śl.12 is identical to SR śl.46. The ŚG contains *-kaca-snāyu-śiro-* and *śukram ityādi* instead of *-kacāḥ snāyu-sirā-* and *śuklam ityādyāḥ* of the SR.

By the procreatory winds, [i.e.] by the winds called *Apāna*. **Through the cleft (*chidra*) of the organ**, [i.e.] through the opening/hole (*randhra*) of the vagina. **Of him** (*tasya*), [i.e.] of the born one. **"The soft": the soft** [ones] [namely] [the ones] endowed with softness[677]. **"The spleen (*plīhan*)"**: the word "spleen (*plīhan*)" is indicative of a particular [piece of] flesh situated on the left side, being the seat of life (*jīva*). **"The liver"**: a similar particular [piece of] flesh, situated on the right side. **The cords (*snāyu*), the vessels (*sirā*), the ducts (*dhamanī*)**[678]: **cords (*snāyu*)** are fine tubes (*nāḍī*); **vessels (*sirā*)** are thicker/grosser than them (= *snāyu*); **ducts (*dhamanī*)** are thick like the stem of the *eraṇḍa* (castor-oil) plant. **Firm (*sthira*)**[679] [means] hard (*kaṭhina*).

677 D reads *yuktāḥ nāḍyaḥ* "tubes joined/endowed with softness".

678 For the general information on the terms *snāyu, nāḍī, dhamanī,* and *sirā*, cf. DAS 2003A, pp.584–585; 560; 553; 584.
J. FILLIOZAT 1975, p.129ff. mentions three kinds of tubular vessels in the body, i.e. *nāḍī, dhamani* and *hirā*, which are mentioned in the Saṃhitā-s (cf. his index under *nāḍī, dhamani* and *hirā*).
According to FILLIOZAT (ibid.), *nāḍī* "flute, chalumeau" means the tube carrying air. But it may denote also tubes carrying other fluids, semen, umbilical nurture etc. For example, AV 6,138,4: *yé te nāḍyaú devákr̥te yáyos tíṣṭhati vŕ̥ṣṇyam,* "deux conduits du corps humain qui sont les canaux ou les cordons spermatique."
MAYRHOFER 1996 (Bd.I), p.7 (for *naḍa* "Schilfrohr"), gives as translation for *nāḍī* "Pfeife, Röhre, Flöte, Ader". He refers to the relationship of this term to Iranian *nay* "Flöte" and Vedic *nada.*
According to FILLIOZAT (ibid.), *dhamani* is derived from the root √*dham* "soufler", cf. R̥V 10,135,7 "the flute is blown". In R̥V, 2,11,8 (*indreṣitāṃ dhamanim*), the term *dhamani* means "un soufflé". FILLIOZAT states "Sāyana ecrit que c'est la voix émise par Indra. Tonnerre, voix du nuage."
MAYRHOFER 1992 (Bd.I), p.775 (on *DHAM*), translates √*dham* as "blasen, ein Blasinstrument spielen, anfachen [das Feuer]", mentioning *dhamani* "[Blas]Rohr". He also refers to R̥V 2,11,8.
Thus, *nāḍī* and *dhamanī* might be associated with the wind instrument.

679 The Ānandāśrama ed. (1896) reads *śirāḥ* instead of *sthirāḥ*. This is not noted by the Adyar ed. It would mean, "Ducts (*dhamanī*-s) are hard vessels (*śirā*) thick like the stem of the *eraṇḍa* plant".

He tells the manner of delivery: "**Is made**" (śl.42a; *kriyate*). **By the procreatory winds**, [i.e.] by the winds as the instruments/means (*sādhana*) of delivery. **Whose limbs are paining**, [i.e.] who has an oppressed (*pīḍita*) body. **Through the cleft of the organ**: the meaning is that (*iti*): through the small/minute (*sūkṣma*) hole of the vagina (*yoni*), which has the shape of a fetter (*yantra*).

He tells the permanence of the individual self, through the exertion/proclivity of drinking the mother's milk etc., of him who has just been born: "**Of him who has just been born**" (śl.43a, *jāta-mātrasya*).

He tells the kinds of substances (*bhāva*): "**Substances are**" (śl.44a, *bhāvāḥ syuḥ*). ***Rasa*** is the juice of food (*annarasa*)[680]. ***Sattva*** is a particular internal_instrument/faculty_of_thought (*antaḥkaraṇa*). **Suitablility (*sātmya*)** is a particular recollection/imprint (*saṃskāra*) due to long_familiarity/frequent_repetition (*paricaya*).

He tells the substances born from the mother: "**The soft**" (śl.45a, *mṛdavaḥ*). **The soft** [ones mean] the regions [of the body] associated with softness. **The spleen (*plīhan*) and the liver (*yakṛt*)** are two particular pieces of flesh situated at the heart.[681] The heart as well as the navel, **the heart and the navel**, [namely] a copulative (*dvandva*) compound [called] *samāhāra* (lit. "aggregate"). [This compound has] the state as if [it were] single, [because of the rule] saying (*iti*) "and a copulative (*dvandva*) [compound] of the parts of a living being, a musical group/band[682] (*tūrya*) and an army, [is treated as if it were single]" (Pāṇini, 2,4,2).

680 The term *āhāra-rasa* "the juice of food" is mentioned in SR śl.40b, too. On the differentiation between *rasa* and *āhāra-rasa*, cf. DAS 2003A, p.528; pp.180–187.

681 This does not reflect the real anatomy. This commentator does not know the real anatomy, or *hṛdaya-sthitau* might have a special meaning, like "near the heart".

682 The term *tūrya* usually means "a musical instrument". But in Pāṇini 2,4,2, it means a musical band. KATRE 1989 (to Pāṇini 2,4,2) explains *tūrya* as meaning *mārdaṅgika-pāṇavikam* "a group of *mṛdaṅga* (drum) and *paṇava* (small drum) players".

They (= the wise ones) know the increase of the body[683] (*śarīropacaya*), colour (*varṇa*), growth (*vṛddhi*), sleep (*supti*)[684], power (*bala*), steadiness/steadfastness (*sthiti*), non-greediness (*alolupatva*), incitement/exertion[685] etc. as born from the nutrient fluid (*rasa*)[686]. (śl.47)[687]

683 The term *śarīra* is defined in contrast with *garbha* "embryo" by SU śārīra., 5,6: *śukra-śoṇitaṃ garbhāśayastham ātma-prakṛti-vikāra-sammūrchitaṃ garbha ity ucyate / taṃ cetanāvasthitaṃ vāyur vibhajati, teja enam pacati, āpaḥ kledayanti, pṛthivī saṃhanti, ākāśaṃ vivardhayati, evaṃ vivardhitaḥ sa yadā hasta-pāda-jihva-ghrāṇa-karṇa-nitambādibhir aṅgair upetas tadā śarīra ity sañjñāṃ labhate / tac ca ṣaḍ-aṅgaṃ [...].* Namely, *garbha* is that which is formed of the semen and procreatory fluid, and is coagulated in the uterus; in contrast, *śarīra* denotes the stage in which the body is increased through the five elements and furnished with the limbs (*aṅga*). If according to this, *śarīropacaya* here would be the increase of the body as defined above.

684 The Ānandāśrama ed. (1896) reads *varṇa-vṛddhis tṛptir* instead of *varṇo vṛddhiḥ suptir*. It is not noted by the Adyar ed. The Ānandāśrama's reading *tṛptir* "satisfaction" seems to be correct, because the parallels, CA śārīra., 3,12 and AS śārīra., 5,22, both contain *tṛptir*. Indeed, ŚG 9,13, the identical verse to SR śl.47, contains *varṇo vṛddhis tṛptir*. Therefore, I conclude that the reading *suptir* is a mistake for *tṛptir*.

685 *Utsāhāḥ*. Plural in the original text. The parallel, CA śārīra., 3,12, also contains a plural, while AS śārīra., 5,22 contains a singular. The parallel, SU śārīra., 4,33, does not mention this term.

686 A and B read *anna-rasajān* instead of *rasajān*: "as born from the juice_of_food/nutrient_fluid", though it obviously violates the metre, and though *annarasa* and *rasa* are identical. On the problem of these two terms, cf. DAS 2003A, p.528 (*āhārarasa*).
The term *rasa* here seems to denote the nutrient fluid from the mother, according to CA śārīra., 3,12 which deals with the *rasaja* substances (*bhāva*): *na hi rasād ṛte mātuḥ prāṇa-yātrāpi syāt, kim punar garbha-janma* "Without the nutrient fluid from the mother, neither the circulation of breath, nor the appearance/birth of the foetus is [possible]".

687 SR śl.47 is identical to ŚG 9,13 (*śarīropacitir varṇo vṛddhis tṛptir balaṃ sthitiḥ / alolupatvam utsāha ityādi rājasaṃ viduḥ //*), though the ŚG's reading *rājasam* "arising from *rajas*" instead of *rasajān* is obviously a mistake.
PARALLELS (the expressions parallel to the SR are underlined):
SU śārīra., 4,33: *śarīropacayo balaṃ varnaḥ sthitir hāniś ca rasajāni.*
CA śārīra., 3,12 (on *rasaja*): *śarīrasyābhinirvṛttir abhivṛddhiḥ prāṇānubandhas tṛptiḥ puṣṭir utsāhāś ceti /12/.*

These (*iti*), [namely] desire (*icchā*), dislike (*dveṣa*), pleasure (*sukha*), pain (*duḥkha*), right and wrong (*dharma-adharma*) and [subliminal] _imprint/[the_fancy_of]_conception/reflection (*bhāvanā*), effort (*prayatna*), knowledge/perception (*jñāna*), the lifespan[688] (*āyus*) and the faculties/organs (*indriya*)[689] are deemed born from the self (*ātman*). (śl.48)[690]

AS śārīra., 5,22: *rasajāni, kṛtsnasya dehasya sambhavo vṛttir vṛddhis tṛptir alaulyam puṣṭir utsāhaś ca.*

ANALYSIS:

SU śārīra., 4,33 is the closest to SR śl.47. CA śārīra., 3,12 also shares expressions such as *vṛddhi, tṛpti* and *utsāha* with the SR. Except for the SU and SR, all the consulted texts include *varṇa* in the group of *rasaja* substances, but not in that of *ātmaja* ones (cf. parallels given in my footnote to SR śl.48). The AS's theory is similar to the CA's, but the AS is closer to the SR than the CA is. The terms which the SU has in common with the SR are not contained in the AS, and vice versa: in the SR, the elements from the SU and AS compensate each other. Thus the SR integrates both pieces of information found in the SU and AS. ŚG 9,13, (the identical verse to SR śl.47), reads *tṛptir* instead of *suptir* in the SR. The term *supti* "sleep", which does not fit the context very well, is obviously a mistake for *tṛpti.* This is supported by the fact that *tṛpti* is mentioned by the CA and AS, too. The AH includes *vṛddhi* and *alolatā* in the *rasaja* group, but *bala* in the *sātmyaja* group. The YS, ViṣṇudhP and AgniP do not mention the *rasaja* group.

688 For the relation between the life span and previous karma, cf. W.D. O'FLAHERTY 1980, p.94ff.
The GaruḍaP (KIRFEL 1954, v.71–72) state that the five, *āyuḥ, karma, vittaṃ, vidyā* and *nidhanam* (death), of the child are determined already in the prenatal period.

689 GARBE 1917, p.320, translates *jñāna-* and *karmendriya*-s as "Wahrnehmungs-sinne" and "Tatsinne, Fähigkeiten" respectively.

690 SR śl.48 is identical to ŚG 9,14. But the ŚG contains *cendriyāṇīty evam ātmajāḥ* instead of *cendriyāṇīty ātmajāḥ matāḥ* in the SR.
In SR śl.48cd, the caesura between c and d is not maintained. Actually, it is a latent caesura between the two words *ca* and *indriya*, which is erased through the *sandhi.*
PARALLELS (the expressions parallel to the SR are underlined):
SU śārīra., 4,33: *indriyāṇi jñānaṃ vijñānam āyuḥ sukha-duḥkhādikaṃ cātmajāni.*
CA śārīra., 3,10: *yāni tu khalv asya garbhasyātmajāni, [...] tad yathā, tāsu tāsu yoniṣūtpattir āyur ātma-jñānam mana indriyāṇi prāṇāpānau preraṇaṃ dhāraṇam ākṛti-svara-varṇa-viśeṣāḥ sukha-duḥkhe icchā-dveṣau cetanā dhṛtir buddhiḥ smṛtir ahaṅkāraḥ prayatnaś ceti /10/.*

AS śārīra. 5,20: *ātmajāni nānā-yoniṣūtpattir manaś cetanendriyāṇi prāṇāpānau dhāraṇam ākṛti-svara-varṇa-viśeṣāḥ kāma-krodha-lobha-bhaya-harṣa-dharmādharma-śīlatā-smṛti-buddhīcchā-dveṣa-prayatnāhaṅkāra-sukha-duḥkhāyur-ātmajñānāni ca.*
YS 3,73–74: *indriyāṇi manaḥ prāṇo jñānam āyuḥ sukhaṃ dhṛtiḥ / dhāraṇā preraṇaṃ duḥkham icchāhaṅkāra eva ca /73/ prayatna ākṛtir varṇaḥ svara-dveṣau bhavābhavau / tasyaitad ātmajaṃ sarvam anāder ādim icchataḥ /74/.*
NOTE on the parallel in the YS:
As already shown in my footnote to SR śl.45 which deals with the six groups of substances/components (*bhāva*), the YS does not mention them, except for the *ātmaja* group. Namely, the YS mentions only the *ātmaja* substances/ components in YS 3, śl.73–74, which are the foregoing passages of the description of the embryonic condition in the first month. The theory that lifespan, knowledge, pleasure and pain etc. are already determined in the uterus is also found, for example, in GaruḍaP v.71–72 (cf my footnote 688 on SR śl.48 "lifespan"). Intriguingly the YS mentions *svara*, and the comm. considers it to be "musical tones" (*svaraḥ ṣaḍja-gāndhārādi*). In fact, SR 1,3, śl.82–86 (SHRINGY 1989, vol.II, p.169) deals with the theory that a singer's voice-quality is an inborn quality (called *śārīra* "the quality pertaining to the body").
Agni P. 369,32d: *[...] caivā 'tmajāni tu /32d/ kāma-krodhau bhayaṃ harṣo dharmādharmātmatā tathā / ākṛtiḥ svara-varṇau tu mehanādyaṃ tathā ca yat /33/* (Compared with ViṣṇudhP, *mehanā* is obviously a mistake for *cetanā*).
ViṣṇudhP 2,114,14d: *ātmajāni nibodha me /14d/ kāmaḥ krodho bhayo harṣo dharmādharmātmatā tathā / ākṛtiḥ svara-varṇau ca cetanādyaṃ tathā vayaḥ /15/.*
ANALYSIS:
The AH does not mention the *ātmaja* group. This text mentions, instead of six, eight groups, *mātṛja, paitṛka, caitana, sātmyaja, rasaja, sāttvika, rājasa* and *tāmasa* (cf. AH śārīra., 3,4cd–8c). In contrast, AS śārīra., 5,18–25 mentions the same groups as the SR does, namely, *mātṛja, pitṛja, ātmaja, sātmyaja, rasaja* and *sattvaja*.
According to the parallels in AS śārīra., 5,20 and CA śārīra., 3,10 (see above), the term *jñāna* (cf. SR śl.48) means "the knowledge of the self" (*ātma-jñāna*). But according to SU śārīra., 4,33, it seems to mean simply "knowledge".
Among the consulted texts, the CA and AS both contain the wording closest to the SR. And it is difficult to decide which of the two texts is closer to the SR; The CA lacks *dharmādharma*, while the AS lacks *prayatna*. The YS also contains expressions similar to the SR. Though the SU also accords with the SR, its line is too fragmental; it makes the impression that it shortened an older text which had been longer.
On the other hand, the SU, in another passage listing the qualities (*guṇas*) of the *karma-puruṣa* (SU śārīra., 1,17), mentions terms similar to the *Atmaja* ones mentioned by CA śārīra., 3,10. SU śārīra., 1,17: *tasya sukha-duḥkhe icchā-dveṣau prayatnaḥ prāṇāpānāv unmeṣa-nimeṣau buddhir manaḥ saṅkalpo*

They say that (*iti*) the faculties/organs of knowing/perception are hearing, touching, then seeing, tasting, smelling. Their ranges/objects (*gocara*), however, are [these] five, namely (*iti*), (śl.49)[691]
sound (*śabda*), touch (*sparśa*), then shape (*rūpa*), taste (*rasa*) [and] smell (*gandha*), respectively. (śl.50ab)

However, they call voice/speech (*vāc*)[692], the hands, the feet, the anus [and] the genital[s] (*upasthā*) the faculties/organs of action (*karmendriya*). (śl.50cd)[693]

Their actions (*kriyā*) are speaking (*vacana*), seizing (*ādāna*), going (*gamana*), discharge (*visarga* lit. "emission"), coition (*rati*), respectively. (śl.51abc)[694]

---

vicāraṇā smṛtir vijñānam adhyavasāyo viṣayopalabdhiś ca guṇāḥ. This might have something to do with the statement of CA śārīra., 1,70–72, considering these to be the marks of the supreme self (*liṅgāni paramātmanaḥ*).

691 SR śl.49 is identical to ŚG 9,15.

692 The term *vāc* seems to mean something more concrete, i.e. "mouth" or "voice-making [anatomical] organ". CA śārīra., 1,26 explicitly states that the tongue (*jihvā*) is the faculty/organ of speech (*vāg-indriya*). The YS (3,91) and the YS's two descendants, the ViṣṇudhP (2,115,35), mention the tongue (*jihvā*) instead of speech (*vāc*). If this is right, the tongue functions as both, the *buddhīndriya* ("sense faculty/organ") perceiving tastes (*rasa*) and the *karmendriya* ("faculty/organ of action") pronouncing words/voices (*śabda*). In contrast, the GaruḍaP (KIRFEL 1954, śl.39) mentions *jihvā* as the *buddhīndriya* for *rasa*, but *vāc* as the *karmendriya* for *śabda*.

693 SR śl.50 is identical to ŚG 9,16. But the ŚG contains *-upasthāny āhuḥ* and *hi* instead of *-upasthān āhuḥ* and *tu* of the SR.

694 SR śl.51 is identical to ŚG 9,17. But the third *pāda* (ŚG 9,17d) contains a quite different wording (*buddhir ahaṅkāras tataḥ param*). The SR's reading of the third *pāda* (śl.51d) poses a difficulty of interpretation; maybe it is a corruption (see my observation in the footnote 695 on SR śl.51cd).
The topic of *buddhi-* and *karma-indriya*, and their objects are dealt with in the following texts:
SU śārīra., 1,4: *śrotra-tvak-cakṣur-jihvā-ghrāṇa-vāg-ghastopastha-pāyu-pāda-manāṃsīti, tatra pūrvāṇi pañca buddhīndriyāṇi, itarāṇi pañca karmendriyāṇi, ubhayātmakam manaḥ.*
SU śārīra., 1,5: *tatra buddhīndriyāṇāṃ śabdādayo viṣayāḥ, karmendriyāṇāṃ yathāsaṅkhyaṃ vacanādānānanda-visarga-viharaṇāni.*
CA śārīra., 1,25ff.: *hastau pādau gudopasthaṃ vāg-indriyam athāpi ca / karmendriyāṇi pañcaiva pādau gamana-karmaṇi /25/ pāyūpasthaṃ visargārthaṃ hastau grahaṇa-dhāraṇe / jihvā vāg-indriyaṃ vāk ca satyā jyotis tamo 'nṛtā /26/ mahābhūtāni khaṃ vāyur agnir āpaḥ kṣitis tathā / śabdaḥ*

The two internal_instruments/faculties_of_thought (*antaḥkaraṇa*) are as follows (*iti*): the mind (*manas*) and the intellect (*buddhi*).[695] (śl.51cd)

---

sparśaś ca rūpaṃ ca raso gandhaś ca tad-guṇāḥ /27/ In v.31, the objects like *śabda* etc. are called *artha, viṣaya* and *gocara*.

CA śārīra., 7,7: *pañca buddhīndriyāṇi, tad yathā, sparśanaṃ rasanaṃ ghrāṇaṃ darśanaṃ śrotram iti / pañca karmendriyāṇi tad yathā, hastau pādau pāyuḥ upasthaḥ jihvā ceti /7/.* Also see CA śārīra., 7,7.

AS śārīra., 5,55ff. (275ff.): *pañca buddhīndriyāṇi, śrotraṃ sparśanaṃ darśanaṃ rasanaṃ ghrāṇaṃ ca /55/ teṣāṃ sabhāgatayā kramād viṣayāḥ śabda-sparśa-rūpa-rasa-gandhāḥ /56/ pañca buddhīndriyādhiṣṭhānāni karṇau tvag-akṣiṇī jihvā nāsike ca /57/ pañca karmendriyāṇi vāk pāyūpastha-pāṇi-pāda-sañjñakāni /58/ tāny api ca vacanotsarga-harṣādāna-gamanārthāni /59/.*

YS 3,91: *gandha-rūpa-rasa-sparśa-śabdāś ca viṣayāḥ smṛtāḥ / nāsikā locane jihvā tvak śrotraṃ cendriyāṇi ca /91/ hastau pāyur upasthaṃ ca jihvā pādau ca pañca vai / karmendriyāṇi jānīyān manaś caivobhayātmakam /92/.*

ViṣṇudhP 2,115,32: *buddhīndriyāni pañcātra śrotraṃ ghrāṇaṃ ca cakṣuṣī / tvak tathā rasanā caiva mahābhūtāśrayāni tu /32/ indriyārthās tathā pañca teṣāṃ nāmāni me śṛṇu / śrotrasya śabdaḥ kathito gandho ghrāṇasya pārthiva /33/ rūpaṃ ca cakṣuṣo jñeyaṃ tvak ca saṃsparśanaṃ tathā / rasanasya rasaś caiva mahābhūtāśrayās tu te /34/ karmendriyāṇi pañcātra teṣāṃ nāmāni me śṛṇu / pāyūpasthaṃ hasta-pādau jihvā caivātra pañcamī /35/ teṣām arthās tathā pañca tān pravakṣyāmy ataḥ param / utsargaṃ ca tathānanda ādāna-gamane tathā /36/ indriyāni daśaitāni teṣāṃ vai nāyakam manaḥ / pañca karmendriyāṇy atra pañca buddhīndriyāṇi ca /37/ indriyārthāś ca pañcātra mahābhūtāni pañca ca /38ab/.*

AgniP 370,3: *pañca karmendriyāṇy atra pañca buddhīndriyāṇi ca / indriyārthāś caiva pañcaiva mahābhūtā manodhipāḥ /3/.*

GaruḍaP (KIRFEL 1954) v.39: *śrotraṃ tvak cakṣuṣī jihvā nāsā buddhīndriyāṇi ca / pāṇī pādau gudaṃ vāk ca guhyaṃ karmendriyāṇi ca /39/.*

ANALYSIS

This is the theory found also in the Sāṅkhya or Vaiśeṣika theory, cf. GARBE 1917, p.320. The SU and AS seem to be the closest to the SR. Though the CA's theory is also almost the same, it allocates the genitals (*upastha*) to the action of discharge (*visarga*), in contrast to the SU and AS which allocate it to the action of the joy (*ānanda, harṣa*) of coitus.

The AH does not mention this topic. But the YS, ViṣṇudhP and GaruḍaP mention it, though the ViṣṇudhP does not. For the Nyāya-Vaiśeṣika theory contained in the CA, see COMBA 1987.

695 This topic is further dealt with in the following verses, SR śl.52–53ab.

The statement of SR śl.51cd does not keep with the various schools which accept only *manas*. The usage of the term *buddhi* as denoting a particular mental faculty/organ is found only in the Sāṅkhya school, cf. GARBE 1917,

p.310. But the Sāṅkhya school counts three internal instruments (*antaḥkaraṇa*), namely, *manas, buddhi* and *ahaṅkāra*, cf. GARBE ibid., p.307. The SR does not mention the third one, *ahaṅkāra*.
According to MÜLLER 1952, p.291, the CA's system, which mentions *buddhi* and *manas*, does not belong to any particular school of philosophy.
The comm. S. states that the statement of SR śl.51 is according to the Vedānta school. But, according to DEUSSEN 1883, p.357, Śaṅkara acknowledges only *manas* which is identified with *buddhi*. However, there seems to have been another theory which distinguishes the one from the other, according to DEUSSEN, ibid., p.338, quoting a statement from the Brahmasūtra.
SR śl.51cd is identical to ŚG 9,17cd, as already remarked. The SR does not have a parallel to the next verse (ŚG 9,18) which consists of a half verse (ab). ŚG 9,17cd–18ab runs as follows: *kriyās teṣām mano buddhir ahaṅkāras tataḥ param /17cd/ antaḥkaraṇam ity āhuś cittaṃ ceti catuṣṭayam /18/.*
The topic of *antaḥkaraṇa* is dealt with, for example, by the texts mentioned below.
SU śārīra., 1,4: *śrotra-tvak-cakṣur-jihvā-ghrāṇa-vāg-ghastopastha-pāyu-pāda-manāṃsīti, tatra pūrvāṇi pañca buddhīndriyāṇi, itarāṇi pañca karmendriyāṇi, ubhayātmakam manaḥ.*
CA śārīra., 1,56ab: *karaṇāni mano buddhir buddhi-karmendriyāṇi ca.*
AS śārīra., 5,60 (280) states that the mind (*manas*) is above all the other organs or faculties (*indriya*), because the latter and their objects are connected through it, and because the mind is at the same time a sense-organ (*buddhi[-indriya]*) and an action-organ: *atīndriyaṃ tu manaḥ sarvārthair anvayāt tad-yogena pañcendriyāṇām artha-pravṛtteḥ / buddhi-karmendriyobhayātmakatvāc ca /60/.*
YS 3,92d: *manaś caivobhayātmakam /92/.* On the notion of *manas* in the YS, cf. YAMASHITA 2001/2002, pp.106–107.
AgniP 370,3: *pañca karmendriyāṇy atra pañca buddhīndriyāṇi ca / indriyārthāś ca pañcaiva mahābhūtā manodhipāḥ /3/ ātmā 'vyaktaś caturviṃśat tattvāni puruṣaḥ paraḥ / saṃyuktaś ca viyuktaś ca yathā matsyodake ubhe /4/ avyaktam āśritānīha rajaḥ-sattva-tamāṃsi ca / āntaraḥ puruṣo jīvaḥ sa param brahma kāraṇam /5/.*
ViṣṇudhP 2,115,38cd: *indriyebhyaḥ parāḥ proktāś catvāro, bhṛgunandana /38/ mano buddhis tathaivātmā avyaktaś ca, mahābhuja / tad-āśrayāṇīndriyāṇi indriyebhyaḥ parāś ca te /39/ tattvāny etāni jānīhi caturviṃśati-saṅkhyayā / yeṣām aikyam mahābhāga, puruṣaḥ pañcaviṃśakaḥ /40/ saṃyuktair viyuktaś ca tathā matsyodake ubhe / manaḥ saṃśayakṛn nityaṃ tathā buddhir vivecanī /41/ ātmā jīvaḥ smṛto, rāma yo bhoktā sukha-duḥkhayoḥ / avyakto miśritānīha rajaḥ-sattva-tamāṃsi ca /42/ puruṣas tv aparo jñeyo yaś ca sarvagato mahān /42ab/.*
ANALYSIS
The SU and AS mention only one mental organ/faculty, i.e. *manas* which has both characters of *buddhi-* and *karma-indriya*. The *manas* as well as the *buddhi-* and *karma-indriya*-s are included in the twenty-four *tattva*-s of Sāṅkhya (cf. SU śārīra. 1,4). In contrast, the CA (śārīra., 1,56ab) mentions two

Comm. K on SR śl.47–74[696]

The text beginning with "**increase of the body**" (śl.47a, *śarīropacayaḥ*) [and] ending with "**[we], afraid of [too great] expansion**[697] **of the text**" (śl.74d, *grantha-vistara-kātarāḥ*), is explained through [mere] recitation/mention (*nigada-vyākhyātaḥ*).

Comm. S on SR śl.47–51

He tells those born from the nutrient fluid (*rasa*) and those born from the self (*ātman*): "**The body**" (śl.47a, *śarīra-*). **Increase** [means] being swollen. **Growth** [means] the superabundance of the measure/quantity (*pramāṇa*). (Comm. S on SR śl.47–48)

He divides/enumerates separately (*vibhajate*) the faculties/organs of knowing/perception and their objects (*viṣaya*): "**The faculties/organs of knowing/perception**" (śl.49a, *jñānendriyāṇi*). **Touching** (*sparśana*): the faculty/organ skin. **Seeing** (*darśana*): the eye. **Tasting** (*rasana*): the tongue. (Comm. S on SR śl.49–50)

He tells the faculties/organs of action and their functions: "**Voice/speech, hands**". (śl.50c, *vāk-kara-*). (Comm. S on SR śl.50–51)

He tells the two internal_instruments (= faculties of thought),

---

mental organs/faculties, *manas* and *buddhi*; these two are included in *karaṇa*-s, to which the *buddhi-* and *karma-indriya*-s belong, too. On the other hand, CA śārīra., 1,32–34 explains *buddhi* to be produced from the contact of the self, *indriya*-s, *manas* and objects. The YS mentions the dual character of *manas*, like the SU and AS. The AgniP and ViṣṇudhP, which have many expressions in common to each other, are in this point not parallel to the YS. The ViṣṇudhP seems to contain many insertions in comparison to its parallel text of the AgniP. But there are discrepancies between the respective statements in the AgniP and ViṣṇudhP. Although some corruptions in the text make comprehension difficult, the AgniP seems to acknowledge only *manas*, as the integrator of the *buddhi-indriya*-s and *karma-indriya*-s, and mentions, besides that, the transcendent (*para-*) *puruṣa* which is called *ātman, avyakta* etc. The theory of the ViṣṇudhP, mentioning *manas* and *buddhi*, is totally different from that of the AgniP.

696 Comm. K does not explain the following verses up to SR 1,2, śl.74.

697 The Ānandāśrama ed. (1896) reads *vistāra* instead of *vistara*. This is not noted by the Adyar ed.

having recourse to the opinion of the Vedāntin-s[698]: "**The mind and the intellect**" (śl.51c, *mano buddhiḥ*). In the Vācaspatya[699], it is said: "Yoga is the dependence of the two instruments/faculties, [namely] the intellect and the mind, on the manifestation of the perception [of] 'I'." (*buddhi-manasoś ca karaṇayor aham iti prakhyāna-pratibhāsālambanatvaṃ yogaḥ*)[700]. (Comm. S on SR śl.51)

SR śl.52–53

Pleasure and pain are to be known as the two objects of the mind (*manas*). (śl.52ab)

The action (*kriyā*) [of] remembrance, confusion/hesitation (*bhrānti*)[701], manifoldness/doubt (*vikalpa*) etc. is considered the business (*vyavasiti*[702]) of the intellect (*dhī* = *buddhi*).[703] (śl.52cd)[704]

698 Śaṅkara does not distinguish *buddhi* from *manas*, but he mentions the different view of others that *manas* is the faculty of reflexion (*saṃśaya, vikalpa*) while *buddhi*, the faculty of decision (*niścaya, adhyavasāya*) (DEUSSEN 1912, p.330).

699 A and B read *vācaspatau* "in the Vācaspati". The Vācaspatya is the title of a Sanskrit dictionary.

700 This is not clear to me.

701 The Ānandāśrama ed. (1896) reads *bhīti* "fear" instead of *bhrānti*; this is not noted by the Adyar ed. D reads *smṛtiḥ prītir vibhrānti* "remembrance, pleasure/love, confusion" instead of *smṛti-bhrānti-*. But this violates the metre.

702 *Vyavasiti* "ascertainment". But according to the commentary, it means something like "business" (*vyāpāra*). In fact, the term *vyavasāya* means "business" in later Sanskrit and New Indo-Aryan.
The Ānandāśrama ed. (1896) reads *dhiyo 'dhyavasitir* instead of *dhiyo vyavasitir;* this is not noted by the Adyar ed. I prefer this reading *adhyavasiti* to *vyavasiti*, because, according to Sāṅkhya, the function of *buddhi* is called *adhyavasāya* "effort/determination/resolution" (GARBE 1917, p.307; Sāṅkhyakārikā 22–23).

703 My translation follows the comm. S. The comm. S interprets it, as if *manas* perceives objects (*viṣaya*), in contrast with *buddhi* which makes *kriyā*. Thus, the pair, *manas* and *buddhi*, corresponds to the pair, *jñānendriya* and *karmendriya*.
But Sāṅkhya allots the functions *saṅkalpa-vikalpau* "die Funktionen des Wünschens und des zweifelnden Überlegens" to *manas* (GARBE 1917, p. 315), and the function *adhyavasāya* to *buddhi* (ibid., p.307). Therefore, if according to Sāṅkhya, this verse could be analysed as follows: *sukhaṃ duḥkhaṃ ca viṣayau vijñeyau manasaḥ; [manasaḥ] kriyā smṛti-bhrānti-vikalpādyā* [*matā*]*; dhiyo* [*kriyā*] *vyavasitir matā* "Pleasure and pain are to be known as the two objects of the mind. The action [of the mind] is deemed remembrance,

The faculties/organs (*indriya*) are such whose origin is Brahman[705]. Others chanted/claimed [the faculties/organs as being] material[706]. (śl.53ab)[707]

---

confusion, manifoldness etc. [The function] of the intellect is ascertainment (*adhyavasiti*)."

704 SR śl.52 is identical to ŚG 9,19. But ŚG 9,19 contains many variants. The theory presented by ŚG 9,17cd–20ab is different from that of SR śl.51cd–53ab. I quote these ŚG verses in the following. ŚG 9,17: (*kriyās teṣām*) *mano buddhir ahaṅkāras tataḥ param /17cd/ antaḥkaraṇam ity āhuś cittaṃ ceti catuṣṭayam /18/ sukhaṃ duḥkhaṃ ca viṣayau manasaḥ kriyā / smṛti-bhīti-vikalpādyā buddhiḥ syān niścayātmikā /19/ aham mametiy ahaṅkāraś cittaṃ cetayate yataḥ /20ab/.*

The ŚG mentions four internal instruments (*antaḥkaraṇa*), namely, *manas, buddhi, ahaṅkāra* and *cetas* (cf. DEUSSEN 1883, p.357). The functions of the two, *ahaṅkāra* and *citta*, are mentioned in ŚG 9,20ab.

But I suspect that this theory of the ŚG originally did not belong to the source text. The reason is the following: ŚG 9,18 consists of only two quarters (*ab*), while ŚG 9,20 unproportionally consists of six (*abcdef*). This fact suggests that either ŚG 18ab or 9,20ab is a secondary interpolation. Namely, it makes the impression that the compiler of the ŚG added *ahaṅkāra* and *citta* into the source text, in which only two *antaḥkaraṇa*-s, namely *manas* and *buddhi,* had been mentioned.

Besides, ŚG 9,19cd, identical to SR śl.52cd, contains an unignorable variant, *buddhiḥ syān niścayātmikā* instead of *dhiyo vyavasitir matā* in the SR. With the ŚG's variant, the verse becomes easier to understand. ŚG 9,19cd gives a too orderly impression in comparison to SR śl.52cd. Therefore, the SR perhaps preserves a more original version than the ŚG.

705 That means, Brahman is the origin of the faculties/organs.

706 I.e. not permanent.

707 The ŚG does not contain a parallel to SR śl.53ab. This suggests that SR śl.53ab is probably an interpolation.

The comm. S explains that these two contradicting theories are those of Sāṅkhya and Vaiśeṣika. In fact the theory that the sense organs are of material nature is contained in the Vaiśeṣikasūtra (jambuvijaya) 3,2,2, according to GARBE 1917, p.322. GARBE, discussing this topic, states that the Sāṅkhya system derives the *indriya*-s from *ahaṅkāra*, which is immaterial. But whether this interpretation of the comm. S. of the SR is actually in accord with the SR, is uncertain.

Also the Nyāyāyika-s who consider the *indriya*-s to be material argue against the Sāṅkhya theorists, cf. PREISENDANZ 1994, vol.2, pp.468–470; p.660. Cf. Nyāyasūtra 1,1,12: *ghrāṇa-rasana-cakṣus-tvak-śrotrāṇīndriyāṇi bhūtebhyaḥ.*

The internal_instrument/faculty_of_thought (*antaḥkaraṇa*) called "*sattva*"[708] is deemed of three kinds due to the difference of the qualities (*guṇa*)[709]. (śl.53cd)[710]

708 In the foregoing verses, SR śl.51d–53ab, the internal instrument (*antaḥkaraṇa*) is divided into two, *manas* and *buddhi*, according to function (*kriyā*). SR śl.53 mentions another aspect of the same internal instrument (*antaḥkaraṇa*), namely, *sattva*.
CA sūtra., 8,4 states that *sattva* is another appellation of *manas*, cf. MÜLLER 1952, p.285. MÜLLER (op.cit., pp.280–281) explains the difference of the two notions, "Im vergleich mit einer Bewegung oder Tätigkeitsäußerung durch *manas* als 'Denken' [...], zielt *sattva* auf den Begriff eines 'Wesens' in einem eigenschaftlichen Zustande." For the threefoldness of *sattva*, i.e., *sāttvika, rājasa* and *tāmasa*, cf. WINDISCH 1908, p.54.
The internal instrument called *sattva* plays a crucial role at the moment of conception, cf. DOSSI 1998, p.133; CA śārīra., 3,13; CA śārīra., 3,3; AH śārīra., 1,1.

709 ROŞU 1978, p.183, states "la prèdominance de l'une sur deux autres marque le psychisme à un moment donné". That means, the three types of the *antaḥkaraṇa* called *sattva*, i.e. *sāttvika, tāmasa,* and *rājasa*, are not static conditions, but dynamic and transitory ones. One's mental condition can occasionally change from one type to another. According to CA śārīra., 3,13, the mental condition (*sattva*) is changeable from time to time in a single person, but the predominant one is considered to be the character of the person (*nānā-vidhāni khalu sattvāni, tāni sarvāṇy eka-puruṣe bhavanti, na ca bhavanty eka-kālaṃ, ekaṃ tu prāyovṛtyāha*), cf. ROŞU ibid., p.187.

710 SR śl.53cd is identical to ŚG 9,20ab.
SR śl.51d–52 considers the *antaḥkaraṇa* to be twofold, i.e. *manas* and *buddhi*, with regard to its action/function (*kriyā*). Now, SR śl.53 observes the same *antaḥkaraṇa* in another aspect, considering it to be threefold, namely, consisting of the three *guṇa*-s, with regard to constitution or disposition.
SR śl.49–55 lists the substances/constituents (*bhāva*) derived from the *sattva*. The *sattvaja* substances/constituents are actually those mentioned in SR śl.54bcd–55, namely *sāttvika-bhāva*-s, *rājasa-bhāva*-s and *tāmasa-bhāva*-s. They are mentioned by the medical texts, too (e.g. AS śārīra., 5,23f.).
The foregoing verses, SR śl.49–53ab, merely serve as the introduction to the notion *antaḥkaraṇa*, in keeping with Sāṅkhya. That means, those mentioned in SR śl.43–53ab, i.e., the five *jñānendriya*-s, five *karmendriya*-s, *manas* and *buddhi*, do not belong to the *sattvaja bhāva*-s. The medical texts do not mention the *jñānendriya, karmendriya, manas* or *buddhi* at all in the passages on the *sattvaja bhāva*-s (e.g., AS śārīra., 5,23f.).

Comm. S on SR śl.52–53

He tells the particular function[s], for the sake of the distinction between the intellect and the mind: "**Pleasure**" (śl.52a, *sukhaṃ*). And from that [follows] the distinction that (*iti*) **the mind** is the instrument/means for the perception of **pleasure and pain**, **the intellect** the instrument/means for **remembrance, confusion/hesitation, manifoldness/doubt etc.**[711] **Business (*vyavasiti*)** [means] function (*vyāpāra*). (Comm. S on SR śl.52)

He observes/investigates the cause of the faculties/organs according to a difference of opinion: "**Brahman**" (śl.53a, *brahma-*). Vedāntin-s, however, relate that (*iti*) the faculties/organs are such whose cause is Brahman, because of the whole manifest world being an illusory_form/transformation (*vivarta*)[712] of Brahman. Or else, the scholars of the Sāṅkhya school deem [them] **such whose origin is Brahman**, [i.e.] immaterial. **Others**, [i.e.] the scholars of the Vaiśeṣika school, say: "[They are] **material**, produced of the [gross] elements (*bhūta*), [i.e.] earth etc."

He tells the third internal_instrument/faculty_of_thought: "**called *sattva***" (śl.53c, *sattvākhyam*). **Called ["]*sattva*["]**: the meaning is: called ["]the nature/natural_condition/disposition["] (*svabhāva*). **Due to the differentiation/difference of the qualities**, [i.e.] due to the difference of the quality of purity (*sattva*), the quality of turbidity (*rajas*) and the quality of darkness (*tamas*). (Comm. S on SR śl.53)

SR śl.54–55

The qualities (*guṇa*) are *sattva* (purity), *rajas* (turbidity) and *tamas* (darkness). (śl.54a)[713]

From the nature (*sattva*)[714] characterised by purity (*sattva*), however, belief that there is [a(n) afterlife/creator] (*āstikya*), a proclivity

711 The term *sādhana* does not belong to the *mūla* text, although the Adyar edition uses boldface for it.

712 B reads *parivartatvāt* "because of being the [repetitive] revolution of Brahman".

713 The types of disposition derived from the three qualities are mentioned in SU śārīra., 1,18.

714 I.e. the internal instrument (*antaḥkaraṇa*) called *sattva*, mentioned in SR śl.53.

solely (*ekaruci*[715]) for the pure *dharma* etc. are held [to arise], (śl.54bcd) [716]
from the nature (*sattva*) characterised by *rajas*, however, the [mental] conditions (*bhāva*)[717], desire, anger, passion/intoxication (*mada*) etc. [are held to arise],[718] (śl.55ab)

715 *Eka-ruci.* same as *eka-rasa*, "of one tenor", "solely devoted to the sole one".

716 The Ānandāśrama edition (1896) reads *śukladharmaikya-*, and the manuscript D reads *vuddhidharmaikya-*, instead of *śuddhadharmaika-*. With the former, the verse would mean "[...] a proclivity for oneness with bright/lucid (*śukla*) *dharma* [...]". With the latter, it would mean "[...] a proclivity by the notion of the belief that there is [a(n) afterlife/creator] (*āstikyabuddhi; v=b*), for oneness with *dharma* [...]". The Adyar edition contains *sāttvikāt*, while the Ānandāśrama edition contains *sāttvikāḥ*. The Adyar edition contains *sattvāt tu rājasād* (SR śl.55a) and *tāmasāḥ* (SR śl.55d) instead of the Ānandāśrama edition's (1896) readings, *rajaso rājasā bhāvāḥ* and *tāmasāt*, respectively. Comm. S contains *sattvād iti* in its explanation to SR śl.55a, cf. Adyar ed., p.47, the fourth line of comm. S.
SR śl.54 is identical to ŚG 9,20cd–21ab. But the ŚG contains *sāttvikāḥ* and *śukla-dharmaika-mati-prakr̥tayo,* instead of *sāttvikāt* and *śuddha-dharmaika-ruci-prabhr̥tayo* of the SR. For the parallelism between ŚG *śukra-dharma-mati-* and AS śārīra., 5,23 *śukla-vartma-rucir*, cf. my footnote 721 on SR śl.55cd.
ŚG 9,20d contains the same reading, *sāttvikāḥ*, as the Ānandāśrama edition (1896) of the SR. ŚG 9,21cd, identical to SR śl.55ab, contains (in *pāda* 21c) the same reading *rajaso rājasā bhāvāḥ*, as the Ānandāśrama edition. In the same manner, ŚG 9,22ab, identical to SR śl.55cd, contains the same *tāmasāḥ*. Therefore, I conclude that these readings in the Ānandāśrama edition of the SR (*sāttvikāḥ, rajaso rājasā* and *tāmasāḥ*) are correct.
With the correct readings of the Ānandāśrama edition (1896) which are confirmed by the ŚG, SR śl.54bcd–55 would be translated as follows:
"From [the quality (*guṇa*) of] *sattva*, [the conditions (*bhāva*)] characterised *sattva* (*sāttvikā*), [namely] belief that there is, a proclivity solely for the pure *dharma* etc. are held [to arise]. (śl.54bcd) From [the quality of] *rajas*, the conditions (*bhāva*) characterised by *rajas* (*rājasa*), [namely] desire, anger, passion etc., are held [to arise]. (śl.55ab) [The conditions] characterised by *tamas* [are held to be], however, sleepiness, sloth, carelessness, distress, deceit etc. (śl.55cd)"

717 This word occurs in SR śl.27, too. There, I translated it as "characteristics" but it may be discussed whether "[mental] condition" fits there, too.
On the other hand, the term *bhāva* here seems to denote the same thing as the *bhāva*-s of *pitr̥ja, mātr̥ja* etc. If so, the term here should be translated as "substance/constituents", too.

718 The Ānandāśrama edition (1896) reads *rajaso rājasā bhāvāḥ* instead of *sattvāt tu rājasād bhāvāḥ*.

from [that] characterised by *tamas*[719], however, sleepiness, sloth, carelessness (*pramāda*), distress (*ārti*)[720], deceit etc. [are held to arise]. (śl.55cd)[721]

719 The Ānandāśrama edition (1896) reads *tāmasāḥ* instead of *tāmasāt*.

720 D reads *-pramādādi* instead of *-pramādārti-*: "[...] carelessness etc., deceit etc. [...]".

721 Cf. AS śārīra., 5,23–25. AS śārīra., 5,23 contains two expressions which are parallel to the SR, i.e. *āstikatva* and *ruci*, among the characteristics of the *sāttvika sattva* (which is called *śuddha-sattva* by the AS).
The expression *śukla-vartma-rucir* (ROŞU 1978 translates it as "l'inclination pour une conduite pure") acoustically resembles *śuddha-dharmaika-ruci-* of the SR. As a matter of fact, ŚG 9,21ab, identical to SR śl.54cd, reads *śukla-dharmaika-mati-*. It seems that these three expressions in the SR, ŚG and AS are parallel to each other. ROŞU 1978, p.186, also points at the resemblance between the AS's *śukra-vartma-rucir* and the AH's (śārīra., 3,7–8) *śukla-dharma-ruci* "inclination pour les choses pures". Thus, it seems that *śuddha-dharmaika-ruci* in SR śl.54cd should be read as constituting one unit like "*eka-ruci* for *śuddha-dharma*".
AS śārīra., 5,24 (*rājasa-sattva*) mentions *kāma* and *krodha*, but not *mada*. In AS śārīra., 5,25, among the characteristics of *tāmasa-sattva,* parallel expressions to the SR text are contained, i.e. *pramāda, nidrā, ālasya*, in this order.
PARALLELS
SU śārīra., 4,33, enumerating the six groups *pitṛja* etc., announces a discussion of the *sattvaja-* ones separately somewhere afterwards. According to Ḍalhaṇa, this seems to refer to SU śārīra., 5,81f, which is, however, parallel to SR śl.72–74. The SU does not have a parallel to SR śl.53ab–55.
CA śārīra., 3,13 calls the three classes *śuddha*, *rājasa* and *tāmasa*. In this passage, the CA mentions little about the respective components of the three classes, but announces a discussion later, namely in CA śārīra. 4,36f. But the contents of CA śārīra., 4,36f. are parallel to those in SR śl.72–74 like in the case of SU śārīra., 5,81f.
AS śārīra., 5,23–25 mentions, as examined above, *āstikatvaṃ, śuklavartma* and *rucir* among the *śuddha-sattvaja-* ones; *kāma* and *krodha* among the *rājasa-* ones; *nidrā, ālasyaṃ, kṣudh-, tṛṣṇā* and *śākā* among the *tāmasa-* ones.
AH śārīra., 3,7–8ab: *sāttvikaṃ śaucam <u>āstikyaṃ śukla-dharma-rucir</u> matiḥ / rājasam bahu-bhāṣitvam māna-<u>krud</u>-dambha-matsaram /7/ tāmasam bhayam ajñānaṃ <u>nidrālasyaṃ</u> viṣāditā /8ab/.*
AgniP 369,34: *tāmasāni tathā jñānam <u>pramādālasya-tṛ</u>ṭ-kṣudhāḥ / moha-mātsarya-vaiguṇya-śokāyāsa-bhayāni ca /34/ <u>kāma-krodhau</u> tathā śauryaṃ yajñepsā bahu-bhāṣitā / ahaṅkāraḥ parāvajñā rājasāni, mahā-mune /35/ dharmepsā mokṣa-kāmitvam parā bhaktiś ca keśave / dākṣiṇyaṃ vyavasāyitvaṃ sāttvikāni vinirdiśet /36/.*

Distinguishing the qualities (*guṇa*), he tells those [characteristics] born from them [one by one]: "***Sattva***" (śl.54a, *sattvaṃ*).

He tells the workings (*kārya*) of the internal_instrument/faculty_of_thought characterised by purity (*sattva*): "**From the nature characterised by purity (*sattva*)**" (śl.54b, *sattvāt*). **Belief that there is (*āstikya*)** [means] the confidence (*śraddhā*) that (*iti*) there is (*asti*) the other world. **A proclivity**, [namely] delight (*prīti*), **solely** with regard to that, [namely] ***dharma*** which is **pure**, unsoiled.[722] Through the word ["]etc.["] (*prabhṛti*), abstaining from non-*dharma* is mentioned. (Comm. S on SR śl.54)

He tells the workings of the internal_instrument/faculty_of_thought characterised by *rajas*: "**From the nature**" (śl.55a, *sattvāt*). **From the nature (*sattva*)**, [namely] from the internal_instrument/faculty_of_thought called "*sattva*", **characterised by *rajas***, in which the quality (*guṇa*) of *rajas* is predominant.

He tells the workings of the internal_instrument/faculty_of_thought characterised by *tamas*: "**sleepiness**" (śl.55cd, *nidrā-*). **Sleepiness** [means] the cessation of the functions with regard to external objects (*viṣaya*), [namely] sleep (*supti*) or deep sleep (*suṣupti*). **Sloth** [means] inactivity even in the functions [which are] the means for [achieving] that desired. **Carelessness** [means] the state without attention. **Deceit** [means] cheating (*pratāraṇa*, lit. "carrying over"). (Comm. S on SR śl.55)

VisṇudhP 2,114,16: *tāmasāni tathā jñāna-pramādālasya-tṛṭ-kṣudhaḥ / moha-mātsarya-vaiguṇya-śokāyāsa-bhayāni ca /16/ kāma-krodhau tathā śauryaṃ yajñepsā bahu-bhāṣitā / ahaṅkāraḥ parāvajñā rājasāni, mahābhuja /17/ dharmecchā mokṣa-kāmitvam parā-bhaktiś ca keśave dākṣiṇyaṃ vyavasāyaś ca sāttvikāni vinirdiśet /18/.*

ANALYSIS

The *sāttvika sattva* is called *śuddha sattva* by the CA and AS. The SU and CA do not deal with this topic except in the parallel passages to SR śl.72–74, on the body (*vigraha, kāya*) of *sāttvika* etc. In these two points, the AH, AgniP and VisṇudhP are the closest to the SR. The AgniP and VisṇudhP are parallel to each other. The YS does not deal with this topic.

722 The manuscript A reads *tatra dānaśuddho* instead of *śraddhā śuddho*: "[...] solely with regard to that [namely] *dharma* which is pure through giving/donation with regard to that, [i.e. which is] unsoiled."

The state of having tranquil faculties/organs (*prasannendriyatā*), diseaselessness, non-sloth etc. are, however, born from suitability (*sātmya*)[723]. (śl.56ab)[724]

The body (*deha*) whose essence is the [gross] elements (*bhūta*)[725]

723 The term *sātmya* "suitability" is mentioned in SR śl.44, too.

724 SR śl.56 is identical to ŚG 9,22cd–23ab. But the ŚG contains *sattvajāḥ*, which is obviously a mistake, instead of *sātmyajāḥ* of the SR (compare it with the parallels in the medical texts given below). Besides, the ŚG contains *mātrātmakās* instead of *bhūtātmakas*.
PARALLELS:
SU śārīra., 4,33: *vīryam ārogyam bala-varṇau medhā ca sātmyajāni.*
CA śārīra., 3,11: *[...] ārogyam anālasyam alolupatvam indriya-prasādaḥ svara-varṇa-bīja-sampat praharṣabhūyastvaṃ ceti.* (In the SR, *alolupatva* are included in the *rasaja* group, while, in the CA, it belongs to the *sātmyaja* group).
AS śārīra., 5,21: *sātmyajāny āyur ārogyam anālasyālolupatvam indriya-prasāda-svara-varṇaujaḥ-sampat-praśastatā-praharṣa-bhūyastvam medhābalaṃ ca.*
AH śārīra., 3,6ab: *sātmyajaṃ tv āyur ārogyam anālasyam prabhā balam.*
Neither the YS, AgniP nor ViṣṇudhP mentions the *sātmyaja* group.
ANALYSIS
The CA and AS, whose passages contain expressions almost identical to each other, are most closely related to the SR. The AS contains both elements from the CA and SU (e.g. *bala* and *medhā* are not mentioned by the CA, but by the SU). The Purāṇa-s do not deal with this topic.

725 CA śārīra., 4,8 states that the consciousness (*cetanā*) of a newly formed embryo takes up (*upādatte*) the gross elements, cf. DAS 2003A, p.556. On the YS's description of the *ātman* receiving the five elements, cf. YAMASHITA 2001/2002, pp.89–90.
In some texts, this topic is dealt with in the description of the embryo's assimilation of nourishment from the mother, cf. the following, and GaruḍaP (KIRFEL 1954) verse 31.
There are two ways of the embryo's assimilation of the gross elements, according to AH śārīra., 1,2. At the time of its entrance into the uterus, the soul is accompanied (*sattvānugaiḥ*) by the gross elements in a germ-like (*bījātmakaiḥ*), minute (*sūkṣma*) state. The gross elements are assimilated by the embryo in the early stage. Later, the embryo assimilates the gross elements through the nutrient fluid (*āhārarasajaiḥ*) from the mother.
WINDISCH 1908, p.49, note 3, states "Die *mahābhūta* [...] kommen zwiefach in Betracht, in Samen und Blut verwandelt (*mahābhūtaiḥ [...] śukrārtavarūpataḥ pariṇataiḥ*) und in den Saft der Speise der Mutter verwandelt (*mātur āhārarasātmanā pariṇataiḥ*)."

receives these [following] qualities (*guṇa*)[726] of it (i.e. = of each [gross] element) from it (i.e. from each [gross] element):[727] (śl.56cd)

For the primitive belief in the Milindapañha, that what a woman eats and drinks forms an embryo, cf. WINDISCH, ibid., p.22.

726 Similar usage of the term *guṇa* is also found in CA śārīra., 4,8 and GaruḍaP (KIRFEL 1954) v.33f. In the YS 3,76–78, dealing with the same topic, the term *guṇa* is not used. The usage of the term *guṇa* in AgniP 369,28f. and ViṣṇudhP 2, 115,2f. is different.

727 Cf. CA śārīra., 4,6. This idea is associated with the theory on the correspondence between micro- and macrocosm, cf. CA śārīra., 4,13: *evam ayaṃ lokasammitaḥ puruṣaḥ. yāvanto hi loke mūrtimanto viśeṣās tāvantaḥ puruṣe, yāvantaḥ puruṣe tāvanto loke iti.*
PARALLELS:
SU śārīra., 1,19: *āntarikṣā śabdaḥ śabdendriyaṃ sarva-cchidra-samūho viviktatā ca, vāyavyās tu sparśaḥ sparśendriyaṃ sarva-ceṣṭā-samūhaḥ sarva-śarīra-spandanaṃ laghutā ca, taijasās tu rūpaṃ rūpendriyaṃ varṇaḥ santāpo bhrājiṣṇutā paktir amarṣas taikṣṇyaṃ śauryaṃ ca, āpyās tu raso rasanendriyaṃ sarva-drava-samūho gurutā śaityaṃ sneho retaś ca, pārthivās tu gandho gandhendriyaṃ sarva-mūrta-samūho gurutā ceti /19/.* Peculiarly *gurutā* is twice mentioned, both among the derivatives of water and those of earth. I wonder if the former might be a mistake for something like *mṛdutā*.
CA śārīra., 4,6; 4,8; 4,12. The list of the respective qualities which the embryo receives from the five gross elements is given in śārīra., 4,12. 4,12: *[...] tatrāsyākāśātmakaṃ śabdaḥ śrotraṃ lāghavaṃ saukṣmyaṃ vivekaś ca, vāyvātmakaṃ sparśaḥ sparśanaṃ raukṣyam preraṇaṃ dhātu-vyūhanaṃ ceṣṭāś ca śārīryaḥ, agnyātmakaṃ rūpaṃ darśanam prakāśaḥ, paktirraukṣṇyaṃ ca, abātmakaṃ raso rasanaṃ śaityam mārdavaṃ snehaḥ kledaś ca, pṛthivyātmakaṃ gandho ghrāṇam sthairyam mūrtiś ceti /12/.*
AS śārīra., 5,2–17. 5,13: *tatrākāśajāni, śrotraṃ śabdaḥ sarvaśrotāṃsi viviktatā ca /13/ vāyavīyāni, sparśanam sparśaḥ praśvāsocchvāsādi-parispandāni lāghavaṃ ca /14/ āgneyāni, darśanaṃ rūpam pittam ūṣmā paktiḥ santāpo medhā varṇo bhās tejaḥ śauryaṃ ca /15/ āmbhasāni, rasanaṃ rasaḥ sveda-kleda-vasāsṛk-śukra-mūtrādi-drava-samūhaḥ śaityaṃ snehaś ca /16/ pārthivāni, ghrāṇaṃ gandhaḥ keśa-nakhāsthy-ādi-mūrta-samūho dhairyaṃ sthairyaṃ ca /17/.*
AH śārīra., 3,2: *śabdaḥ sparśaś ca rūpaṃ ca raso gandhaḥ kramād guṇāḥ / khānilāgny-ab-bhuvām eka-guṇa-vṛddhy-anvayaḥ pare /2/ tatra khāt khāni dehe 'smin śrotraṃ śabdo viviktatā / vātāt sparśatvam ucchvāsā vahner dṛg rūpa-paktayaḥ /3/ āpyā jihvā-rasa-kledā ghrāṇa-gandhāsthi pārthivam /*
YS 3,76–78: *ākāśāl lāghavaṃ saukṣmyaṃ śabdaṃ śrotram balādikaṃ / vāyoś ca sparśanaṃ ceṣṭāṃ vyūhanaṃ raukṣyam eva ca /76/ pittāt tu darśanam paktim auṣṇyaṃ rūpam prakāśitām / rasāt tu rasana śaityaṃ snehaṃ kledaṃ samārdavam /77/ bhūmer gandhaṃ tathā ghrāṇam gauravam mūrtim eva ca / ātmā gṛhṇāty ajaḥ sarvaṃ tṛtīye spandate tataḥ /78/* (This topic is situated

between the descriptions of the second and third month. The reading *balādikam* in śl.76b might be a mistake for *bilādikam* ("opening etc."). The Trivandrum edition, with the Bālakrīḍā commentary, reading *tathā balam,* is no better. Anyway, *bala* "power" or "semen" would not be suitable for a quality of space.
AgniP 369,28: *kha-śabdāt kṣudra-śrotāṃsi dehe srotraṃ viviktatā / śvāsocchvāsau gatir vāyor vakra-saṃsparśanaṃ tathā /28/ agne rūpaṃ darśane syād ūṣmā paṅktiś ca pittakam / medhā varṇam balaṃ chāyā tejaḥ śauryaṃ śarīrake /29/ jalāt svedaś ca rasanaṃ dehe vai samprajāyate / kledo vasā rasā raktaṃ śukra-mūtra-kaphādikam /30/ bhūmer ghrāṇaṃ keśa-nakhaṃ roma ca śirasas tathā /31ab/.*
ViṣṇudhP 2,115,2: *bhūmiḥ pañca-guṇā jñeyājalaṃ jñeyaṃ catur-guṇam / tejas tu tri-guṇaṃ, rāma, pavano dviguṇo mataḥ /2/ tatraika-guṇam ākāśaṃ nityaṃ jñeyam manīṣibhiḥ / śabdaḥ sparśaś ca rūpaṃ ca raso gandhaś ca pañcamaḥ /3/ bhūmer guṇaṃ vijānīyād upānte gandha-varjitāḥ / rasa-gandha-vihīnās tu tejasaḥ parikīrtitāḥ /4/ gandho rasas tathā rūpaṃ nāsti vāyor, bhṛgūttama / gandho rasas tathā rūpaṃ sparśaḥ khe na ca vidyate /5/ raso gandhas tathā rūpaṃ sparśanaṃ śabda eva ca / bhūmy-ādīnāṃ guṇāḥ proktāḥ pradhānā, bhṛgunandana /6/ ākāśajāni srotāṃsi tathā śrotraṃ viviktatā / śvāsocchvāsau parispando vāk ca saṃsparśanaṃ tathā /7/ vāyavīyāni jānīyāt sarvāṇy etāni paṇḍitaḥ / rūpaṃ sandarśanam paktim pittam ūṣmāṇam eva ca /8/ medhā varṇam balaṃ chāyā tejaḥ śauryaṃ tathaiva ca / sarvāṇy etāni jānīyāt taijasāni śarīriṇām /9/ ambhasānīha rasanaṃ svedaḥ kledo vasā tathā / rasāsṛk-chukra-mūtrādi dehe drava-cayas tathā /10/ śaityaṃ snehaś ca, dharmajña, tathā śleṣmāṇam eva ca / pārthivānīha jānīhi ghrāṇa-keśa-nakhādi ca /11/ asthnām samūho dhairyaṃ ca gauravaṃ sthiratā tathā /12ab/.*
GaruḍaP (KIRFEL 1954) v.31 (of version 1): *puṣṭo nāḍyāḥ suṣumnāyā yoṣid-garbha-sthitas tvaran /31 version 1/ kṣitir vāri havir-bhoktā pavanākāśam eva ca / ebhir bhūtaiḥ piṇḍitais tu nibaddhaḥ snāyu-bandhanaiḥ /32/ mūla-bhūtā ime proktāḥ sapta-nāḍyante sthitāḥ /32* of version 1/ Verse 33: *tvacāsthi nāḍayo roma māṃsaṃ caivātra pañcamam / ete pañca guṇāḥ proktā mayā bhūmeḥ, khageśvara /33/ yathā pañca-guṇāś cāpas tathā tac chṛṇu, kaśyapa / lālā mūtraṃ tathā śukram majjā raktaṃ ca pañcamam / apām pañca guṇāḥ proktā jñātavyās te prayatnataḥ /34/ kṣudhā tṛṣā tathā nidrālasyaṃ kāntir eva ca / tejaḥ pañca-guṇam proktaṃ, tārkṣya, sarvatra yogibhiḥ /35/ rāga-dveṣau tathā lajjā bhayam mohas tathaiva ca / ity etat kathitaṃ, tārkṣya, vāyujaṃ guṇa-pañcakam /36/ dhāvanaṃ cālanaṃ caiva ākuñcana-prasāraṇaṃ / nirodhaḥ pañcamaḥ prokto vāyoḥ pañca guṇāḥ smṛtāḥ /37/ ghoṣaś cintā ca gāmbhīryaṃ śravaṇaṃ satya-saṅkramaḥ / ākāśasya guṇāḥ pañca jñātavyās, tārkṣya, yatnataḥ /38/.*
Hastyāyurveda p.415 (sth.3, adhy.8), v.125ff.

ANALYSIS

The classical medical texts and the Purāṇa-s both contain similar theories in accordance with Sāṅkhya, except for the GaruḍaP whose theory sometimes deviates from Sāṅkhya. The descriptions of this topic in the classical medical texts are usually brief, but the AS contains some details which are not found in

Sound (*śabda*), the faculty/organ of hearing (*śrotra*), having spaces/hollows (*suṣiratā*), distinctness (*vaiviktya*), [the state] of being a perceiver of the minute/subtle, (śl.57ab)
and the opening (*bila*)[728] from the space, (śl.57c)
and from the wind, touch and the faculty/organ of touching. (śl.57d)[729]

Comm. S on SR śl.56–57

He tells the substances (*bhāva*)[730] born from suitability (*sātmya*): "**Tranquil**" (śl.56a, *prasanna-*). **Diseaselessness** [means] being devoid of disease (*nīrogatva*).

He distinguishes the qualities of the [gross] elements in the body: "**The body**" (śl.56c, *deho*). **Whose essence is the [gross] elements**, [namely] commenced by/from the five gross elements (*mahābhūta*). Even though the body's being commenced by just single (*ekaika*) elements [is true], nevertheless the supporting/strengthening (*upaṣṭambhakatva*) of (= by) the other elements indeed (*eva*) exists. (Comm. S on SR śl.56)

Having spaces/hollows [means] being with a cavity/cavities.

---

the other classical medical texts. The SR's verses (śl.60–68ab) on the ten kind of vital winds (*prāṇa*) and their functions are parallel to the YY (Yogayājñavalkya) 4,47–71, and seem to be a secondary insertion (cf. *Situating the text §2.3.3.*).

The AS, AgniP and ViṣṇudhP mention, among the qualities of the element earth, *keśa, nakha* and *asthi*, i.e. terms like in SR śl.71ab. The AgniP and ViṣṇudhP are parallel to each other, but not parallel to the YS. The text of the ViṣṇudhP seems to contain more insertions than the AgniP does.

Though all the texts that I consulted deal with this topic, it does not seem that the SR is very closely related to any of them.

728 The Ānandāśrama ed. (1896) reads *balaṃ* "power", but this would be unsuitable for a quality of space, for the space element usually has qualities which have something to do with space, dryness, voids including those inside the body etc.

729 SR śl.57 is identical to ŚG 9,23cd–24ab. But the ŚG contains *śabda-śrotra-mukharatā vaicitryaṃ sūkṣma-vāg dhṛtiḥ* and *balaṃ*, instead of *śabdaṃ śrotraṃ śuṣiratāṃ vaiviktyaṃ sūkṣma-bodhṛtām* and *bilaṃ* in the SR.

730 For the translation of the term *bhāva* "substance/constituent" or "[mental] condition", cf. my footnote 717 on SR śl.55ab (*bhāva*). Also cf. *bhāva* "substance" in SR śl.44a.

Distinctness[731] [means] the state of that separated, [namely] the situation/position apart (*pṛthak*) through being the provider[732] of the nature of the body (*deha-svabhāva-prāpakatva*).

Being a perceiver of the minute [means] thorough cognisance (*parijñāna*) without effort of even (*api*) an object difficult to perceive. The difference is that (*iti*) the **opening (*bila*)** is a large/gross (*sthūla*) cavity but a hollow/perforation (*suṣira*) is a small one.

He tells the qualities of the wind in the body: "**From the wind**" (śl.57c, *vāyoḥ*). **Touch (*sparśa*)** is a quality. **The faculty/organ of touching**[733] [means] the faculty/organ skin. (Comm. S on SR śl.57)

SR śl.58–60ab

[From the wind] [the body] obtains[734] throwing upwards, throwing down, contracting, further, going and stretching out[735], these five actions (*karman*) enumerated (*iti*),

731 The quality *vaiviktya* = *pṛthaktva*, is one of the twenty-four *guṇa*-s in Vaiśeṣika (cf. FADDEGON 1969, p.129). Also see POTTER 1977 (p.52; also *separatedness* in the register). I was, however, not able to find a relevant discussion on *deha-svabhāva-prāpakatva*.

732 A and B contain a variant *prakaṭatvena* "through being manifest".

733 Obviously *sparśanendriya* is a misprint of *sparśanendriyam*.

734 "Obtains" (*gṛhṇāti*) is contained in SR śl.60b.
The term *gṛhṇāti* means "to receive [the qualities of the gross elements]" like *ādatte* (SR śl.56d). The term *grahaṇa* "reception, perception" occurs in CA śārīra., 4,8, which the commentator Cakrapāṇi explains as meaning *gṛhṇāti bhūtāni* (cf. ROŞU 1978, p.161). ROŞU associates this term with the philosophical expressions derived from the root √*grah*, namely, the expressions for perception such as *grāhaka/grāhya, graha, atigraha*, *grahītṛ* etc.
CA śārīra., 4,8, contains *upādatte* (*guṇopādāna-kāle*), too, which is close to *ādatte* of SR śl.56d. In this passage, the CA associates the embryo's acquisition of the qualities [of the gross elements] (*guṇopādāna*) with cosmogonical creation. According to the CA (ibid.), the acquisition of the qualities (*guṇopādāna*) is accomplished in a very short time (*aṇunā kālena*).

735 The five actions (*karman*) are mentioned in the Vaiśeṣika theory, cf. Vaiśeṣikasūtra 1,1,7: *utkṣepanam avakṣepanam ākuñcanam prasāraṇaṃ gamanam iti karmāṇi*. SU nidāna., 1,17 states that the *Vyāna* makes [things] move in fivefold manners (*pañcadhā ceṣṭayati*). The commentator Ḍalhaṇa considers the fivefold movement to be the five actions (*prasāraṇākuñcana-vinamanonnamana-tiryaggamanāni pañca ceṣṭāḥ*) mentioned above. These terms are found e.g. in SU śārīra., 7,3 among others, where *ākuñcana*

dryness[736], (śl.58)[737]
further (*tathā*) these (*iti*) ten alterations of wind[738]: the *Prāṇa*[739],

---

(contraction) and *prasāraṇa* (expansion) are considered to be the characteristics (*viśeṣa*) of the vessels (*sirā*).
The five *karman*-s might be the movements of the body. In the embryological passages in Kauṇḍinya's Pañcārthabhāṣya, a commentary on the Pāśupatasūtra, the five actions (*karman*) of Vaiśeṣika are obviously associated with the foetus' movements in the uterus.
HARA 1980, p.147, translates these passages (the terms in question have been underlined), "[...] suffering in the womb (*garbha*) is as follows: when an individual soul (*puruṣa*) is placed in his mother's womb (*udara*), like a man whose limbs are cramped into a broken cart, he (*puṃs*) necessarily experiences the annoyances of confinement; he has no room, no sufficient space **to bend, stretch out etc.**, but is obstructed in all his motions, like a prisoner stupefied (*mūḍha*) in a dark and doorless chamber (*advāraka andha-tamas*)." The original expression for "to bend, stretch out" is *ākuñcana-prasāraṇa*. HARA (ibid.) refers to Vaiśeṣikasūtra 1,1,7.
HARA (ibid., p.149) also mentions Viṣṇupurāṇa 6,5,1f., "Incapable of extending (*prasāraṇa*) or contracting (*ākuñcana*) his own limbs and reposing amidst a mud of faeces and urine, he [= the embryo] is every way incommoded."
In the Hastyāyurveda 3,9,14 (p.438), *prasāraṇa, ākuñcana* etc. are considered to be the movements (*ceṣṭā*) of the elephant foetus (*prasāraṇākuñcanādyāś ceṣṭāḥ kābhiś ca ceṣṭate*). Hastyāyurveda 3,9,109, contains *saṅkucana* instead of *ākuñcana* (Hastyāyurveda 3,9,109, *prasāraṇaṃ ca gātrāṇāṃ tathā saṅkucanaṃ ca yat / catvāriṃśat sirās tasya viniyuktā hi dhāraṇe*).

736 D reads *jajñire* instead of *rūkṣatām*. "[They] know throwing upwards, [...] these five actions enumerated. Further, [the body] obtains these ten alterations of wind: [...]."

737 SR śl.58 is identical to ŚG 9,24cd–25ab. But the ŚG reads *vāyutaḥ* "from the wind" instead of *rūkṣatām*.

738 The term *prāṇa* here denotes the first of the ten vital winds (*vāyu*), cf. GEENENS, p.80.
Among the texts I consulted, only the SR and GaruḍaP (KIRFEL 1954, p.353, v.42) mention the subsidiary winds like *Nāga* etc.
According to WOODROFFE 1990, p. 44, these subsidiary winds are mentioned in the Śāradātilaka: "The Minor *vāyu*-s are nāga, kṛkarā, devadatta, dhanañjayā, producing hiccup, closing and opening eyes, assistance to digestion, yawning, and distention".
Dhyānabindūpaniṣad (in the Yogopaniṣad, SASTRI, MAHADEVA 1920) v.56cd–58ab (p.199): *prāṇo 'pānaḥ samānaś codāno vyānas tathaiva ca /56/ nāgaḥ kūrmaḥ kṛkarako devadatto dhanañjayaḥ / prāṇādyāḥ pañca vikhyātā nāgādyāḥ pañca vāyavaḥ /57/ ete nāḍī-sahasreṣu vartante jīvarūpiṇaḥ /60ab/.*
Buddhist Tantra texts also mention them, cf. RINPOCHE 1986, pp.121–122. According to ZYSK 1993, p.211, the five subsidiary winds, the *Nāga* etc.,

Apāna, further those whose names are the *Vyāna, Samāna* [and] *Udāna, Nāga* ("snake"), *Kūrma* ("tortoise"), and *Kṛkara* ("a kind of partridge"), *Devadatta* ("god-given") [and] *Dhanañjaya*, (śl.59)[740] [and] lightness. (śl.60ab)[741]

Comm. S on SR śl.58–60ab

Throwing upwards (*utkṣepa*) is the action producing[742] union with the region of the upper space (*ākāśa*). **Throwing down (*avakṣepa*)** is the action producing union with the region of the lower space.

---

mentioned in Yoga in addition to the five main winds, the *Prāṇa* etc., are not mentioned in the classical medical texts. Further, ZYSK (ibid.) states that the classical medical texts contain no reference to the respiratory techniques in relationship to the five vital winds. In contrast to that, Yoga emphasises the quasi-medical respiratory techniques for purifying the vital winds and maintaining their proper circulation in the vessels in the body.

These passages, i.e. the SR's description of the subsidiary winds, which are parallels to the YY, seem secondary.

739 For the theory of respiration and the vital wind, cf. ZYSK 1993.

740 SR śl.59 is identical to ŚG 9,25cd–26ab. But the ŚG has all the nouns in the nominative case, and contains *kṛkalo* instead of *kṛkaraṃ*. Further, this verse is parallel to YY 4,47: *prāṇo 'pānaḥ samānaś ca udāno vyāna eva ca / nāgaḥ kūrmo 'tha kṛkaro devadatto dhanañjayaḥ //.*

The verses, SR śl.60cd–68ab, which are also contained in the ŚG, are parallel to YY 4,41cd–71ab, as studied in *Situating the text §2.3.3* of this thesis.

741 SR śl.60ab is identical to ŚG 9,26cd.

PARALLELS

AgniP 369,28: *śvāsocchvāsau gatir vāyor vakra-saṃsparśanaṃ tathā. Vakra* seems a mistake (maybe for *vaktra*?).

ANALYSIS

SR śl.60cd–68ab mentions the respective actions of the five vital winds. The prototype of the theory in which every single role is allotted to each vital wind (*Apāna, Prāṇa, Samāna, Vyāna* and *Udāna*) is found in PraśnaUp 3,1–12; MaitrīUp 2,6, cf. ZYSK 1993, p.205, left column. The five vital winds are mentioned in the CA and SU, e.g. CA cikitsā., 28,5–12; SU nidāna., 1, 12–20a, cf.ZYSK 1993, p.207. ZYSK gives a list containing the respective places and actions of the vital winds, according to the CA and SU.

In summary, the theory of the five vital winds in the SR often deviates from those in the two medical texts, although it contains similarities to the two medical texts (the details are examined in the footnotes on SR śl.60cd–68ab: footnotes 747, 750, 753, 756, 767, 768).

742 *Saṃyoga-janaṭakaṃ* is obviously a mistake and should be read *saṃyoga-janakaṃ*.

Contracting (*ākuñcana*) is bringing toward oneself. **Going (*gamana*)** is the action producing union with the region of space not separated from a particular direction (*dig-viśeṣa*). **Stretching out (*prasāraṇa*)** [is union with what is] farther (*paratas*). **Dryness** is lack of unctuousness (*snigdhatva*). (Comm. S on SR śl.58)

The *Prāṇa, Apāna* etc. are **the ten** transformations of the wind. ***Nayanam***[743] ("leading"?). **Lightness** is lack of heaviness. The relation [of words] is that (*iti*) the body **obtains** these qualities from the wind. (Comm. S on SR śl.59–60a)

SR śl.60cd–65

The foremost of them, the *Prāṇa*, situated below the navel-bulb (*nābhi-kanda*), (śl.60cd)
moves in the mouth, in the two nostrils, in the navel [and] in the lotus of the heart, (śl.61ab)[744]
[and is] the cause of the articulation of sound/word[s][745] (*śabda*), expiration (*niḥśvāsa* "sighing out"), exhalation (*ucchvāsa* "sighing up")[746], coughing etc.[747] (śl.61cd)[748]

743 The word *nayanam* is not contained in the *mūla* text. It is puzzling, most probably a mistake.

744 SR śl.60cd–61ab is identical to ŚG 9,27. But the ŚG contains *mukhyataraḥ, nābhekaṇṭhād* and *caraty asau*, instead of *mukhyatamaḥ, nābhi-kandād* and *caraty āsye* of the SR, respectively. This verse is parallel to YY 4,49cd–50ab: *prāṇa evaitayor mukhyaḥ sa sarva-prāṇa-bhṛt sadā /49/ āsya-nāsikayor madhye hṛn-madhye nābhi-madhyame /50ab/.* But the YY does not make the statement that the *Prāṇa* is situated below the bulb of the navel.

745 The term *śabda* denotes the object of the sense organ/faculty (*jñānendriya*) of hearing and the object of the faculty/organ of action (*karmendriya*) of speaking.

746 The terms *ussāsa* (Skt. *ucchvāsa*) and *nissāsa* (Skt. *niḥśvāsa*) occur in the Tandulaveyāliya (verse 7–8) meaning "inhalation" and "exhalation", respectively.

747 According to CA cikitsā., 28,6, the places of the *Prāṇa* are the head, chest, throat, tongue, mouth and nose. The statement of the SR is partly inconsistent with that of the CA, in that the SR does not mention the head or chest, but the navel and heart. The SU (nidāna., 1,13) mentions only the mouth as a place for the *Prāṇa*.
The term *kanda* "bulb" is again mentioned in SR śl.147a which is parallel to YY 4,16.
The actions allotted to the *Prāṇa* also vary in all three texts. According to CA cikitsā., 28,6, its actions are spitting, sneezing, belching, respiration (*śvāsa*) and digestion. But according to SU nidāna., 1,13, the *Prāṇa* sustains the body

The *Apāna*, however, is situated in the anus/rectum (*guda*), in the penis (*meḍhra*), and in the hips (*kaṭī*), shanks (*jaṅghā*) and belly,
in the navel-bulb[749], in the groins (*vaṅkṣaṇa*), in the thighs and knees. (śl.62)[750]

(*dehadhṛk*), lets the body absorb food/nutrition (*annam praveśayaty*), and supports the other vital winds (*prāṇāṃś cāpy avalambate*). The SR accords with the CA only in mentioning respiration. The SR is inconsistent with the SU. The action of speech is allotted to the *Udāna* in the CA and SU. With the SR's mention of cough (*kāsa*), SU nidāna., 1,14ab might be compared, stating that the *Prāṇa*, being excited (*duṣṭa*), causes hiccup and difficult breathing (*hikkāśvāsādi*).

In summary, the CA is closer to the SR than the SU is, but there are still some inconsistencies between the CA and SR.

748 SR śl.61cd is identical to ŚG 9,28 (consisting of two *pāda*-s). But the ŚG contains the variants *[...]-niḥśvāsa-* and *[...]-cchvāsāder api kāraṇam*, instead of *[...]-niḥśvāsa-* and *[...]-cchvāsa-kāsādi kāraṇam* of the SR. Further, this verse is parallel to YY 4,66ab, *niḥśvāsocchvāsa-kāsāś ca prāṇa-karmeti kīrtitāḥ*. The YY, however, does not mention *śabdoccāraṇa*.

749 C.E. (= Calcutta Edition ?) contains *nābhau skandhatra* instead of *nābhi-kande*: "in the navel, in the shoulders, ..."

750 SR śl.62–63ab is identical to ŚG 9,29abcdef. ŚG 9,29ab contains a variant *kaṭi-jaṅghodareṣv api* instead of *kaṭī-jaṅghodareṣu ca* of SR śl.62ab. ŚG 9,29cd contains the variants, *-kaṇṭhe, vṛṣaṇayor* and *-jānuṣu*, instead of the SR's readings, *-kande, vaṅkṣaṇayor* and *-jānuni* (SR śl.62cd), respectively. ŚG 9,29ef contains the variants, *tasya* and *kīrtitam*, instead of the SR's readings, *asya* and *kīrtitaḥ* (SR śl.63ab), respectively.

The Ānandāśrama edition (1896) of the SR contains *-ūrujānuṣu* (SR śl.62d) instead of *-ūrujānuni* of the Adyar edition. According to a footnote in the Adyar edition, C.E. has the same reading as the Adyar edition. Similarly, the ŚG contains *-ūrujānuṣv*, as shown above. The reading *-rujānuṣu* is probably correct. Grammatically, a *dvandva* compound of the body-limbs should be singular, but, in the SR, plural also occurs for *kaṭījaṅghodareṣu* in the *pāda* c.

The Ānandāśrama edition (1896) reads *kīrtitam* (SR śl.63b) instead of *kīrtitaḥ*, of the Adyar edition. The Adyar edition (p.48, corrigendum no. 7) notes that C.E. also reads *kīrtitam*. The ŚG also contains *kīrtitam*.

Further, SR śl.62 and śl.63cd are parallel to YY 4,53 (*apāna-nilaye kecid guda-meḍhroru-jānuṣu / udare vaṅkṣaṇe kaṭyāṃ jaghane tau vadanti hi //*) and 4,66cd (*apāna-vāyoḥ karmaitad viṇ-mūtrādi-visarjanam*), respectively.

The places and actions of the *Apāna* are mentioned in CA cikitsā., 28,10; SU nidāna., 1,19. According to the CA, its places are the testicles, bladder, penis, navel, thighs, groins and anus. The bladder is not mentioned in the SR. The SR allots the navel to the *Prāṇa*. The SU mentions only the stomach (*pakvāśaya*) as the *Apāna*'s place, but also states that the *Apāna*, being excited, causes diseases

Emission of urine, faeces etc. is proclaimed as its action. (śl.63ab)[751]

The *Vyāna* is situated in the eyes, ears, and ankles (*gulpha*), in the hip and in the nose. (śl.63cd)[752]

And its action is holding, letting go and seizing etc. of the *Prāṇa* [and] *Apāna*. (śl.64ab)[753]

---

in the bladder and anus. Neither the CA nor SU mentions the hips, shanks, belly, navel-bulb, groin, thighs or knees, though the CA states that the *Apāna* is situated also in the intestines (*antra*). The CA allots the groins (*vaṅkṣaṇa*) to the *Apāna*.

The CA and SU both state that the actions of the *Apāna* are releasing semen, urine, faeces, menstrual discharge and the foetus. The SR's statement is in concordance with this.

In summary, it is only in a few points that the CA's and SU's statements on the *Apāna*'s places accord with the SR's. In contrast, the statements on its actions are in concordance in all three texts.

751 C.E. contains *-visargakarmakīrtitam* instead of *-visargaḥ karma kīrtitaḥ* of the Adyar edition. I interpret it as *-visarga-karma kīrtitam*, "Its action of emission of [...] is proclaimed". If the Ānandāśrama edition's reading *kīrtitam* is to be taken instead of *kīrtitaḥ*, then the translation is: "Its action is proclaimed as emission of urine, faeces etc."
Kīrtitam is contained also in the ŚG (see above).

752 SR śl.63cd is identical to ŚG 9,30ab. But the ŚG contains *jihvā ghrāṇeṣu* instead of *kaṭyāṃ ghrāṇe ca* of the SR. In fact, the Ānandāśrama edition (1896) also reads *ghrāṇeṣu*.
Further, this verse is parallel to YY 4,52 (Trivandrum edition: *vyānaḥ (śrotrā)kṣi-madhye ca kṛkaṭyāṃ gulphayor api / ghrāṇe gale sphijoddeśe tiṣṭhaty atra na saṃśayaḥ //*). The list in the YY is more complete than that in the SR.

753 SR śl.64ab is identical to ŚG 9,30cd. But the ŚG contains *prāṇāyāma-* instead of *prāṇāpāna-* of the SR. The ŚG's *prāṇāyāma* seems a mistake.
This verse is furthermore parallel to YY 4,67ab (Trivandrum edition: *hānopādāna-ceṣṭādi vyāna-karmeti ceṣyate*). The actions, *hāna* and *upādāna*, respectively correspond to *tyāga* and *grahaṇa* of the SR. The list of actions in the SR is a more complete version than that in the YY in contrast to the list of places where the case is opposite (cf. the foregoing footnote).
The places and actions of the *Vyāna* are mentioned in CA cikitsā., 28,9; SU nidāna., 1,17. Both, the CA and SU, state that it pervades all over the body. The places listed in the SR are not mentioned by the two medical texts. According to CA cikitsā., 28,9, the actions of the *Vyāna* are movement, extension, contraction and blinking. To these, SU nidāna., 1,17, adds carrying of the nutrient fluid (*rasa*), sweating and carrying of blood. If we compare SR śl.20 with SU nidāna., 1,17, we find that *tyāga* and *grahaṇa* should be separated from

The *Samāna*, pervading the whole body together with the [internal] fire, (śl.64cd)
roaming about through the 72,000 apertures of tubes (*nāḍī*)[754],
properly carrying [to their destination] the juices/saps of what is eaten and drunk, is the maker of nourishment of the body.[755] (śl.65)[756]

---

prāṇāpānadhṛti. In this case, the translation is: "holding (*dhṛti*) of the *Prāṇa* and *Apāna*, letting go (*tyāga*), and seizing (*grahaṇa*)". The CA and SU do not mention the *Prāṇa* or *Apāna* in describing the action of the *Vyāna*. In summary, the SR is inconsistent with the two medical texts.

754 The term *nāḍī* may be translated as "tubular vessel" or "tube", cf. DAS 2003A, p.560. DAS prefers "tube", in order to make the distinction of the terms, *nāḍī, sirā* and *dhamanī*, clear. JOLLY 1901, p.44 and p.109, translates "Röhre, Ader". The term *nāḍī* might mean "pulse" (cf. JOLLY 1901, p.22, *nāḍī-parīkṣā*), but JOLLY remarks that this usage is of a later period.
In the SR, the term *nāḍī* occurs only in the verses parallel to the YY. The unique exception is SR śl.39c and 40a, *nāḍī* denoting the umbilical cord. The classical medical texts (SU, AS, AH, CA) also call the umbilical cord *nāḍī*, cf. my footnote 629 on SR śl.39c. Except for this case, the term *nāḍī* does not often occur in the śārīrasthāna-s of these texts. Usually the terms such as *sirā, dhamanī* and *srotas* are used to denote a "tubular vessel". These terms occur also in the SR's anatomical description, SR śl.86cd–119 (śl.101ff *sirā*; śl.111 *dhamanī*; śl.87, śl.112 *srotas*).
The term *nāḍī* in SR śl.65b has a Haṭhayogic background. The 72,000 *nāḍī*-s radiating from the heart are mentioned in YS 3,108, in its description of Haṭhayoga. The 72,000 *nāḍī*-s are mentioned also in Haṭhayogic texts like the Haṭhayogapradīpikā etc.

755 The vital wind's participation in digestion is dealt with by GaruḍaP (KIRFEL 1954) v.43–47.

756 SR śl.64cd–65 is identical to ŚG 9,31–32ab. Further, this verse is parallel to YY 4,54cd–56 (Trivandrum ed.: *samānaḥ sarva-gātreṣu sarvam prāpya vyavasthitaḥ /54cd/ bhukta-sarva-rasaṃ gātre vyāpayan vahninā saha / dvi-saptati-sahasreṣu nāḍī-mārgeṣu sañcaran /55/ samāna-vāyur evaikaḥ sāgnir vyāpya vyavasthitaḥ / agnibhiḥ saha sarvatra sāṅgopāṅga-kalebare /56/*). The content of YY 4,56 is a repetition of that of YY 4,54cd–55. The redundancies like *vahninā saha, sāgnir, agnibhiḥ* etc. give rise to the suspicion that the YY was enlarged through adding these to an originally more compact text. It should be noted that *prāpya* (YY 4,54d) is merely an alternation of *vyāpya* (YY 4,56b) which is the same wording as in SR śl.64b. In comparison with the YY, the SR is more compact. Therefore, the SR seems to preserve an older version of the text. SR śl.65d *deha-puṣṭikṛt* is parallel to YY 4,68ab (Trivandrum ed.: *poṣaṇādi samānasya śarīre karma kīrtitam*).
The places and actions of the *Samāna* are mentioned in CA cikitsā., 28,8; SU nidāna., 1,16. According to the CA, the *Samāna*'s places are the channels

He relates the place[s] and action[s] of the chief one, [namely] the *Prāṇa*: "**Of them**" (śl.60c, *teṣām*). **Foremost (*mukhyatama*)**, [namely] being the main (*pradhāna*), because of the others', [namely] the *Apāna* etc.', following it. **Below the navel-bulb**, [i.e.] in the region of the base (*Ādhāra*)[757], [it] **moves (*carati*)**, [i.e.] wanders about (*vicarati*). (Comm. S on SR śl.60–61)

He relates the place[s] and action[s] of the *Apāna*: "**The *Apāna*, however**" (śl.62a, *apānas tu*). **In the hips, shanks and belly**: Here "in the five" (*pañcasu*) is to be supplied (*adhyāhārya*), because, otherwise, of attainment of the state of being like one (i.e. a singular) according to (*iti*) [the rule] ["]And the *dvandva* of the parts of a living being, a musical group and an army [is a singular]["] (Pāṇini 2,4,2).[758]

---

carrying sweat, morbific entities (*doṣa*) and water. The CA also states that the *Samāna* is situated beside the internal fire (*antaragni*). The SU states that the *Samāna* moves in the *āmāśaya* and *pakvāśaya* ("the receptacles of undigested and digested food", cf. DAS 2003A, p.130), and is connected with the internal fire. The CA and SU both state that the *Samāna*'s action is digestion (CA *agni-bala-pradaḥ*; SU *annam pacati*). The SR does not accord with the two classical medical texts except for stating that the *Samāna* moves together with the internal fire. In contrast to the SR, the CA and SU allot the actions of pervading the whole body and carrying the nutrient fluid (*rasa*) to the *Vyāna*. The mention of 72,000 tubes (*nāḍī*) is obviously according to the Haṭhayogic theory (cf. YS 3,108: *dvāsaptati-sahasrāṇi hṛdayād abhiniḥsṛtāḥ / hitāhitā nāma nāḍyas*; Haṭhayogapradīpikā 1,39: *dvāsaptatisahasrāṇi nāḍīnām mala-śodhanam*; ibid., 3,123). Besides, AgniP 370,40ab and ViṣṇudhP 115,91cd mention seventy-four crore (*koṭi*) of openings.

757 SR 2, śl.60cd–68ab is parallel to the YY, as already mentioned. That means that the expression *nābhi-kanda* is most probably according to the Haṭhayoga theory. The YY deals with an old theory of Haṭhayoga which does not use the denomination *cakra* for the centres of the *kuṇḍalinī*, except for that in the navel. In this theory, only the centre of the navel is called *Nābhi-cakra*. Therefore, the term *ādhāra* in the YY cannot denote a *cakra* (*Ādhāra-cakra*), but simply the lowermost region of the body (cf. my footnote 1096 on SR śl.145c, *ādhāra*). The commentator Siṃhabhūpāla (comm. S) was perhaps aware that this statement in the SR is according to the old theory, for he properly explains this term as meaning *ādhāra-pradeśe* "in the region of the *ādhāra*".

758 Cf. WACKERNAGEL 1957 (Bd.II,1), p.163ff. (§69) on *klassisches Singulardvandva.*
Pāṇini 2,4,2 is also referred to by comm. S on SR śl.24d (Adyar ed., p.38, l.3 from the bottom, *prāṇy-aṅgatvād dvandvaikavad-bhāvaḥ*).

But in [the case of] the employment of a numeral (*saṅkhyopādāna*), the state of being like one (*ekavadbhāva*, i.e. a singular) does not obtain, because of the following (*iti*) being mentioned in the Vicāracintāmaṇi:

["]If [you ask] why, [we answer] [']ten teeth and lips['] (*daśa dantoṣṭhāḥ*, plural *dvandva*)" is, here/in_this_case, an adjective/qualifier expressing number (*saṅkhyā-vācī iha bhedakaḥ*)["]. [759] (Comm. S on SR śl.62)

"**Urine, faeces etc.**" (śl.63a, *mūtra-purīṣādi-*). Here by the word ["]etc.["] (*ādi*), mention/inclusion (*grahaṇa*) of semen (*vīrya*), impurity of the eyes (*dūṣikā*) etc. [is intended].

He relates the place[s] and action[s] of the *Vyāna*: "**The *Vyāna***" (śl.63, *vyāno*). **The holding (*dhṛti*)** [is] retention (*dhāraṇa*), of the two winds, [namely] **of the *Prāṇa* and *Apāna***; that is to say (*iti yāvat*) *kumbhana*[760]. **Letting go (*tyāga*)** [is] making slip away outside. **Seizing (*grahaṇa*)** is making enter inside. By the word ["]etc.["] (*ādi*), the action of the mongoose (*nakula-karma*)[761] etc., well known in Yogaśāstra-s, [is intended]. (Comm. S on SR śl.64ab)

He relates the place[s] and action[s] of the *Samāna*: "**The *Samāna***" (śl.64c, *samāno*). The relation [of words] is that (*iti*)[762] **roaming about together**[763] **with the fire**, [i.e.] the abdominal one, [which is] the cause of digestion of what is eaten and drunk, **carrying the juices/saps of what is eaten and drunk into the apertures of**

759 I was not able to identify this sentence. But this grammatical rule is prescribed in Pāṇini 2,4,15. The instance *daśa dantoṣṭhāḥ* is mentioned in its commentary. Cf. WACKERNAGEL 1959 (Bd. II,1), p.163 (§69b).

760 The Yoga technique of breath control (*prāṇāyāma*) consists of inhaling (*pūraka*), holding (*kumbhaka*) and exhaling (*recaka*), cf. Haṭhayogapradīpikā (WALTER 1893), intro. xix; xxvi.

761 Unfortunately, I was not able to identify this term. Tāraḷekara's Marathi translation (1975, see under "Saṅgītaratnākara" in the bibliography) explains it as meaning *prāṇāyāma-viṣayaka karmeṃ*, "the action whose object/scope is the restraining of the vital wind", which is not very helpful.
"Mongoose" (*nakula*) is not clear. The term *nakula* might be a mystic name for the sound *h*. So, *nakula-karma* might mean "action/pronunciation of the sound *h*".

762 Adyar ed., p.49, l.15, *karotiti* is obviously a mistake and should be read *karotīti*.

763 A and B read *sa* instead of *saha*: "it (= *samāna*), roaming about with the fire [...]".

tubes (*nāḍī*),[764] it makes (= brings about) **nourishment of the body**. (Comm. S on SR śl.64cd–65)

SR śl.66–68

The *Udāna* sits in the feet, in the hands [and] in the joints of the limbs.

Raising (*unnayana*), ascension (*utkramaṇa*) etc. of/in the body (*deha*)[765] are proclaimed as its action(s) (*karman*).[766] (śl.66)[767]

The five [winds], [namely] the *Nāga* etc., are situated depending on the elements (*dhātu*), skin etc.[768] (śl.67ab)

764 This explanation is different from the translation of SR śl.64cd-65 above, in which *nāḍī-randhreṣu* is taken to be connected with *sañcaran* situated next to it, and not *ānayan*.

765 The comm. S interprets *deha-unnayana* as "leading the body upwards" and *utkramaṇa* as "death" (i.e. ascension of the soul).

766 Instead of *karmāsya dehonnayanotkra*, the manuscript D reads *dehasya karma nayanotkra°*, and the Ānandāśrama edition (1896) reads *karmāsya dehonnamanotkra°*. Using the first variant, which I interpret as *dehasya karma nayano°*, the translation would be: "The action of the body, [namely] leading, ascension etc., is proclaimed." Meanwhile the translation of the latter variant is the same as the Adyar edition which uses *unnayana.*
But the parallels in the ŚG and YY (see below) support *dehonnayana*.

767 SR śl.66 is identical to ŚG 9,32cd–33ab. Furthermore, SR śl.66ab is parallel to YY 4,54ab (Trivandrum ed.: *udānaḥ sarva-sandhi-sthaḥ pādayor hastayor api*). SR śl.66cd is identical to YY 4,67cd (*udāna-karma tat proktaṃ dehasyonnayanādi yat*).
The places and actions of the *Udāna* are mentioned in CA cikitsā., 28,7; SU nidāna., 1,14. The *Udāna*'s places are, according to the CA, the navel and throat. But the SU merely mentions the *Udāna*'s upward movement in the body, although it states that the *Udāna*, being excited/disturbed, causes diseases in the area over the clavicle (*jatru*). In contrast, the SR allots the navel to the *Prāṇa*. The SR allots the feet/legs (*pāda*) to the *Udāna*, while the CA allots the parts of the legs, i.e. the thighs and groins, to the *Apāna*. The *Apāna*'s actions are, according to the CA, speech, effort, energy, strength and complexion; but, according to the SU, speech and song. The CA and SU both allot the movements of the body to the *Vyāna*. To put it in a nutshell, the SR's statement on the places and actions of the *Apāna* is different from that of the two classical medical texts. The comm. S interprets *utkramaṇa* as meaning "death" (see above). But the original meaning simply seems to be "upward moving of the body".

768 These five subsidiary vital winds are not mentioned in the classical medical texts.

Ejection[769] etc., winking etc., and sneezing etc. respectively, lassitude[770] etc. [and] swelling etc. are proclaimed as their action[s] (*karma*). (śl.67cd–68ab)[771]

From fire (*agni*), however, [the body] receives[772] the eye (= the faculty/organ of seeing), shape (*rūpa*), bile (*pitta*), digestion (*pāka*)[773], brightness (*prakāśatā*). (śl.68cd)[774]

Comm. S on SR śl.66–68cd

He relates the place[s] and action[s] of the *Udāna*: "***Udāna***" (śl.66a). **Raising (*unnayana*) of the body** is leading upwards (*ūrdhva-nayana*). **Ascension** is death. By the word ["]etc.["] (*ādi*), hiccough etc. [are intended]. (Comm. S on SR śl.66)

He relates the place[s] and the action[s] of the five winds, [namely] the *Nāga* etc: "**Skin etc.**" (śl.67a, *tvag-ādi*). The locus (*avasthāna*) of **the *Nāga*** is in **the skin**; of **the *Kūrma***, in **the blood**;

The expression *tvag-ādi-dhātūn* is remarkable. The SR begins the metabolic chain of the seven elements (*dhātu*) with skin (cf. SR śl.79). In contrast, the classical medical texts most often mention the nutrient fluid (*rasa*) as the first link of the chain, although these texts contain a few passages mentioning skin as the first link (cf. DAS 2003A, p.547ff., on *tvac*).
For the chain of the seven elements (*dhātu*), cf. JAMISON 1986; DAS 2003A, p.553ff, on the term *dhātu*. According to JAMISON (ibid.), the chain beginning with skin is more archaic than the one beginning with the nutrient fluid (*rasa*).

769 The term *udgāra* may mean various actions of discharging from the mouth like spitting, vomiting, belching etc.

770 The manuscript D reads *tandrī* instead of *tandrā*.

771 SR śl.67–68ab is identical to ŚG 9,33cd–34. But the ŚG contains *kṣut-pipāsādikaṃ, tandrī-* and *śokādi,* instead of *kṣuta-prabhṛti ca* (SR śl.67d), *tandrā* and *śophādi* (SR śl.68ab), respectively. The manuscript D of the SR reads *tandrī*, too.
Further, these verses are parallel to YY 4,57ab (*nāgādi-vāyavaḥ pañca tvag-asthy-ādiṣu saṃsthitāḥ*), and YY 4,68cd–70ab (*udgārādi-guṇo yas tu nāga-karmeti cocyate /68cd/ nimīlanādi kūrmasya kṣutaṃ vai kṛkarasya ca / devadattasya viprendre tandrī karmeti kīrtitam /69/ dhanañjayasya śophādi sarva-karma prakīrtitam /70ab/*). YY 4,69c reads *tandrī* as the manuscript D of the SR and the ŚG do.

772 The verb *ādatte* ("receives") is contained in SR śl.69.

773 The manuscript D reads *pakva* instead of *pākaṃ*.

774 SR śl.68cd is identical to ŚG 9,35ab. But the ŚG contains *rocakaṃ* and *dīptaṃ* instead of *locanaṃ* and *pittaṃ*, respectively.

of **the *Kṛkara***, in **the flesh**; of **the *Devadatta***, in **the fat (*medas*)**; of **the *Dhanañjaya***, in **the bone**. (Comm. S on SR śl.67)

Lassitude (*tandrā*) is sluggishness (*ālasya*). By the word ["]etc.["] (*prabhṛti*), yawning etc. [are intended].

He relates the qualities of [the gross element] fire (*tejas*) in the body: "**From fire, however**" (śl.68c, *agnes tu*). **The eye (*locana*)** is the faculty/organ eye (*cakṣus*). **Bile** is a particular *doṣa*.[775] **Digestion (*pāka*)** is digesting/cooking (*pacana*) of the juice of food (*anna-rasa*)[776]. **Brilliance (*prakāśatā*)** is the state of having fiery energy (*tejas*).

SR śl.69–74

non-endurance (*amarṣa*), sharpness/severity (*taikṣṇya*), warmth, the vital fluid (*ojas*), and fiery energy (*tejas*), valour, further (*tathā*)[777] having a good memory (*medhāvitā*); (śl. 69abc)[778]

775 *Doṣa*. DAS 2003A, p.548, translates it as "morbific entity".
JOLLY 1901 explains it as follows: wind, bile and phlegm, which are called *doṣa*-s, are the three kinds of principal energy which decide the constitution and character of an individual and dynamically influence his corporeal condition. Health is considered to be the normal condition or equilibrium of the three *doṣa*-s, while sickness is considered a disorder of this equilibrium (JOLLY ibid., p.40–41).
The term *doṣa* means etymologically "fault" or "source of faults", cf. DAS (ibid.). That means that this term originally denoted the disorder of the principal energy, namely an excess of wind, bile or phlegm, which causes sickness. Also cf. R.F.G. MÜLLER 1961, p.105.

776 DAS 2003A, p.578ff. (on the term *rasa*), explains this term as follows: the term *anna-rasa* "juice of food" is often a synonym of *āhāra-rasa* which is usually taken to be an intermediate between food and the first element (*dhātu*), *rasa* ("chyle" or "nutrient fluid"). But its actual identification in the old texts is not easy, since *āhāra-rasa/anna-rasa* can also simply mean "*rasa* of food" which is terminologically not differentiated from *rasa*.

777 The manuscript D reads *samādatte* "[the body] receives" instead of *tathādatte*. The manuscripts *ka., kha.* and *ga.* read *tadādatte* "then [...] [the body] receives".

778 For the derivatives of fire, the AS (śārīra., 5,15) shows a good parallelism to the SR (*āgneyāni, darśanaṃ rūpam pittam ūṣmā paktiḥ santāpo medhā varṇo bhās tejaḥ śauryaṃ ca /5,15/*). But the AS does not mention *prakāśatā, amarṣa, taikṣṇya* or *tejas*. The SU (śārīra., 1,19) mentions *taikṣṇya* and *amarṣa*. The CA (śārīra., 4,12) mentions *prakāśa*.

from water, however, [the faculty/organ of] tasting (*rasana*), taste (*rasa*), (śl.69d)[779]
coolness, unctuousness (*sneha*), fluid/flowing (*drava*), sweat, urine etc.[780], also softness; (śl. 70ab)[781]
from earth, the faculty/organ of smelling (*ghrāṇa*), smell, fixedness (*sthairya*), and firmness/perseverance (*dhairya*)[782], heaviness/dignity (*gaurava*), (śl.70cd)[783]
the beard, hair, and nails, teeth, bones etc. and anything else hard. (śl.71ab)[784]

[There are] the nature (*prakṛti*) of the elements (*dhātu*), [namely] wind (*vāta*) etc.[785], [and] further (*tathā*) the nature of space etc.

779 SR śl.69 is identical to ŚG 9,35cd–36ab. But the ŚG contains *tīkṣṇa-sūkṣmāṇām* instead of *taikṣṇyam ūṣmāṇam.*

780 The Ānandāśrama edition (1896) reads *drava-sveda-mūtrādi* instead of *dravaṃ svedam mūtrādi*. The meaning is slightly modified with this variant, taking all three, i.e. fluid, sweat and urine, in one compound.

781 In the list of the derivatives of water, the AS (śārīra., 5,16) shows a good parallelism to the SR (*āmbhasāni, rasanaṃ rasaḥ sveda-kleda-vasāsṛk-śukra-mūtrādi-drava-samūhaḥ śaityaṃ snehaś ca*). But the AS does not mention *mṛdutā*. CA śārīra., 4,12 mentions *mārdava*.

782 The Ānandāśrama edition (1896) reads *sthairya-dhairye* instead of *sthairyaṃ dhairyaṃ*.

783 SR śl.70 is identical to ŚG 9,36cd–37ab. But the ŚG contains *gātrāṇi* instead of *mūtrādi*.

784 SR śl.71ab has no parallel in the ŚG.
For the derivatives of earth, the parallels in the AS (śārīra., 5,17) and ViṣṇudhP (2,115,11–12ab) mention the nails, hair and bones. The AgniP (369,31ab) also mentions the nails and hair.
Among the parallels, AS śārīra., 5,17 is the closest to SR 70cd–71ab (AS śārīra., 5,17: *pārthivāni, ghrāṇaṃ gandhaḥ keśa-nakhāsthy-ādi-mūrta-samūho dhairyaṃ sthairyaṃ ca /17/*). But the AS does not contain an equivalent to *gaurava* of the SR, while the SU (śārīra., 1,19) mentions *gurutā*.

785 ŚG 9,43 makes a similar statement: *vāta-pitta-kaphās tatra dhātavaḥ parikīrtitāḥ*, "With regard to that (*tatra*), wind, bile and phlegm are proclaimed to be elements".
This topic is dealt with in detail by SU śārīra., 4,62–80. It mentions seven natures (*prakṛti*), i.e. mental and physical constitutions, due to the mixture of the three morbific entities (*doṣa*), cf. SU śārīra., 4,62. Nature (*prakṛti*) is unchangeable (SU śārīra., 4,78). Although it is derived from the morbific entities (*doṣa*), it does not cause abnormality (SU śārīra., 4,79). SU śārīra. 4,80, presents the theory of another school (*kecid āhuḥ*) which deals with *bhautikī prakṛti*, namely, the natures (*prakṛti*) derived from the five gross elements

(*mahābhūta*). The SU (ibid.) states that the first three of them, namely, the natures (*prakṛti*) derived from wind, fire and water, respectively correspond to the natures (*prakṛti*) derived from the three morbific entities (*doṣa*), i.e. wind (*vāta*), bile (*pitta*), and phlegm (*śleṣma*). In this way, the three morbific entities (*doṣa*) are associated with the three gross elements (*mahābhūta*) (also cf. KIRFEL 1951). But, in these passages of the SU, the three morbific entities, wind (*vāta*) etc., are not called *dhātu* "element". The following verses (SU śārīra., 4,81–98) deal with the types of mental constitution or disposition called *kāya*, which is the same topic as in SR śl.72–74.
CA śārīra., 4,34, mentions only the three morbific entities (*doṣa*) which cause abnormality or sickness in the mind and body (*sattva-śarīra*). The CA does not mention the types of nature (*prakṛti*) derived form the morbific entities (*doṣa*). The following passages (CA śārīra., 4,36–40), in contrast, deal with the same topic as in SR śl.72–74, associating the three types of *antaḥkaraṇa* called *sattva* (i.e. *śuddha*), *rājasa* and *tāmasa*, with various kinds of beings like deities, animals and plants.
The AS has no parallel.
The AH śārīra., 3,83–104, mentions seven natures (*prakṛti*) derived from the morbific entities (*doṣa*), like the SU. AH 3,83, states that nature (*prakṛti*) is determined by the most prominent morbific entity (*doṣa*) of the semen, procreatory discharge, pregnant mother, mother's food, her activity, uterus, and season. In this passage, wind (*vāta*), bile (*pitta*) and phlegm (*kapha*) are respectively accompanied by the appositions, wind, fire and *soma* (here probably "water" or "fluid"), though the relation between the three morbific entities (*doṣa*) and the three gross elements (*mahābhūta*) is not explicitly discussed. The AH does not mention the types of mental constitution associated with various beings.
The YS does not deal with this topic.
In contrast to the YS, the AgniP (369,37–39) and ViṣṇudhP (115,19–21) contain descriptions of the three morbific entities' (*doṣa*) influence on disposition and dream, which are the topics belonging to the theory of the natures (*prakṛti*) derived from the morbific entities (*doṣa*).
In summary, the SU is the closest to the SR. The CA and AH, in some points, contain similarities to the SR.
In SR śl.72, the three morbific entities (*doṣa*), wind (*vāta*) etc., are called *dhātu* "elements". This *dhātu* is obviously something else than the seven *dhātu*-s, i.e. skin etc. mentioned in SR śl.67 and śl.79.
JOLLY 1901, p.39 and p.41, mentions *dhātu* as another denomination for the three morbific entities (*doṣa*). According to JOLLY, *dhātu* is a neutral denomination in contrast to *doṣa* which etymologically means something negative like "source of faults".
According to DAS 2003A, p.553ff. (on *dhātu*), the problem seems to be more complicated. DAS gives the examples, in which the three morbific entities (*doṣa*) and the seven elements (*rasa* etc.) are all included in the same category

(śl.71cd)

[And that which] pertains to *sattva* (= *sattva-guṇa*) is sevenfold, [namely] the body (*vigraha*) of Brahman, Indra and Yama[786], and [that] pertaining to Varuṇa, further [that] pertaining to Kubera, [that] pertaining to the sages (*ṛṣi*), [and] the body (*vigraha*) of the *gandharva*-s. (śl.72ab)

The body (*deha*) pertaining to *rajas* is six fold, [i.e. that] pertaining to the *piśāca*-s ("goblins"), further [that] pertaining to the *rakṣas* ("demons"), [that] pertaining to the *asura*-s ("devils"), [that] pertaining to the birds, [that] pertaining to the snakes, further, the body (*deha*) of the ghosts (*preta*) as the last one (*para*). (śl.73)[787]

The body (*deha*)[788] pertaining to *tamas*, threefold, is that whose form[s] are beasts/animals, fish [and] trees. (śl.74ab)[789]

---

called *dhātu*. He also gives an example, in which the gross elements (*mahābhūta*) are called *dhātu*.

786 *Brahmendriyamavigrahaḥ* is obviously a misprint for *brahmendra-yama-vigrahaḥ*.

787 The manuscript D reads *yasya* instead of *dehaḥ*. It would be translated, "[The body] pertaining to *rajas* is six fold, of which (*yasya*) [there] are [that] pertaining to the *piśāca*-s [...]".
The Ānandāśrama edition (1896) reads *yaś ca*. With this reading, it would be translated, "[And that which] pertains to *rajas* is six fold, [namely that] pertaining to the *piśāca*-s [...]".
The manuscript *gha.* reads *yakṣapaiśāco* instead of *dehaḥ paiśāco*. This seems to be grammatically unsuitable, because the noun *yakṣa* does not have a *vṛddhi* in contrast to the other ones. As a matter of fact, *yakṣa* is not mentioned by any parallels in the medical texts (see below).

788 The Ānandāśrama edition (1896) again reads *yaś ca* instead of *dehaḥ*, which is the same as śl.73ab. This is not noted by the Adyar edition. It would be translated, "[And that which] pertains to *tamas* is threefold, [namely that] whose form[s] [are] beasts/animals, fish [and] trees."

789 The term *vigraha* (lit. "body") seems to mean the type of psyche or mentality, according to the parallels in the SU and CA.
Parallels are found in SU śārīra., 5,81–98; CA śārīra., 4,36–40.
The SU calls it *kāya* "body" like the SR, but in SU śārīra., 5,86 and 5,87, the same thing is called *sattva* "psyche". The commentator Ḍalhaṇa (on SU śārīra., 5,81–87) explains that they are the types of psyche (*citta-prakṛti* lit. "the nature of the psyche"). The types mentioned by the SU are (1) *sāttvika kāya*-s, namely, *brahman, mahendra, varuṇa, kubera, gandharva, yama* and *ṛṣi*, (2) *rājasa kāya*-s, namely, *asura*, snake (*sarpa*), bird (*śakuna*), *rākṣasa*, *piśāca* and *preta*; (3) *tāmasa kāya*-s, namely, beast (*paśu*), fish (*matsya*) and tree (*vanaspati*).

Afraid of [too great] expansion of the text,[790] we do not tell their characteristics. (śl.74cd)

Comm. S on SR śl.69–74

Non-endurance (*amarṣa*) is wrath (*kopa*). **Sharpness/severity (*taikṣṇya*)** is intolerance. **Warmth (*ūṣman*)** is a particular heat producing sweat. **The vital fluid** is already mentioned. **Valour (*śūratā*)** is the state without fear. **Having a good memory (*medhāvitā*)** is non-forgetting of the experienced.

He relates the qualities of [the gross element] water: "**From water, however**" (śl.69d, *jalāt tu*). **[The faculty/organ of] tasting (*rasana*)** is the faculty/organ of tasting (*rasanendriya*). (Comm. S on SR śl.69)

Unctuousness (*sneha*) is a particular quality of voice/speech (*vāc*), [which has] become (= which is) the cause of the usage (*vyavahāra*)[791] of saying (*iti*) "tender/charming" (*snigdha*).[792]

These types are enumerated as the types of *sattva*, or psyche, by the CA. The types mentioned by the CA are (1) *sāttvika* ones, namely, *brahman, ṛṣi, indra, yama, varuṇa, kubera* and *gandharva*, (2) *rājasa* ones, namely, *asura, rākṣasa, piśāca, sarpa, preta* and bird, (3) *tāmasa* ones, namely, beast, fish and tree (*vanaspati*).

The SU, CA and SR are all in concordance. The SR calls the trees *"aṅghripa"* instead of *"vanaspati"* in the SU and CA.

The AS has no parallel, except for the discussion on the substances derived from the mind (*sattva*). It lists the substances which are *śuddhasattvaja, rājasa* and *tāmasa*, but does not mention the notion of "body".

790 The manuscript *gha.* and C.E. read *kātarāt* instead of *kātarāḥ*. It is, however, grammatically unsuitable, because *kātara* is an adjective.

791 The term *vyavahāra* means the everyday usage of language, or verbal expression. This statement here means that the expression *snigdha* "tender/charming" is derived from *sneha*. For this term, also cf. my footnote 364 on SR śl.2c *"usage"*.

792 The term *snigdha* denotes a property of the body in the medical theory (cf. JOLLY 1901, p.18, *snigdha, picchila*). This term in the *mūla* text has nothing to do with music. But the comm. S makes up an interpretation that *snigdha* denotes the sweetness of voice. This kind of association could be compared with the musicological theory contained in the musicological text, Mataṅga's Bṛhaddeśī (anuccheda 4), that the qualities of voice like sweetness, hoarseness etc. are determined by the three morbific entities (*doṣa*); the unctuousness (*snigdha*) of voice is caused by phlegm.

He relates the qualities of [the gross element] earth: "**From earth**" (śl.70c, *bhūmer*). **Fixedness** is not being hasty. **Firmness/perseverance** is the state of not being bewildered even in a calamity. **Heaviness**[793] is the quality which is (lit. having become) the non-inherent cause of falling (*patana-asamavāyi-kāraṇa*)[794]. (Comm. S on SR śl.70)

He relates the other divisions/differences of the body: "**Wind etc.**" (śl.71c, *vāta-ādi*). **Wind_etc.** are wind, bile and phlegm. **Space etc.** is space, wind, fire (*tejas*), water [and] earth. (Comm. S on SR śl.71)

[It is said] that (*iti*): **sevenfold**, of seven kinds, is the body **which pertains to *sattva*** of **Brahman** etc., **six fold** is the body **pertaining to *rajas*** of **the *piśāca*-s** etc., **threefold** is **the body pertaining to *tamas*** of **beasts/animals** etc.

Thinking (*iti*): ["]For what (*kim iti*) is their characteristic not mentioned ?["], hence he says: "**Their**" (śl.74c, *teṣām*).

SR śl.75–78

They relate six limbs (*aṅga*)[795] of the body (*piṇḍa*)[796], namely (*iti*) the head, feet/legs, further the hands and the waist. Now (*atha*), the

793 According to the comm. S, *gaurava* does not mean "dignity" but "heaviness".

794 "Non-inherent cause" (*asamavāyikāraṇa*) is a technical term of the Vaiśeṣika school. "An individual which is not inhered in by the effect, but which is "closely related" (*pratyāsanna*) to the inherence cause, may function as a noninherence" (POTTER 1977, p.56). Also cf. ibid., p.280; p.288; pp.652–654. Because of this, the meaning of the sentence "Heaviness is the quality which is the non-inherent cause of falling" can be interpreted as follows. When a substance falls, the substance is the inherent cause of the motion of falling. Heaviness inheres in the substance. That means, heaviness is closely related to the inherent cause of falling. So, heaviness is the non-inherent cause of falling.

795 PARALLELS

SU śārīra., 5,3: *[…] evaṃ vivardhitaḥ sa yadā hasta-pāda-jihvā-ghrāṇa-karṇa-nitambādibhir aṅgair upetas tadā "śarīram" iti saṃjñāṃ labhate / tac ca ṣaḍaṅgaṃ: śākhāś catasro, madhyam pañcamaṃ, ṣaṣṭhaṃ śira iti /3/ ataḥ param pratyaṅgāni vakṣyante […] /4/.*

CA śārīra., 7,5*: tatrāyaṃ śarīrasyāṅga-vibhāgaḥ, tad yathā, dvau bāhū, dve sakthinī, śirogrīvam, antarādhiḥ, iti ṣaḍaṅgam aṅgam.*

CA śārīra., 7,11: *ṣaṭpañcāśat pratyaṅgāni ṣaṭsv aṅgeṣūpanibaddhāni, yāny aparisaṅkhyātāni pūrvam aṅgeṣu parisaṅkhyāyamāneṣu, tāny anyaiḥ paryāyair iha prakāśyāni bhavanti* etc.

secondary appendages (*pratyaṅga*) in their entirety too will be mentioned.[797] (śl.75)

Seven [layers of] skin[798] [and] seven *kalā*-s[799] [are] covered with

---

AS śārīra., 5,2: *[...] ṣaḍ-aṅgaṃ śirontarādhir dvau bāhū sakthinī ca netra-nābhi-pāṇi-pādādīni tv asya pratyaṅgāni.*
AH śārīra., 3,1: *śiro 'ntarādhir dvau bāhū sakthinīti samāsataḥ / ṣaḍ-aṅgam aṅgam pratyaṅgaṃ tasyākṣi-hṛdayādikam.*
YS 3,84 mentions *ṣaḍaṅgāni*, but does not explain what they actually are.
On YS 3,84, cf. YAMASHITA 2001/2002, pp.104–105.
AgniP 369,42cd: *ṣaḍ-aṅgaṃ sakthinī bāhu-mūrdhā-jaṭharam īritam.*
ViṣṇudhP 2,114,25: *ṣaḍ aṅgāni pradhānāni kathayiṣyāmi te śṛṇu / dvau bāhū sakthinī dve ca mūrdhā jaṭharam eva ca.*
ANALYSIS
All the texts contain the same theory. They are also similar in wording, except for the SU. The SU's very brief remark is probably an abbreviated version of an older text. The AgniP and ViṣṇudhP contain a passage parallel to each other, but this passage is not contained in the YS.
The mentioning of body limbs is already found in the Śatapathabrāhmaṇa. ŚB 10,2,3,5 (WEBER's ed., p.769, l.1): *dáśa vā́ ime púruṣe prāṇā́ś catvāry áṅgāny anyātmā́ pañcadaśá*, "Ten vital winds are in the man. Four limbs and the trunk. [Total makes] fifteen".

796 The term *piṇḍa* may mean the embryo in the second month, too (cf. my footnote 535 on SR śl.23c). In fact, SR śl.24cd–25 states that, in the third month, there are already the buds of the body-limbs and secondary appendages. The Tantric text, Śāradātilaka, in its description of birth, considers the child to be born *piṇḍita-śarīra* "the one whose body is balled". So, the term *piṇḍa* here seems to be associated with the embryonic formation of the body.
But the parallels to SR śl.75, i.e. SU śārīra., 5,3 and CA śārīra., 7,5, both contain the term *śarīra* instead of *piṇḍa*, although the wording of these two texts is the closest to that of SR śl.75 among the parallels. SU śārīra., 5,3 especially resembles SR śl.75, in mentioning the limbs (*aṅga*) and secondary appendages (*pratyaṅga*) in relationship to the foetus. Therefore the *piṇḍa* in SR śl.75 seems to mean the same thing as *śarīra*.
The term *piṇḍa* in SR śl.75 does not necessarily mean the foetus, because SR śl.78 mentions the faeces and semen. The foetus does not have semen, as stated in Hastyāyurveda sthāna 3, adhyāya 8, śl.96ab (p.413), *śukraṃ rajo malaś caiva garbhasthasya na jāyate* "Semen, procreatory-menstrual fluid, impurities (sg. in the text) of the one situated in the womb do not arise".

797 The same expression is also found in SU śārīra., 5,4: *ataḥ param <u>pratyaṅgāni vakṣyante</u>*.
According to YAMASHITA 2001/2002, p.108, the secondary appendages are referred to as *sthāna*-s in the YS 3,96–99ab.

798 PARALLELS

SU śārīra., 5,6: *tvacaḥ sapta, kalāḥ sapta, āśayāḥ sapta, dhātavaḥ sapta [...]*. The names of the seven layers of skin are listed in SU śārīra. 4,4, beginning with the innermost one: *avabhāsinī* (*sidhma-padma-kaṇṭakādhiṣṭhānā*), *lohitā* (*tilakālakanyaccha-vyaṅgādhiṣṭhānā*), *śvetā* (*carmada-lāja-gallī-maṣakādhiṣṭhānā*), *tāmrā* (*vividha-kilāsa-kuṣṭhādhiṣṭhānā*), *vedinī* (*kuṣṭha-visarpādhiṣṭhānā*), *kohiṇī* (*granthy-apacy-arbuda-ślīpada-gala-gaṇḍādhiṣṭhānā*), and *māṃsadharā* (*bhagandara-vidradhy-arśo'dhiṣṭhānā*). Certain diseases arise in a particular layer of skin, as quoted in the parentheses. The seven *kalā*-s are mentioned in SU śārīra., 4,5–20: *māṃsadharā, raktadharā, medodharā, śleṣmadharā, purīṣadharā, pittadharā,* and *śukradharā.*
CA śārīra., 7,4, mentions only six layers of skin: *udakadharā, asṛgdharā, sidhma-kilāsa-sambhavādhiṣṭhānā, dadru-kuṣṭha-sambhavādhiṣṭhānā, alajī-vidradhi-sambhavādhiṣṭhānā,* and the sixth skin called "the seat of boils" (*arus*) (*yāṃ cāpy adhiṣṭhāyārūṃṣi jāyante*). The CA does not mention the *kalā*-s.
AS śārīra., 5,29 lists the six skin layers, but also mentions another theory of seven skin layers (5,37). The names of the six layers of skin are the same as mentioned by the CA, except for the sixth which is called *prāṇadharā.* The sixth layer of skin is called *prāṇadharā* in the AgniP and ViṣṇudhP, too. The CA mentions the same names of the *kalā*-s as the SU.
AH śārīra., 3,8: *tatra sapta tvaco 'sṛjaḥ* /8d/ *pacyamānāt prajāyante kṣīrāt santānikā iva.* The seven *kalā*-s are mentioned in AH śārīra., 3,10, though some of them are missing.
YS 3,84: *tasya ṣoḍhā śarīrāṇi ṣaṭ tvaco dhārayanti ca / ṣaḍ aṅgāni tathā 'sthnāṃ ca saha ṣaṣṭyā śata-trayam.* This text mentions only six skin layers, cf. YAMASHITA 2001/2002, p.105. The comm. Mitākṣara explains it like this: the *dhātu*-s are six (*rakta-māṃsa-medo'sthi-majjā-śukrākhyā ṣaḍ dhātava eva*); therefore, the skin layers, being situated between the *dhātu*-s (*rambhā-stambha-tvag iva bāhyāntara-rūpeṇa sthitāḥ*), are six. The six limbs are mentioned in SR śl.75. The 360 bones are mentioned in SR śl.90.
AgniP 369,43: *ṣaṭ tvacā bāhyato yadvad anyā rudhira-dhārikā / kilāsa-dhāriṇī cānyā caturthī kuṇḍa-dhāriṇī /43/ pañcamīm indriya-sthānaṃ ṣaṣṭhī prāṇa-dharā matā / kalā saptamī māṃsa-dharā dvitīyā rakta-dhāriṇī /44/ yakṛt-plīhāśrayā cānyā medo-dharā 'sthi-dhāriṇī / majjā-śleṣma-purīṣāṇāṃ dharā pakvāśayāsthitā /45/ ṣaṣṭhī pitta-dharā śukra-dharā śukrāśayā 'parā /46ab/.* AgniP 369,43 mentions six layers of skin, but it mentions only five names among the six, namely, *rudhiradhāriṇī, kilāsadhāriṇī, kuṇḍadhāriṇī, indriyasthānaṃ* and *prāṇadharā.* The expression *kalā saptamī* "the seventh *kalā*" in śl.44c indicates that this text considers the *kalā*-s to be seven. Actually in the following verses, seven names are listed: *māṃsa-dharā, raktadhāriṇī, medodharā, asthidhāriṇī, majjā-śleṣma-purīṣāṇāṃ dharā, pitta-dharā* and *śukradharā.* The contents of the verses 45 and 46ab, mentioning the receptacles (*āśaya*), corresponds to that of SR 2,80cd–81.
ViṣṇudhP 2,115,26: *ṣaṭ tvacaś ca śarīre 'smin kīrtyamānā nibodha me / bāhyato hy adharā, rāma, tvacā rudhira-dhāriṇī /26/ vilāsa-kāriṇī cānyā caturthī*

cords (*snāyu*)[800], phlegm [and] the *jarāyu*[801]. Ripened/digested by the

---

kuṣṭha-kāriṇī / pañcamī vidradhi-sthānaṃ ṣaṣṭhī-prāṇa-dharā matā /27/ kalāḥ satya (misreading for *sapta*) *smṛtā dehe tāsāṃ vakṣyāmi lakṣaṇam / ekā māṃsa-dharā nāma dhamanyo yatra saṃsthitāḥ /28/ asṛg-dharā dvitīyā tu yakṛt-plīhāśrayā matā / medo-dharā tṛtīyā syāt sūkṣma-sthūlāśrayā tu yā /29/ majjāśrayā caturthī tu tathā śleṣmaparā matā /28/ purīṣa-dhāriṇī cānyā yathā pakvāśaye sthiteḥ* (misreading for *sthite*) */30/ ṣaṣṭhī pitta-dharā nāma jaṭharāgnau samāśritā / śukrāśayā śukra-dharā tathā jñeyā ca saptamī /31/.* ViṣṇudhP 115,27 is parallel to the AgniP, although there are some discrepancies. The names of the skin layers are *adharā, rudhiradhāriṇī, vilāsakāriṇī, kuṣṭhakāriṇī, vidradhasthāna* and *prāṇadharā.* The *kalā*-s are *māṃsadharā, asṛgdharā, medodharā, majjāśrayā, purīṣadhāriṇī, pittadharā* and *śukradharā.*

ANALYSIS

The SU mentions seven skin layers, while the other texts mention six. The AS mentions both theories of the CA and SU, namely, six and seven layers of skin. The AH mentions the seven skin layers in concordance with the SU. The CA's theory of six skin layers is in concordance with that of the AgniP and ViṣṇudhP. The SU's naming of the skin layers seems to be based on the colour of each layer. The names of the diseases, *kilāsa, kuṣṭha, vidradha* etc., which the SU allots to the respective skin layers, are also mentioned for the six layers of skin by the CA, AgniP and ViṣṇudhP. That means that, in contrast with the SU's theory which names the seven layers of skin according to their colours, the theory of six skin layers of the CA, AgniP and ViṣṇudhP names these layers according to the substances (*māṃsa, rakta* etc.) which they carry or to the diseases which arise in them.

Concerning the theory of the seven *kalā*-s, all the texts are in concordance, except for the CA which does not mention it in the śārīrasthāna.

The SR mentions the seven layers of skin, as do the SU and AS.

799 *Kalā.* The *kalā* is the boundary line between the elements (*dhātu*) (cf. R.F.G. MÜLLER 1961, p.90–91). The seven *kalā*-s between the seven layers of the elements (*dhātu*) consist of the liquid extract from the essence of the elements (*dhātu-sāra*) which is compared to the sap of wood (cf. SR śl.77b). Their respective names indicate that they contain or hold flesh, blood, fat, phlegm, faeces, bile and semen. Cf. JOLLY 1901, p.43. "Visceral part between the receptacles of the elements (*dhātu*)" (DAS 2003A, p.536ff.).

800 *Snāyu* "cord", "sinew", cf. DAS 2003A, p.584ff. JOLLY 1901, p.43, explains that it denotes "Sehnen und Nerven, eigentlich Bänder". Also cf. R.F.G MÜLLER 1961, p.149.

801 Here, the term *jarāyu* means a membraneous covering, not necessarily connected with a foetus, cf. DAS 2003A, p.546ff. DAS refers to SU śārīra., 4,7, which is parallel to SR śl.76. I further discuss the matter below (footnote 804).

fires of the *kośa*[802] ("sheath"), they are respectably between (*antarā-*

802 The term *kośa-agni* occurs twice in the text (SR śl.76c and śl.80c). In the parallels, AH śārīra., 3,62, and AS śārīra., 6,83, different expressions are used, namely, *agni* "fire" and *dhātv-agni* "the fire of the elements". Basing on these parallels, I infer that *kośāgni* denotes "the heat of the elements (*dhātu*)". SHRINGY 1999, p.64, interprets it in the same manner, translating "the internal heat of the tissues".
The term *kośa* does not occur anywhere in the śārīrasthāna-s of the medical texts consulted (SU, CA, AS, AH).
But the two commentators of the YS, Vīramitrodaya and Mitākṣara (on YS 3,84), mention *kośa*. Vīramitrodaya states that, according to the SU, the skin layers have the nature of six *kośa*-s (*ṣaṭ-kośātmikā tvaco*). But actually such a theory is not dealt with at all in the description of the skin layers in the SU's śārīrasthana. On the other hand, Mitākṣara states the following: the nutrient fluid (*anna-rasa*), being cooked by the abdominal/digestive fire (*jāṭharāgni*), becomes blood; the blood, being cooked by the fire which is situated in its own *kośa* (*sva-kośa-sthenāgninā*), becomes flesh; the flesh, being cooked by the fire of its own *kośa* (*sva-kośānala*), becomes fat; the fat, becomes bone, in the same way, then, the bone becomes marrow. The last member of the chain of elements (*dhātu*), namely semen, is called "the final element" (*carama-dhātu*). The "final element" is considered to be the first *kośa* of the self, because it does not change further into another element (*carama-dhātos tu pariṇatir nāstīti sa evātmanaḥ prathamaḥ kośaḥ*). Therefore, what the term *kośa* here seems to mean something like "sheath". The seven elements (*dhātu*) are compared to the sheaths or layers which cover the self. Although the commentator Mitākṣara considers this theory to be well known in Āyurveda, such a theory is actually not dealt with in the śārīrasthāna-s of the classical medical texts such as the SU, CA, AS or AH.
WOODROFFE 1990, p.44, mentions the term *kośa*, which seems to denote the same thing as in SR śl.76 and the commentary of Mitākṣara on the YS. WOODROFFE states, "In the gross body (*śarīra-kośa*), there are six external *kośa*-s, viz., hair, blood, flesh, which come from the mother, and bone, muscle, marrow, from the father". That means that the *kośa*-s here denote the seven elements (*dhātu*). Regrettably, he gives no reference.
This theory is contained in, e.g., Śāradātilaka 1,47cd–48ab: *snāyv-asthi-majjānaḥ śukrāt, tvaṅ-māṃsāsrāṇi śoṇitāt // sāṭ-kauśikam idam proktaṃ sarva-deheṣu dehinām.*
Indeed, DAS 2003A, §10.8 (p.276), points to the confusion in Tantric texts, caused by two different lists, one being the list of seven *dhātu*-s, but with skin in the place of *rasa* (like in SR śl.79ab!). The other is a list of six substances, which are the same as in the list of the seven *dhatu*-s, but without semen; these six substances are found in six "sheaths" (*kośa*).
The term *kośa* might denote "embryo", too. But this meaning does not fit the context of SR śl.76c. Cf. MārkaṇḍeyaP 11,5: *nārikela-phalaṃ yadvat sa kośaṃ*

antarā) the elements (*dhātu*), (śl.76)[803]
and, as (*bhūta*) the borders of the elements (*dhātu*), are deemed to be similar to the pith/essence of wood.[804] (śl.77ab)

---

vṛddhim ṛcchati / tadvat prayāty asau vṛddhiṃ sa kośo 'dhomukhaḥ sthitaḥ /11,6/ "[...] the *kośa* (= embryo) grows, being upside down." This Purāṇa (11,10) states that the embryo becomes hard through the fire (*kāṭhinyam agnināyāti*).

803 The C.E. edition reads *svasvakośāgnibhiḥ pakvās te tridhā*. "Ripened in three ways by the respective fires of the *kośa*-s." In this case, *kośāgnibhiḥ* might be an apposition to *snāyu-śleṣma-jarāyubhiḥ*. Compare it with SR śl.80c, *svasvakośāgninā* (but the Ānandāśrama edition reads *svasvakośādinā*).

804 The meaning here might be as follows. The *kalā*-s are the visceral parts which look like border lines between the elements (*dhātu*) (and the receptacles (*āśaya*), according to AH śārīra., 3,9). They are covered with cords (*snāyu*), phlegm and a membraneous covering (*jarāyu*). They are heated by the heat of the body and are transformed sevenfold. They are compared to the pith of wood (*kāṣṭha-sāra*). The first *kalā* is the one holding flesh. Over the flesh of this *kalā*, vessels, ducts, cords and channels spread like lotus roots.
In SU śārīra., 4,7, a parallel to SR śl.76ab, DAS considers the term *jarāyu* to denote a membraneous covering, not necessarily connected with a foetus (cf. DAS 2003A, p.546ff. I follow him in the interpretation of *jarāyu* in SR śl.76b. AH śārīra., 3,9cd–10ab, parallel to this passage, obviously misunderstands the original meaning, and substitutes *jarāyu* with *aparā* ("*ulba*, afterbirth, the outer_skin/membrane of the foetus"). The commentator Ḍalhaṇa, on SU śārīra., 4,7, explains that living beings are born covered in the *ulba*, and the membraneous covering of the *kalā* looks like the *ulba* (*jarayur ulbākāro yena veṣṭitāḥ prāṇino jāyante, kalā-veṣṭako 'pi tadvad eva*).
PARALLELS
SU śārīra., 4,5–9:*kalāḥ khalv api sapta bhavanti dhātvāśayantara-maryādāḥ /5/ yathā hi sāraḥ kāṣṭheṣu chidyamāneṣu dṛśyate / tathā hi dhātu-māṃseṣu chidyamāneṣu dṛśyate /6/ snāyubhiś ca praticchannān santatāṃś ca jarāyuṇā / śleṣmaṇā veṣṭitāṃś cāpi kalābhāgāṃs tu tān viduḥ /7/ tāsām prathamā māṃsadharā, yasyām māṃse sirā-snāyu-dhamanī-srotasām pratānā bhavanti /8/ yathā bisa-mṛṇālāni vivardhante samantataḥ / bhmau paṅkodaka-sthāni tathā māṃse sirādayaḥ /9/*. The commentator Ḍalhaṇa, explaining the term *maryādāḥ* in SU 4,5d, remarkably uses the term *sīmā-bhūtāḥ*, which is the same expression as in SR śl.77a. The SU's description is very close to that of the SR. But it does not mention the fire (*kośāgni*) which heats and transforms the *kalā*-s.
AS śārīra., 5,45–47: *yas tu dhātv-āśayāntareṣu kledo 'vatiṣṭhate sa yathāsvam ūṣmabhir vipakvaḥ snāyu-śleṣma-jarāyu-cchannaḥ kāṣṭha iva sāro dhātu-sāra-śeṣo rasa-śeṣo 'lpatvāt kalā-sañjñaḥ /45/ tā dhātv-āśayāntara-maryādāḥ sapta kalāḥ /46/ tāsām prathama māṃsadharā nāma, yasyām māṃse sirā-snāyu-dhamanī-srotasām bhumāv iva paṅkodakena bisa-mṛṇālānām pratānāni*

The first is the one holding flesh[805]. Vessels (*sirā*), ducts (*dhamanī*), cords (*snāyu*) and channels (*srotas*)[806] grow in the flesh like the bulbous roots of the lotus in the mud. (śl.77cd–78ab)

The others are the ones holding blood, fat, phlegm, faeces, bile and semen.[807] (śl.78cd)

---

bhavanti /47/. The expressions are similar to those of the SU. Some compounds of the AS are found split up in the SU (e.g. *snāyu-śleṣma-jarāyu-cchannaḥ*). The commentator Indu explains *ūṣman* as *dhātūṣman* "the heat of an element (*dhātu*)".

AH śārīra., 3,9cd–10ab, which is quoted by the commentary K of the SR. Its wordings are parallel to those in AS śārīra., 5,45.

ViṣṇudhP 2,115,28: *ekā māṃsa-dharā nāma dhamanyo yatra saṃsthitāḥ.* Compare this passage with SR śl.77. The same passage is found in the AgniP, too, but there *dhamanī* is not mentioned.

ANALYSIS

The SU (śārīra., 4,5–9), AS (śārīra., 5,45–47) and AH (śārīra., 3,9cd–10ab) are parallel to the SR śl.76–78. The AS's wording is the closest to that of the SR. The SU's wording gives the impression that an older, terser version of the text was supplemented and enlarged by insertions, while, in the AS, the original version is better preserved. These texts, however, do not contain a counterpart for *kośāgni*, which seems to denote the heat contained in each *dhātu*, according to the commentator Indu on AS śārīra., 5,45. The commentator Ḍalhaṇa's (on SU śārīra., 4,5d) explanation of *maryādā* as *sīmā-bhūta*, i.e. the same expression as contained in SR śl.77a, might suggest that Ḍalhaṇa referred to a text which contained this expression.

805 C.E. reads *māṃsodbhavā* "born from the flesh". It seems to be unsuitable, because the parallels in the medical texts call the same thing *māṃsadharā*.

806 For the term *sirā* "vessel", cf. DAS 2003A, p.584. The translation "Ader" by JOLLY 1901, p.43, is improper, because it denotes not only the blood vessels, but also the vessels for wind, bile, phlegm etc., cf. R.F.G. MÜLLER 1961, p.146 (229).

For *dhamanī* "duct", cf. DAS 2003A, p.553. Also see R.F.G. MÜLLER 1961, p.106 (106); JOLLY 1901, p.41 and p.44.

For *snāyu* "cord", cf. my footnote 800 on SR śl.76b.

For *srotas* "aperture", "channel", cf. DAS 2003A, p.585. Also see R.F.G. MÜLLER 1961, p.149 (240).

807 I have already discussed these in my footnote 798 on SR śl.76a, ("seven [layers of] skin"). The medical texts which I consulted contain the synonym *purīṣa* instead of *śakṛt*.

Having mentioned the six kinds of substances (*bhāva*) of the born [child] [which are] derived from the mother etc.[808], and the qualities and material causes of the respective elements (SR śl.67f) of the material body, he relates the manifestation/appearance (*prapañca*) of the body, [namely] limbs (*aṅga*), secondary appendages (*pratyaṅga*), elements (*dhātu*), ducts (*dhamanī*), *cakra*-s etc. by means of: ["]**They relate [...] of the body**["] etc. (śl.75a, *piṇḍasyāhuḥ*). **The secondary appendages** are skin [layer]s, *kalā*-s etc. **Will be mentioned**: The meaning is: [they will be mentioned] just immediately after. "**Seven [layers of] skin**" (śl.76a, *tvacaḥ sapta*): **The seven [layers of] skin** are *Bhāsinī* etc., being produced from blood being cooked/ripened. As mentioned in Āyurveda:

> Thus the body is constituted of the [gross] elements (*bhūta*)[809]. In it, seven skin [layer]s arise from blood being cooked/ripened, like the skin [of cream] (*santānikā*) from milk.[810]
> The skin [layer]s are (lit. the skin [layer] is[811]) *Bhāsinī*, *Lohinī*[812], *Śvetā*, *Tāmrā*, further *Vedinī*, [and] *Rohinī*.[813] The seventh[814] is renowned as *Māṃsadharā*.[815]

808 The Ānandāśrama edition (1896) reads *mātṛjādi-ṣaḍvidhān* instead of *mātṛjādīn ṣaḍvidhān*.

809 I.e. of the five elements, earth, water, fire, wind and space. That means that the body is material.

810 The Ānandāśrama edition (1896) reads *jīvāḥ sāntānikā* (correcting *sāntātikā*) instead of *kṣīrāt santānikā*. The manuscript D similarly reads *jīvāḥ santānikā*. Comparing it with the original text of the AH, I conclude that this is unsuitable.

811 *Syāt*. Optative. Cf. Introduction to the *English translation*.

812 The Ānandāśrama edition (1896) reads *lohitā* instead of *lohinī*. But, in the medical texts, it is usually called *lohinī*.

813 The Ānandāśrama edition (1896) reads *tvak tāmrā vedinī* instead of *tāmrā tvag vedinī*.

814 The Ānandāśrama edition (1896) reads *saptadhā* instead of *saptamī*. With this, it is translated, "The skin is renowned [to be] sevenfold (*saptadhā*), *Bhāsinī [...]*".

815 The first and second lines are quoted from AH śārīra., 3,8cd–9ab. But the third and fourth lines are not contained in the edition of the AH which I consulted. The names listed in the third and fourth lines are in concordance with the theory of the SU and AS. The SU calls the first skin *Avabhāsinī*, while the AS calls it *Bhāsinī*. But, in the same place, the AS states that the colour of this skin layer is *avabhāsinī* "shining, bright".

"**Seven *kalā*-s**" (śl.76a, *kalāḥ sapta*): The moisture of the nutrient fluid (*rasa*), a particular **essence (*sāra*)** of **the elements (*dhātu*)**[816], situated in the interval[817] between an afore-going and a following **element (*dhātu*), covered with cords (*snāyu*), phlegm and a *jarāyu*[818], cooked/ripened by the fire of the *kośa*[819], similar to the pith/essence (*sāra*) of wood[820]**, is called a "*kalā*", because of [its]

The names like *bhāsinī* "bright", *lohinī* (derived from *loha* "iron" or "red"), *śvetā* "white", *tāmrā* "copper-coloured" might be associated with the lustre of metals (gold, iron, silver, copper etc.).

Atharvaveda 10,2, v.11, describes the colours of the blood, among which *lohinī* and *tāmra* are mentioned: "Who disposed in him waters, moving apart, much moving, produced for river-running, strong, ruddy, red, dark, and turbid, upward, downward, crosswise in man?" (tr. by WHITNEY 1987, Vol.I, p.478) (*kó asminn ā́po vyádadhāt viṣūvṛ́taḥ purūvṛ́taḥ sindhusṛ́tyāya jātā́ḥ / tīvrā́ aruṇā́ lóhinīs tāmra-dhūmrā́ ūrdhvā́ ávācīḥ púruṣe tiráścīḥ //*).

PARALLELS to SR 2,76 (seven skin layers):

SU śārīra., 4,4: *tasya khalv evampravṛttasya śukra-śoṇitasyābhipacyamānasya kṣīrasyeva santānikāḥ sapta tvaco bhavanti / tāsām prathamā 'vabhāsinī nāma [...].*

The names of the skin layers in the SU are the same as those in the SR, except for the first one called *Avabhāsinī* instead of *Bhāsinī*.

CA śārīra., 7,4, lists only six skin layers, and the CA's description is totally different from that in the SR and SU. The CA's six skin layers are: 1. *udakadharā*, 2. *asṛgdharā*, 3. *sidhma-kilāsa-sambhavādhiṣṭhānā*, 4. *dadrū-kuṣṭha-sambhavādhiṣṭhānā*, 5. *alajī-vidradhi-sambhvādhiṣṭhānā*, and the sixth skin layer.

816 The manuscript C reads *dhātv-ādhāra viśeṣaḥ*, "a particular base (= receptacle?) of the elements (*dhātu*)" instead of *dhātu-sāra-viśeṣaḥ*. This seems to be unsuitable.
In contrast, the variant of the Ānandāśrama edition (1896), *sāra-śeṣaḥ*, seems to be correct, as I will show below.

817 The Ānandāśrama edition (1896) reads *dhātv-anya* instead of *dhātv-antarāla-sthaḥ*. I consider the resulting sequence to be a pair of compounds, i.e. *pūrvottara-dhātv-anya* "different from an afore-going and a following element (*dhātu*)" and *snāyu-śleṣmajarāyu-cchannaḥ*.

818 As I have already discussed in my footnotes 801 (on SR śl.76b) and 804 (on SR śl.77ab), the term *jarāyu* here denotes a membranous covering.

819 The term *kośa* "sheath/case", here meaning a *dhātu*, is already discussed in my footnote 802 on SR śl.76c.

820 The comparison of the blood to the pith (*rasa*) of wood is found in BṛhadāraṇyakaUp 3,9,28f, which deals with an old theory of the chain of elements (*dhātu*), cf. my footnote 926 on SR śl.103.

having a subtle form.[821] And it is said thus:

"That called a *kalā*, the moisture in the intervals between the receptacles (*āśaya*) of the elements (*dhātu*), fully (*vi-*) ripened/cooked separately/individually (*svaṃ svaṃ*) by heat, covered with phlegm, cords (*snāyu*) and the *aparā* (= *jarāyu*)[822], is like the pith/essence of wood."[823] (AH śārīra., 3,9cd-10ab)

The qualification ["]**being covered with cords (*snāyu*) etc.**["], and so on[824], is to be made known as common to both the skin [layer]s [and] the *kalā*-s. The meaning is: **"They"** (śl.76d, *tāḥ*), the skin [layer]s and the *kalā*-s, **respectively** as (*bhūta*) **the borders between**[825] **the elements (*dhātu*)**, in the respective intervals of the seven elements (*dhātu*), [namely] skin [layer] etc., are producers (*āpādaka*) of non-intermingling of those very elements (*dhātu*).

821 The nutrient fluid (*rasa*) is heated by the heat of the body, and transformed into the seven elements (*dhātu*), one after another.
The meaning of the commentary K, which explains the term *kalā* as having a subtle form, is elucidated by AS śārīra., 5,45: "The residue of the essence of the elements (*dhātu*) [and] the residue of the nutrient fluid is called (lit. is that whose name is) *kalā* ("digit/small_portion") because of [its] littleness/paucity" (*dhātu-sāra-śeṣo rasa-śeṣo 'lpatvāt kalā-sañjñaḥ*). This verse is quoted by Ḍalhaṇa on SU śārīra. 4,7, *dhātu-rasa-śeṣo'lpatvāt kalāsañjñaḥ* ("[...] the residue of the elements (*dhātu*) and nutrient fluid").
The Ānandāśrama edition (1896) reads *sāra-śeṣaḥ* "rest/residue of the essence" instead of *sāra-viśeṣaḥ*. Comparing it with *dhātu-sāra-śeṣa* in the above-quoted AS śārīra., 5,45, I prefer the Ānandāśrama edition's *śeṣaḥ* to the Adyar edition's *viśeṣaḥ*.

822 As already discussed in my footnotes 801 (on SR śl.76b) and 804 (on SR śl.77), the term *jarāyu* in this case means a membranous covering which is not necessarily related to a foetus. The commentary K fails to understand the correct meaning, and falsely explains it as *aparā* "afterbirth, foetal envelope".

823 The Ānandāśrama edition (1896) omits a considerable part. I.e. the whole part from *kāṣṭha-sāropamaḥ* up to *tathācoktam* (which corresponds to the Adyar edition, p.52, ll.12–13), and the words at the beginning of the quoted verse from the AH. Thus, the final part of this line runs: *kośāgni-pakvaḥ svasvam ūṣmaṇā śleṣma-snāyv-aparācchannaḥ kalākhyaḥ. kāṣṭha-sāravad iti.*

824 The Ānandāśrama edition (1896) reads *snāyutvādicchannatvāpādakaṃ* instead of *snāyutvādichannatvādikaṃ*. I suspect that *āpādaka* is a mistake caused by *asaṅkīrṇatāpādakā ity arthaḥ* in the line below.

825 The Ānandāśrama edition (1896) contains *madhye* instead of *madhye madhye*.

"The first" (śl.77c, *ādyā*): **The first** skin [layer] and[826] *kalā* are ["]**the one holding flesh**["] by name. **The other (*parā*)** six skin [layer]s and *kalā*-s are holders of blood etc.[827]

Comm. S on SR śl.75–78

He divides the [main] limbs and the secondary appendages in the body: **"Of the body"** (śl.75a, *piṇḍasya*). **Respectively between the elements (*dhātu*)**: The meaning is "in the respective intervals of the seven elements (*dhātu*)". The meaning is: **The first** skin [layer], [namely] **the one holding flesh**, is hard, **similar to the pith/essence (*sāra*) of wood**, but the others are not hard. Thinking: ["]Where do the vessels (*sirā*) etc. arise?["], he then says: **"In the flesh"** (śl.77d, *māṃse*). **Ducts (*dhamanī*)** are particular vessels (*sirā*). **A cord (*snāyu*)** is a particular binder of flesh, resembling a cord (*tantu*). Thinking: ["]What do the other skins and *kalā*-s hold?["], he says: **"Blood"** (śl.78a, *asṛg*).

SR śl.79–82

The seven elements (*dhātu*)[828] are[829] skin, blood, flesh, fat (*medas*), bone, marrow (*majjan*) and semen.[830] And among them (*tatra*), skin is

826 The Ānandāśrama edition (1896) reads *ca tvak kalā* instead of *tvak kalā ca*.

827 The *kalā* holding blood (*rakta-dharā*) is mentioned by the SU, AS, AH, AgniP and ViṣṇudhP. Among these texts, only the ViṣṇudhP calls this *kalā* "*asṛg-dharā*", adopting the synonym *asṛj* instead of *rakta*. The comm. K of the SR also calls it *asṛg-dharā*. The CA and YS do not mention the *kalā*-s at all.

828 For the term *dhātu*, cf. DAS 2003A, p.553ff.: "In medical texts mostly the seven elements [...] which are formed in a process of successive transformation due to the breaking down of food."
There is a musicological theory which associates the seven elements (*dhātu*) with the seven musical tones. The comm. S on SR 1,3,55cd–56ab (Adyar ed., p.120, l.5) quotes the musicological text, Mataṅga's Bṛhaddeśī (anuccheda 29, cf. P.L. SHARMA 1992, p.44), which associates the seven musical tones with the seven continents and the seven elements (*dhātu*) of the human body. This passage of the Bṛhaddeśī, in its turn, contains a quotation attributed to the SU, *tvag-asṛṅ-māṃsa-medosthimajjā-śuklāni dhātavaḥ*. Unfortunately, I was not able to identify this passage in the SU. Śāradātanaya's BHĀVAPRAKĀŚANA, a musicological text written in about the same period as the SR, in its seventh chapter (G.O.S. XLV, p.186, l.5ff.), deals with the theory that the seven tones originate in the seven elements (*dhātu*) of the human body. Remarkably, the

Bṛhadddeśī's quotation from the SU and the Bhāvaprakāśana both begin the chain of elements (*dhātu*) with the skin, like the SR does.
On the other hand, the classical medical texts usually consider the chain of elements (*dhātu*) to begin with the nutrient fluid (*rasa*). But the chain beginning with the skin is also sporadically found in these texts, cf. DAS 2003A, §10,7 (p.273ff.). For the historical development of the theory of the chain of elements (*dhātu*), cf. JAMISON 1986.

829 Optative *syus*. For the translation of optative, cf. Introduction to the *English translation.*

830 PARALLELS
The SU's śārīrasthāna contains no parallel, although SU śārīra., 5,6 states that the elements (*dhātu*) are seven. The names of the elements (*dhātu*) are listed in SU sūtra., 14,10 (*rasād raktaṃ tato māṃsam māṃsān medaḥ prajāyate / medaso 'sthi tato majjā majjñaḥ śukraṃ tu jāyate* /10/). The chain of elements (*dhātu*) beginning with skin is mentioned in SU nidāna., 1,25–29.
CA's śārīrasthāna does not mention the elements. But they are mentioned in CA sūtra., 28,4.
AS śārīra., 5,54 briefly refers to AS sūtra., 19,16–24, which is said to mention *doṣa, dhātu* and *mala.* In the nineteenth chapter of the AS's sūtrasthāna, the chain of elements (*dhātu*) is said to begin with the skin.
AH śārīra., 3,62: *sāras tu saptabhir bhūyo yathāsvam pacyate 'gnibhiḥ rasād raktaṃ tato māṃsam māṃsān medas tato 'sthi ca /62/ asthno majjā tataḥ śukraṃ śukrād garbhaḥ prajāyate /63ab/.*
AgniP 369,40: *rasas tu prāṇiṇāṃ dehe jīvanaṃ rudhiraṃ tathā / lepanaṃ ca tathā māṃsam meha-sneha-karaṃ tu tat /40/ dhāraṇaṃ tv asthi-majjā syāt pūraṇaṃ vīrya-vardhanam / śukra-vīrya-karaṃ hy ojaḥ prāṇa-kṛj jīva-saṃsthitiḥ /41/ ojaḥ śukrāt sāra-taram āpītaṃ hṛdayopagam /42ab/*
ViṣṇudhP 2,114,22: *rasas tu prāṇino dehe jīvano rudhiras tathā / lepanaṃ ca tathā māṃsam medaḥ snehakaraṃ ca tat /22/ dhāraṇaṃ tv asthi-kathitam majjā bhavati pūraṇī / garbhotpādakaraṃ śukraṃ tathā vīrya-vivardhanam /23/ tejaḥ prāṇakaraṃ nityaṃ tatra jīvo vyavasthitaḥ / śukrād api paraṃ sāram apītaṃ hṛdayopamam /24/* (The reading *apīta* is obviously a mistake and should be read *āpīta*. This text substitutes *tejas* for *ojas* which is contained in the parallel of the AgniP).
GaruḍaP 1ab (KIRFEL 1954): *tvacā-raktaṃ tathā māṃsam medo majjāsthi jīvitaṃ.*
The Hastyāyurveda mentions the elements (*dhātu*) in 3,9 v.201–205 (p.452) and in 3,9 v.37 (p.440).
ANALYSIS
All the medical texts mention the elements (*dhātu*), but solely the AH lists them individually in the śārīrasthāna. While in the SU, CA and AS, the list is contained in the sūtrasthāna. So, the AH is the exception.
Contrastingly the YS does not mention the elements (*dhātu*). The AgniP and ViṣṇudhP, which are parallel to each other, mention the chain of elements

[already] mentioned.[831] Blood arises[832] from the juice of food (*annarasa*) cooked/ripened by the fire pertaining to the stomach.[833] (śl.79–80a)[834]

Thus/In_the_same_manner (*evaṃ*), the other elements (*dhātu*) are born respectively through blood etc., [which are] cooked/ripened by every single one's own fire of sheath (*kośa*)[835]. (śl.80bcd)

(*dhatu*) beginning with the nutrient fluid (*rasa*). The GaruḍaP contains the chain beginning with the skin.

831 The manuscript D reads *coktatvād raktaṃ* instead of *coktā tvag*. "And among them, through/from being mentioned (= as mentioned), blood [...]". This reading is unsuitable, because such a topic is not mentioned in the previous passages.

832 Optative *bhaved*. On the translation of optative, cf. Introduction to the *English translation.*

833 *Kha.* and *ṅa.* read *jaṭhara* instead of *jāṭhara*.
PARALLELS
Only the AH contains a parallel to SR śl.79d–80. AH śārīra., 3,62: *sāras tu saptabhir bhūyo yathāsvam pacyate 'gnibhiḥ / rasād raktaṃ tato māṃsam māṃsān medas tato 'sthi ca /62/ asthno majjā tataḥ śukraṃ śukrād garbhaḥ prajāyate /63ab/.*
Neither the SU, CA nor AS lists the elements (*dhātu*) in the śārīrasthana. In these texts, the list seems to be contained in other sthāna-s than the śārīrasthana (cf. my footnote 830).
The same theory as in SR śl.79d–80 is dealt with in AS śārīra., 6,83:
tatrāhāra-raso vyāna-vikṣipto yathāsvaṃ saptasu dhātv-agniṣu kramāt pacyamānaḥ svātma-bhāva-cyuti-samanantaram eva prāpta-raktādi-dhātu-sañjñakaḥ [...] /83/.
ANALYSIS
According to SR śl.79ab, the chain of the seven elements (*dhātu*) begins with the skin. But in SR śl.79, skin is dealt with separately from the other elements (*dhātu*). In SR śl.79d–80, the chain of elements (*dhātu*) actually seems to begin with *anna-rasa,* which is then transformed into the blood etc. In this point, it resembles the usual classical medical theory, according to which the chain begins with the nutrient fluid (*rasa*). For the terms *rasa* and *ahārarasa* (*annarasa*), cf. DAS 2003A, p.578ff.; p.528.
The term *kośa-agni* has been already mentioned in SR śl.76c.

834 SR 79ab is identical to ŚG 9,37cd. But the ŚG contains *-medau* and *-śukrāṇi* instead of *-medo* and *-śuklāni*, respectively.

835 For the term *kośāgni* "fire of sheath", cf. my footnote 802 on SR śl.76c, *kośa*. The Ānandāśrama edition (1896) reads *svasvakośādinā* "[cooked/ripened] by every single one's own sheath" instead of *svasvakośāgninā*. But if we compare it with SR śl.76, this variant seems to be unsuitable. The manuscript D reads *svasva-kośāgnibhiḥ*.

The seats (*āśraya*) of blood[836], phlegm, unripe [food] and bile, of ripened [food] (*pakva*), further of wind
and of urine are respectively seven, called ["]receptacles["] (*āśaya*). (śl.81)[837]

The receptacle of the embryo (*garbhāśaya*, "the womb"), [namely] the eighth [receptacle], of women is between the receptacles of bile and of cooked/ripened [food]. (śl.82ab)

The heart, whose form is [that of] a lotus, through tranquil

836 For the receptacle of blood (*rakta*), cf. DAS 2003A, pp.577–578. According to DAS (ibid.), this term does not necessarily denote the heart. The receptacle of blood referred to in SR śl.81 does not seem identical to the heart mentioned by SR śl.82.

837 PARALLELS

SU śārīra., 5,8: *āśayās tu, vātāśayaḥ pittāśayaḥ, śleṣmāśayo, raktāśaya, āmāśayaḥ, pakvāśayo, mūtrāśayaḥ, strīṇāṃ garbhāśayo 'ṣṭama iti.*

SU śārīra., 5,39: *[...] pitta-pakvāśayor madhye garbhaśayyā yatra garbhas tiṣṭhati.*

The CA's śārīrasthāna does not list the receptacles (*āśaya*), although CA śārīra., 7,10 mentions *purīṣādhāra, āmāśaya* and *pakvāśaya* in the list of the *koṣṭhāṅga*-s.

AS śārīra., 5,61: *saptāśayāḥ, kramād asṛk-kaphāma-pitta-pakva-vāyu-mūtrādhārāḥ /61/ strīṇām pitta-pakvāśayor-madhye garbhāśayo 'ṣṭamaḥ /62/.*

The AH's śārīrasthāna does not contain the list of the receptacles.

AgniP 370 ("śārīrāvayava"), 6cd: *saptāśayāḥ smṛtā dehe rudhirasyaika āśayaḥ /6/ śleṣmaṇaś cāma-pittābhyām pakvāśayas tu pañcamaḥ / vāyu-mūtrāśayaḥ sapta strīṇāṃ garbhāśayo 'ṣṭamaḥ /7/ pittāt pakvāśayo 'gneḥ syād yonir vikāśitā dyutau / padmavad garbhāśayaḥ syāt tatra dhatte sa-raktakam /8/.*

ViṣṇudhP 2,114,49: *saptāśayāḥ smṛtā dehe śṛṇu tān api, bhārgava / āśayo rudhirasyaikaḥ kaphasya ca tathā paraḥ /49/ āma-pittāśayau cānyau jñeyaḥ pakvāśayo 'paraḥ / vāyu-mātrāśayau* (*mātra* is obviously a mistake for *mūtra*) *cānyau āśayāḥ sapta kīrtitaḥ /50/ strīṇāṃ garbhāśayo, rāma, pitta-pakvāśayāntare / aṣṭamaḥ sa bhavet tāsāṃ yatra garbhaḥ sa tiṣṭhati /51/.*

ANALYSIS

The statements of the SR, SU, AS, AgniP and ViṣṇudhP are all in concordance. The CA and AH, unexpectedly, do not contain a parallel in the śārīrasthāna. The YS does not mention the receptacles (*āśaya*), though it mentions *purīṣādhāna* and *āmāśaya* among various entrails (YS 3,94–95). The position of the *garbhāśaya* between the *pitta°* and *pakva-āśaya* is mentioned by the SU, AS and ViṣṇudhP. This position is implied by the AgniP, too. R.F.G. MÜLLER 1941–1942, p.163, refers to Mahābhārata 7,253,11 and 7,333,24, which mention this position, too (also see MārkaṇḍeyaP 10,5–6).

phlegm and blood[838], (śl.82cd)
is[839] hollow (*suṣira*)[840], facing downward, situated between the liver and the spleen.[841] (śl.83ab)[842]

838 When phlegm and blood are tranquil (cf. optative *syād*).

839 Optative *syād*. For the translation of optative, cf. Introduction to the *English translation*.

840 C.E. reads *snāyuvaṃ śyam adho°* instead of *suṣiraṃ syād adho°*. This could be interpreted as *snāyu-vaṃśyam adho°* "[The heart] is [like] a [bamboo] joint of cords (*snāyu*)".

841 D reads *yaccadehāntare (yac ca dehāntare)* instead of *yakṛt-plīhāntara-*. This could mean "[...] and which is situated in the inside of the body", but seems unsuitable.

842 PARALLELS
SU śārīra., 4,31–32: *śoṇita-kapha-prasādajaṃ hṛdayaṃ, yadāśrayā dhamanyaḥ prāṇavahāḥ, tasyādho vāmataḥ plīhā phupphusaś ca, dakṣiṇato yakṛt kloma ca, tad viśeṣeṇa cetanā-sthānam atas tasmiṃs tamasāvṛte sarva-prāṇinaḥ svapanti /31/ bhavati cātra: puṇḍarīkeṇa sadṛśaṃ hṛdayaṃ syād adhomukham / jāgratas tad vikasati svapataś ca nimīlati /32/.* Also cf. 4,34: *hṛdayaṃ cetanāsthānam uktaṃ, suśruta, dehinām / tamo'bhibhūte tasmiṃs tu nidrā viśati dehinām.*
CA śārīra., 7,8: *hṛdayaṃ cetanādhiṣṭhānam ekam.* Except for this passage, I did not find any parallel in the śārīrasthāna.
AS śārīra., 5,71: *hṛdayam punaḥ śleṣma-rakta-prasādāt sambhavati padma-kośa-saṅkāśaṃ suṣiram adhomukhaṃ, tad viśeṣeṇa cetanāyāḥ sthānaṃ sarva-bhāvānāṃ ca cetanānugatānām, tasya vāma-pārśve plīhā puḥpusaś ca / dakṣiṇato yakṛt kloma ca /71/.*
The AH's śārīrasthāna contains no parallel.
The YS contains no parallel. The heart is mentioned in the list of the entrails in YS 3,95. Though the structure of the heart is described in YS 3,108–111, it accords with Haṭhayogic theory, but not with the classical medical theory.
AgniP 370,16: *kapha-prasārād bhavati hṛdayam padma-sannibham / adhomukhaṃ tac chuṣiraṃ yatra jīvo vyavasthitaḥ /16/ caitanyānugatā bhāvāḥ sarve tatra vyavasthitāḥ / tasya vāme yathā plīhā dakṣiṇe ca tathā yakṛt /17/ dakṣiṇe ca tathā kloma padmasyaivam prakīrtitam //18ab//* (17ab is parallel to SR śl.83ab) The reading *kapha-prasārād* in v.16a is obviously a mistake for *kapha-prasādād*.
ViṣṇudhP 2,115, 60: *kapha-prasādād bhavati hṛdayam padma-sannibham / adhomukhaṃ tat-suṣiraṃ yatra jīvo vyavasthitaḥ /60/ caitanyānugatā bhāvāḥ sarve tatra vyavasthitāḥ / tasya vāme tathā plīhā dakṣiṇe ca tathā yakṛt /61/ dakṣiṇe ca tathā kloma padmasyaiva prakīrtitam /62ab/.*
ANALYSIS
The SU, AS, AgniP and ViṣṇudhP make statements similar to that of the SR. The SU is the most closely related to the SR, but it does not contain the term

Comm. K on SR śl.79–83ab

With regard to (*uddiśya*) the elements (*dhātu*), he shows their appearance/arising, through "**skin, blood**" etc. (śl.79a, *tvag-asṛg*). "**And among them, skin is [already] mentioned**" (śl.80cd, *tatra coktā tvag*). The word "and" (*ca-kāra*) is of deviating [syntactical] order (*bhinnakrama*): [it should be] "And skin" (*tvak ca*). This is the meaning, thus (*iti*): Here, even the external skin, counted as an element (*dhātu*)[843], is one whose characteristic is mentioned through the very characteristic of skin which is (lit. having become) the border between elements (*dhātu*). [844] "**Blood, phlegm**". (śl.81a, *rakta-śleṣma-*): Here, through the word ["]blood["], the blood differentiated from the element (*dhātu*) is desired to mention (= intended).

Comm. S on SR śl.79–83ab

He divides the elements (*dhātu*): "**Skin**" (śl.79a, *tvag-*). He expresses the manner of appearance/arising of blood etc: "**Blood**" (śl.79d, *rakta-*). **Unripe [food] (*āma*)** is unripe juice [of food] (*rasa*)[845]. He relates the extra receptacle (*āśaya*) of women: "**The receptacle of the embryo**" (śl.82a, *garbhāśaya*). He relates the manner of appearance/ arising, the particular form and the place of the heart: "**Through tranquil**" (śl.82c. *prasannābhyām*).

SR śl.83cd–85

And this is the place of consciousness[846] (*cetana*). Then, when this is covered with darkness, when [it (=lotus)] is closing, the self sleeps,

---

suṣira (SR śl.83a), while the AS, AgniP and ViṣṇudhP do contain it. Only the SU describes the process of sleep, like the SR śl.83d–84ab does.

843 Literally: "with/through *dhātu*-hood".

844 This commentator here is obviously considering the chain of elements (*dhātu*) as begining with the nutrient fluid (*rasa*), according to the classical medical theory.

845 The term *rasa* denotes "nutrient fluid", cf. DAS 2003A, p.578ff.

846 Ḍalhaṇa on SU śārīra., 4,31, considers the heart to be specially important for perception (*cetanā*), although the whole body is the place of perception (*sakala-śarīram eva cetanāsthānam*), cf. ROŞU 1978, p.207. For the early Buddhist theory which designates the heart as the chief centre of consciousness, cf. WAYMAN 1982, p.631. This theory even goes back to the Ṛg Veda, cf. DAS 2003A, p.592ff. Also cf. YAMASHITA 2001/2002, pp.106–107.

also when [it] is opening up, [the self] awakens.[847] (śl.83cd–84ab)
Sleep[848] (*svāpa*) is twofold[849] through dream-sleep (*svapna*) and deep-sleep (*suṣupti*)[850]. (śl.84cd)

If the outer faculties/organs are dissolved in the heart, [but] the consciousness/intellect[851] awakens/wakes, that is called ["]dream-sleep["].[852] (śl.85ab)

If the mind is dissolved[853] in the vital wind (*prāṇa*), that is[854] the ["]deep-sleep["] of the self. (śl.85cd)

Gone/dissolved (*apīta*)[855] into itself (*svam*), [namely] the supreme self[856], the self sleeps (*svapiti*): thus (*iti*) it is considered from this.[857] (śl.86ab)

847 According to the comm. S, *nimīlati* and *vikasati* are past participles, locative sg. A parallel is contained in SU śārīra., 4,31–32: *tadviśeṣeṇa cetanā-sthānam atas tasmiṃs tamasāvṛte sarvaprāṇinaḥ svapanti /31/ bhavati cātra, puṇḍarīkeṇa sadṛśaṃ hṛdayaṃ syād adhomukham / jāgratas tad vikasati svapataś ca nimīlati /32/.*
If I follow the SU's parallel, *vikasati* and *nimīlati* of the SR also could be taken to be a third person sg. present. However, then it would be difficult to translate SR śl.84ab, especially the quarter 84b. As it is, the SU has the other two correspondences to the SR, namely *svapiti* and *jāgarti* as participles, which parallels the construction of the SR, except that the pairs of finite verbs and participles are exchanged.

848 MITTWEDE 1993 studies the theory of sleep in CA sūtra., 21,35 and followings. (*yadā tu manasi klānte karmātmanaḥ klamānvitāḥ viṣayebhyo nirvartante tadā svapiti mānavaḥ*). But this theory of the CA differs from that of the SU and SR. MITTWEDE reports that the Bṛhadāraṇyakopaniṣad contains an early form of the SU's statement that sleep is caused by *tamas*, when it prevails in the heart.

849 The Ānandāśrama edition (1896) reads *dvidhā* instead of *dvedhā*.

850 These two divisions of sleep is according to the Upaniṣadic thought, cf. Bṛhadāraṇyakopaniṣad 2,1,18–19 (RADHAKRISHNAN 1953, pp.189–190); Śaṅkara's comentary on the Brahmasūtra 3,2,1–9 (DEUSSEN 1883, pp.369–380); Kaivalyopaniṣad 17; Brahmabindūpaniṣad 11. ŚG 10,58 mentions the three states, *jāgrat, svapna* and *suṣupti*.
The SU contains no parallel. Instead, SU śārīra., 2,33 makes a different manner of classification of sleep (*nidrā*), i.e. *tāmasī*, *svābhāvikī* and *vaikārikī*.

851 The term *citta* here seems to denote the same thing as *cetana*.

852 C.E. reads *svapnam* instead of *svapnas*, though it is grammatically false.

853 D reads *manas saṃlīyate* "the mind sticks/lies_down/melts_away" instead of *manaś cel līyate*.

854 Optative *syāt*.

855 *Cha.* reads *apītaṃ* instead of *apītaḥ*.

856 C reads *paramātmānaṃ* instead of *parātmānaṃ*.

"**And this is the place of consciousness**" (śl.83c, *etac ca cetana-sthānam*): The meaning is thus (*iti*): **This** is the lotus of the heart, **whose form is [that of] a lotus, facing downward, hollow (*suṣira*).** Here, the word ["]and["] (*ca*) has the meaning of a collection (*samuccaya*) of the other places which are (lit.: having become) the cause[858] of the manifestation of consciousness. The other places are, however, the vital wind (*prāṇa*)[859] etc., mentioned[860] in the *śruti*, "Brahman is the vital wind (*prāṇa*), Brahman is *ka*, Brahman is *kha*/space" etc. (Cāndogya-Up. 4,10,5). **The place of consciousness** is the place of consciousness, [namely] of the supreme Brahman (*kūṭastha-brahman*; lit. "Brahman situated on the peak"[861]), the place of manifestation. Here, the word ["]consciousness["] (*cetana*) is not serving_ for/synonymous_with (*para*) ["]the individual self (*jīva*)["]. Why? Because of contrariness to the *śruti*: "Now, this which is the city of Brahman" etc. (Ātmaprabodha 1: *atha yad idam brahma-puram idam puṇḍarīkaṃ veśma*). For the *śruti* through (*iti*) ["]Whose abode/form (*sanniveśa*) is the white lotus (*puṇḍarīka*)["][862] here expresses the white lotus' (= the heart's) being the place of Brahman. For thus (*tathā hi*), due to the word *dahara*'s effecting synonymity with [the word] ["]Brahman["] after the refutation of the synonymity with [the words] "space" (*ākāśa*)[863] and "individual self", through the important/extensive collection [of references] (*mahatā sandarbhena*), [namely]:

857 The etymological explanation in SR śl.86ab is based on ChāndogyaUp 6,8,1, as stated by the comm. S. ChāndogyaUp 6,8,1 is discussed in Śaṅkara's Bhāṣya, cf. DEUSSEN 1883, p.263 (XX,2). Also cf. ROŞU 1978, p.210; BṛhadāraṇyakaUp 2,1,17–19.

858 The Ānandāśrama edition (1896) lacks *hetu*. This is not noted by the Adyar edition.

859 For the vital wind (*prāṇa*) as the seat of the self (*ātman*), cf. ZYSK 1993, p.204.

860 The Ānandāśrama edition (1896) reads *uktādīni* instead of *uktāni*. This may be a mistake caused by the following *prāṇādīni*.

861 For the attribute *kūṭastha*, cf. Śaṅkara's Bhāṣya on sūtra 1,3,19 (Ānandāśrama series, p.273): *kūṭastha-nityadṛk-svarūpam ātmānam [...].*

862 *Puṇḍarīka-sanniveśam* means the same thing as *puṇḍarīkaṃ veśma* in Ātmaprabodha 1, according to JACOB 1985, p.550.

863 The Ānandāśrama edition (1896) reads *ākāro* instead of *ākāśa*. The editor marks it with a question mark.

> Even[864] at the time of sleep, this whole world of knowing and unknowing living beings (*jīva-loka*), indeed, though having attained Brahman's world called "the space/cavity of *dahara*" (*dahara-ākāśa*), whose seat is the white lotus of the heart, does not know: "I have (lit. am one who has) gained the absorption_into/identification_with (*bhūya*) Brahman", because of their sight being covered with the coating of the darkness of beginningless ignorance

etc.,
– after [first] beginning with:

> Now this which is, in this regard (*asmin*), the city of the Brahman, the fine (*dahara*) dwelling place, resembling a minute cave, whose form/abode is the white lotus (= heart), is the *dahara*[865]

in the section on the *dahara* in the Bhāmatī[866], [and then] quoting the *śruti* (ChāndogyaUp 8,1,1)[867]:

864 The Ānandāśrama edition (1896) reads *svāpakāle hi* instead of *svāpakālepi.*

865 ChāndogyaUp 8,1,1: *atha yad idam asmin brahmapure daharam puṇḍarīkaṃ veśma daharo 'sminn antarākāśas tasmin yad antas tad anveṣṭavyaṃ tad vāva vijijñāsitavyam.* This passage of the ChāndogyaUp is quoted in Śaṅkara's Bhāṣya commenting 1,3 sūtra 14, *dahara uttarebhyaḥ*. It deals with the space (*ākāśa*) called "*dahara*" which is situated in the lotus of the heart. It discusses the question of whether this space belongs to the gross element of space (*bhūtākāśa*), the individual self (*vijñānātmā*), or the supreme self (*paramātmā*). This question arises, because, in the daily usage of speech (*rūḍha*), the term *ākāśa* usually means the gross element of space. Śaṅkara insists that the space of *dahara* belongs to the supreme self.

866 Vācaspatimiśra's Bhāmatī is the commentary on Śaṅkara's Bhāṣya which is, in its turn, a commentary on Bādarāyaṇa's Brahmasūtra. The text quoted here is from Brahmasūtra 1,3,5 (sūtra 14–19) which is entitled "the section of *dahara*" (*Dahara-adhikaraṇa*).

867 According to Śaṅkara's Bhāṣya (Ānandāśrama series 21; see Śāṅkarabhāṣya in my biblio), it is ChāndogyaUp 8,3,2. Śaṅkara quotes this passage in his commentary on sūtra 1,3,15. Quoting this verse, he states, *tathā hy ahar ahar jīvānāṃ suṣuptāvasthāyām brahmaviṣayaṃ gamanaṃ [...] loke 'pi kila gāḍhaṃ suṣuptam ācakṣate brahmībhūto brahmatāṃ gata iti*, "For example, there is individual selves' going every day to the domain of Brahman [...] In the world (= daily life), indeed, they refer to one sleeping very deeply as 'become Brahman, gone to the state of Brahman'."

"All these creatures, going to Brahman day by day, do not find this world of Brahman." (partial negation) (Or total negation: "None of these creatures [...] finds [...].") –,

here too it should be understood that (*iti*) the consciousness is Brahman alone, because of [its (= consciousness')] being the occupier of the lotus of the heart.

"**Then while this**" etc. (śl.83d, *tad asmin*): **While this** (*asmin*), [namely] the consciousness (*cetana*), [namely] the Brahman, **is covered**, [namely] enclosed, **with darkness**, [namely] with beginningless ignorance, **then**, when that lotus of the heart **closes** (*nimīlati*), [namely] becomes a bud (*mukulī-bhavati*), then **the self**, [namely] the individual self (*jīva*), **sleeps** (*svapiti*), [namely] slumbers (*nidrāti*); [and] then, when the lotus **opens up** (*vikasati*)[868], then the individual self **awakens** (*jāgarti*), [namely] perceives (*budhyate*) [itself as] "I". ["]Too["] (*api*) [is used] in the meaning of ["]and["] (*ca*).[869] This is [already] mentioned thus (*iti*): Through the self's ignorance of [its] own nature, even in the state of an individual self, the individual self's conditions of being awake, dream-sleep and deep-sleep, [namely] the causes of the designations "*anādi-viśva-taijasa-prājña*"[870], arise/appear. "**Twofold**" (śl.84c. *dvedhā*): **Sleep (*svāpa*)** is slumber (*nidrā*), **dream-sleep** and **deep-sleep** [are] its distinction[s]/specie[s].

868 The comm. K seems to take *nimīlati* to be a finite verb, because it uses a relative construction.

869 The Ānandāśrama edition (1896) lacks the part beginning with *apiścārthe*. It concludes with *manaś cel līyata ityādy asmad-viracite 'dhyātma-viveke vīkṣyatām budhair ity antaḥ spaṣṭārthaḥ*.

870 For *anādi-viśva-taijasa-prājña*, cf. OLIVELLE 1992, p.201, note 69 on the NāradaparivrājakaUp, "These are categories of the elaborate cosmological scheme of Advaita Vedānta. The self (*ātman*) in the waking state that thinks it has a physical body and perceives objects through the senses is called by the technical term Viśva. The self in the dream state that thinks it has a subtle body and is engaged in dreaming is called Taijasa. The self in deep sleep, when all cognition ceases but when it is still under the influence of the ignorance [*māyā*] that conceals Brahman, is called Prājña." They are mentioned in the ParamahaṃsaparivrājakaUp, too.

Comm. S on SR śl.83cd–86ab

He says that (*iti*) the very heart is the cause of sleeping[871] and waking: "**And this**" (śl.83c, *etac ca*). **The place of consciousness (*cetana*)**, [namely] of the self. **When that (*asmin*)** is (*sati*) **closing (*nimīlati* locativus absolotus), the self sleeps. When it** is (*sati*) **opening up (*vikasati* locativus absolutus)**, it **awakens.**

Telling of the twofoldness of sleep, he characterises: "**Twofold**" (śl.84c. *dvedhā*). **The consciousness/intellect (*citta*)**, [namely] the mind (*manas*), **awakens**, sees things pertaining to the dream (*svāpnika*).

He etymologises the word "**sleeps**" (*svapiti*): "**Gone/dissolved into itself**" (śl.86a, *svam apītaḥ*). From the *śruti*: "He/It has gone into (*apīta*) himself/itself indeed; therefore they refer to him/it as sleeping (*svapiti*)" (Chāndogya-Up. 6,8,1).

SR śl.86cd–119

Of those characterised by a body, the ears, eyes, nostrils, mouth, anus and penis are (opt. *syuḥ*) the nine channels (*srotas*) carrying impurities outside. (śl.86cd–87ab)

There are[872] three extra ones of women: two of the breasts, [one] of blood in the vagina. (śl.87bc)[873]

871 The term *svapna* here seems to mean sleep in general in contrast to *jāgaraṇa* "waking".

872 Optative. For the translation of optative in technical Sanskrit, cf. Introduction to the *English translation.*

873 PARALLELS
SU śārīra., 5,10: *śravaṇa-nayana-vadana-ghrāṇa-guda-meḍhrāṇi nava srotāṃsi narāṇām bahirmukhāni etāny eva strīṇām aparāṇi ca trīṇi dve stanayor adhastād raktavahaṃ ca.*
CA śārīra., 7,12: *nava mahanti chidrāṇi sapta śirasi, dve cādhaḥ.*
AS śārīra., 6,38–39: *srotāṃsi puṃsāṃ nava, karṇau netre nāsāpuṭau mukham pāyur mūtra-pathaḥ, anyani ca trīṇi strīṇāṃ stanau rakta-pathaś ca.*
AH śārīra., 3,40cd: *srotāṃsi nāsike karṇau netre pāyv-āsya-mehanam /40cd/ stanau rakta-pathaś ceti nārīṇām adhikaṃ trayam /41ab/.*
Neither the YS, AgniP nor ViṣṇudhP mentions the nine channels (*srotas*).
Hastyayurveda 3,9, v.59 and followings (p.441): *haste ca tāluni dve dve mukhe netre kaṭi-dvaye / dve karṇayoś ca pratyekam evam ekādaśaiva tu /59/ stanayoś caiva meḍhre ca nāgasya tu gude tathā / vidyāt pañcadaśaitāni śrotāṃsi (= srotāṃsi) vadatāṃ vara /60/.*

Sixteen "nets/networks" (*jāla*)[874] [are] situated at/in bones, cords (*snāyu*), vessels (*sirā*) and flesh.

Six "bunches" (*kūrca*) are at hands, feet, neck and urinary organ. (śl.88)[875]

ANALYSIS
The SU, AS and AH mention the nine channels (*srotas*), like the SR. The CA briefly mentions nine openings (*chidra*). The Purāṇa-s mention nothing.

874 Cf. SU śārīra., 5,12.
ZYSK 1986, p.699, states "It appears [...] that the *jāla*-s are the networks of veins, arteries, nerves, etc., which enclose the various bundles (*kūrca*s) of muscles, etc."

875 PARALLELS
SU śārīra., 5,12: *māṃsa-sirā-snāyv-asthi-jālāni pratyekaṃ catvāri / pratyekaṃ catvāri, tāni maṇibandha-gulpha-saṃśritāni paraspara-nibaddhāni paraspara-gavākṣitāni ceti, yair gavākṣitam idaṃ śarīram.*
SU śārīra., 5,13: *ṣaṭ kūrcāḥ te hasta-pāda-grīvā-meḍhreṣu.*
SU śārīra., 5,14: *mahatyo māṃsa-rajjavaś catasraḥ: pṛṣṭha-vaṃśam ubhayataḥ / peśī-nibandhanārthaṃ dve bāhye ābhyantare ca dve.*
SU śārīra., 5,15: *sapta sevanyaḥ, sirasi vibhaktāḥ pañca, jihvā-śephasor ekaikā / tāḥ parihartavyāḥ śastreṇa.*
AS śārīra., 5,84: *ṣoḍaśa jālāni, teṣām māṃsa-sirā-snāyv-asthi-jāni / catvāry ekaikatra gulphe maṇibandhe ca parasparagavākṣitāni tāni sthitāni /84/ ṣaṭ kūrcā hasta-pāda-grīvā-meḍhreṣu /85/ catasro māṃsa-rajjavaḥ pṛṣṭha-vaṃśam ubhayataḥ peśī-bandhanārthaṃ tāsāṃ dve bāhye dve cābhyantara-sthite /86/ sapta sīvanyaḥ / tāḥ pañca śirasi jihvā-mehanayor ekaikā / parihāryāś ca śastreṇa.*
AH śārīra., 3,14ab: *jālāni kaṇḍarāś cānge pṛthak ṣoḍaśa nirdiśet/ ṣaṭ kūrcāḥ sapta sīvanyo meḍhra-jihvā-śirogatā /14/ Śastreṇa tāḥ pariharec catasro māṃsarajjavaḥ /15ab/.* (The *kaṇḍarā* is mentioned by SR śl.95b, too.) AgniP 370,23cd: *dehe pādādi-śīrṣānte jālāni caiva ṣoḍaśa /23cd/ māṃsa-snāyu-śirāsthibhyaś catvāraś ca pṛthak pṛthak / maṇibandhana-gulpheṣu nibaddhāni parasparam /24/ ṣaṭ kūrcāni smṛtānīha hastayoḥ pādayoḥ pṛthak / grīvāyāṃ ca tathā meḍhre kathitāni manīṣibhiḥ /25/ pṛṣṭha-vaṃśasyopagatāś catasro māṃsa-rajjavaḥ / tāvatyaś ca tathā peśyas tāsām bandhana-kārikāḥ /26/ sīraṇyaś (= mistake for sīvanyaś) ca tathā sapta pañca mūrdhānam āśritāḥ / ekaikā meḍhra-jihvās tā [...]/27abc/.*
ViṣṇudhP 2,115,72cd: *jalāni* (= mistake for *jālāni*) *ṣoḍaśaivātra vibhāgas teṣu kathyate /72cd/ māṃsa-snāyu-śirāsthibhyaḥ catvāras tu pṛthak-pṛthak / maṇibandhāni* (= mistake for *maṇibandha-*) *gulpheṣu nibaddhāni parasparam /73/.* The *śaṅku* ("dart") mentioned in v.74 seems to denote the same thing as *kūrca* : *śaṅkūni ca smṛtānīha hastayoḥ pādayos tathā / grīvāyāṃ ca tathā meḍhre kathitāni manīṣibhiḥ /74/. V.75: dehe 'smiṃś ca tathā jñeyāś catasro māṃsa-rajjavaḥ / pṛṣṭha-vaṃśobhayagate dve dve tatra prakīrtite /75/ tāvantyaś*

Four tendons/cords of flesh (*māṃsarajju*) are at the two sides of the backbone.

Five "sutures/seams" (*sīvanī*)[876] are on the head. Two are deemed in the tongue and the sexual organ. (śl.89)

The groups of bones (*asthi-rāśi*)[877] are agreed [to be] fourteen or

---

ca tathāpy etās tāsām bandhana-kārakāḥ / sīvanyaś ca tathā sapta pañca mūrdhānam āśritāḥ /76/ ekā meḍhra-gatā caikā tathā jihvā-gatā parā /77ab/.

ANALYSIS

The terms, *jāla, kūrca, māṃsarajju* and *sīvanī*, are respectively translated into "network", "bundle", "large tendon" and "suture (of the body)" by ZYSK 1986, p.700.

(1) *jāla* "net". The SU, AS, AH and the two Purāṇa-s (AgniP and ViṣṇudhP) mention the sixteen nets (*jāla*). According to these texts, the nets (*jāla*) are sixteen, because four nets (*jāla*) are each situated in the flesh, cords (*snāyu*), vessels (*sirā*) and bones. Neither the CA's śārīrasthāna nor the YS mentions them.

(2) *kūrca* "bunch". The same textual condition as in the case of the nets (*jāla*) applies to that of *kūrca*. But the ViṣṇudhP seems to call it *śaṅku* "dart". The SU, AS, AgniP and ViṣṇudhP all state that the bunches (*kūrca*) are situated in the wrists (*maṇibandha*) and ankles (*gulpha*).

(3) *māṃsarajju* "tendon of flesh" (cf. ZYSK 1986, p.699). As above the same applies to *māṃsarajju*. The SU, AS, AgniP and ViṣṇudhP all state that the tendons of flesh (*māṃsarajju*) bind the *peśī*-s ("muscles, pieces of flesh"). But the ViṣṇudhP's statement on this topic is incomplete.

(4) *sīvanī* "suture". This term also has the same textual condition as the nets (*jāla*).

In summary, the SR's theory accords with that of the SU. The SU (śārīra., 5,15), AS (śārīra., 5,87) and AH (3,15ab) warn against cutting the nets (*jāla*), *kaṇḍarā*-s, bunches (*kūrcā*), tendons of flesh (*māṃsarajju*) and sutures (*sīvanī*) with the knife.

876 The Ānandāśrama edition (1896) reads *sevanyaḥ* instead of *sīvanyaḥ*; this is not noted by the Adyar edition.

877 PARALLELS

SU śārīra., 5,16: *caturdaśāsthnāṃ saṅghātāḥ.* 5,17: *ye hy uktā asthi-saṅghātās te khalv aṣṭādaśaikeṣām.*

AS śārīra., 5,88.

AH 3,15c: *caturdaśāsthisaṅghātāḥ.*

ANALYSIS

The group of bones (*asthi-rāśi*) is called *asthnāṃ saṅghāta* or *asthi-saṅghāta* in the SU, AS and AH. The SU mentions two different amounts, fourteen and eighteen, for the groups of bones, as does the SR. The AS (śārīra., 5,89) and AH (śārīra., 3,15d) contain abbreviated versions of the statement of the SU, but the two texts seem to misunderstand the original meaning, attributing the

eighteen. The number of bones in the body is a triad of hundred to which sixty is added (= 360[878]). (śl.90)

---

amount eighteen only to the *sīmanta*-s which are components attached to the groups of bones (*asthi-saṅghāta*). Neither the CA, YS, AgniP nor ViṣṇudhP mentions the groups of bones.

878 PARALLELS (and the passages dealing with the same topic)
SU śārīra., 5,18: *trīṇi saṣaṣṭāny asthi-śatāni veda-vādino bhāṣante, śalya-tantreṣu tu trīṇy eva śatāni* etc: "The propounders of (Āyur)veda speak of 360 bones of the body; but among the followers of the doctrine of surgical medicine (*śalyatantra*), there are only 300 bones. Of them, 120 are in the limbs, 117 in the pelvic region, the sides, the vertebral column, and the chest, and sixty-three above the nape of the neck (i.e. above the clavicles). Thus the total number of bones is 300." (Translated by ZYSK 1986, p.700.)
AS śārīra., 5,90: *trīṇi ṣaṣṭyā'dhikāny asthi-śatāni.*
AH śārīra., 3,16abc.
YS 3,84: *'sthnāṃ ca sa-ṣaṣṭi ca śata-trayam.* The following verses (śl.85–90) contain the list of the places of the bones and their respective amounts, cf. YAMASHITA 2001/2002, pp.105–106.
AgniP 370,27d: *asthi-ṣaṣṭi-śata-trayam.* Its list of the bones (śl.28–34c) is almost identical to that of the YS (3,85–90c).
ViṣṇudhP 115, 77cd: *asthnām atra śatāni syus trīṇi ṣaṣṭy-adhikāni tu.* Its list of the bones (śl.78–83ab) is almost identical to that of the YS (3,85–90c).
GaruḍaP (KIRFEL 1954) v.50: *asthnāṃ hi dvy-adhikam proktaṃ ṣaṣṭy-adhika-śatatrayaṃ.* It states that the bones are 362.
Hastyāyurveda 3,9, v.58 (p.441) states that there are 320 bones, but 366 bone joints (*sandhi*).
ANALYSIS
For the bones, the SR mentions two different numbers, 360 and 300. SR śl.92ab states that the number 300 is according to the school of Dhanvantari. The SU, too, mentions these two numbers, but this text ascribes 360 to the "knowers of the [Āyur-]veda" (*vedavidāṃ*), while it ascribes 300 to the science of surgery (*śalyatantre*). In the following passages listing the names of the bones, the SU adopts the number 300. The AH's statement is similar to that of the SR in considering the number 300 to be according to Dhanvantari's school. The AS mentions only the number 360.
Only the SU adopts the number 300 in enumerating the respective places and numbers of the bones (SU śārīra., 5,17–19). The AS (śārīra., 5,90–94) enumerates 360 in a manner different from the SU. The AH does not list them.
The CA (śārīra., 7,6) presents a system of enumeration of the bones, totally different from that of the SU. This system has similarities to that of the YS.
The statements of the AgniP and ViṣṇudhP seem to be directly borrowed from YS 3,85–90c. The AgniP and ViṣṇudhP very often contain the same deviations from the YS, e.g. *sūkṣmaiḥ* (AgniP śl.28a; ViṣṇudhP śl.78a) instead of *sthālaiḥ* (YS, 3,85a), though sometimes, these two Purāṇa-s are different from each

The sages call those bones fivefold: "*valaya*-s[879] ('circle'), *kapāla*-s[880] ('skull-[bones]'), *rucaka*-s ('teeth'), and *taruṇa*-s[881] ('young' or 'soft'), *nalaka*-s ('long bone'[882])".[883] (śl.91)[884]

---

other; for example, only the ViṣṇudhP (śl.83b) contains *puruṣasyāsthi-saṅgrahaḥ* originated from YS 3,90d.

The SR mentions both the amounts 360 and 300. But the SR's list of the bones (SR śl.91) is identical to that in SU śārīra., 5,27, i.e., it accords with the school of Dhanvantari.

HOERNLE 1907, p.104ff., informs us of the statements on the bones in the ŚB; the number 360 of the bones is associated with that of the bricks in an altar (ŚB 10,5,4,12). HOERNLE (§43, p.112) further discusses the data on the bones from AV 10,2.

For the information on the bones in the AV, cf. AV 9,8, v.15 (ribs); v.20 (*kapāla* of the head); 10,2, v.3.; v.8 (*kapāla*, and the gathering of bones). AV 10,2,26 relates that Atharvan sewed together the head and heart of man. This might perhaps deal with the origin of the sutures in the skull and heart.

879 Hastyāyurveda 3,9,v.7cd (p.440): *valayākṛti-kalpāni saptaṣaṣṭiś ca sandhayaḥ.*

880 Hastyāyurveda 3,9,v.43b (p.440), skull-bones; 3,9,v.53 (441): *kapālāsthy ekam evāhuḥ sarvato jaghanāśritam.*

881 Hastyāyurveda 3,9, v.47ab (p.440): *gala-nāḍyāṃ catuḥṣaṣṭis taruṇāsthīni dantinaḥ.*

882 Its name may derive from it being thin and long like a reed (*nala*).

883 These names are mentioned in SU śārīra., 5,20. The SU first lists the names of the bones according to their positions in the body. The SU next presents another manner of classification, in which the bones are divided into five groups according to their shapes (cf. HOERNLE 1907, pp.76–81, §30). This manner of classification according to the shape is not contained in the CA, AS, AH, YS or the Purāṇa-s. ZYSK 1986, p.701, translates SU śārīra., 5,20 as follows. "Groups and types of bones. These bones are five-fold: *kapāla* (shell-like bones; flat bones), *rucaka* (teeth), *taruṇa* (soft bones; cartilages) *valaya* (round bones), *nalaka* (long bones). Of them, the bones in the knees, buttocks, shoulders, cheeks, palate, temples, and skull are flat bones; the bones of the teeth are teeth; the bones of in the nose, the ears, the neck, and the eyeballs [are] soft] bones (cartilages); [the bones] in the sides, the vertebral column, and the chest [are] round bones; the remaining [bones] are called long bones."

Among the names of the bones mentioned by CA śārīra., 7,11, only *nalaka* (line 15–16a, *ūru-nalaka, bāhu-nalaka* "hollow bones of the thighs and arms" according to HOERNLE p.25) and *kapāla* (line 30. *śiraḥ-kapāla*) are mentioned by the SU's list, too. These two names are also found in the Bheḷasaṃhitā śārīra., adhyāya 7. HOERNLE (ibid., p.38f.) reports that the Bheḷasaṃhitā's information on the bones resembles that of the CA. The YS and its two descendants (AgniP and ViṣṇudhP) mention only *kapāla*. The Hastyāyurveda at

In this regard[885], Dhanvantari mentioned only (*eva*) three hundred bones.

There are[886] here, however, two hundred bone-junctures, augmented by ten (= 210)[887]: (śl.92)[888]

[They] are[889] *koraka*-s ("hole", i.e. "bud-shaped"), *pratara*-s ("crossing", "a raft")[890], *tunnasīvanī*-s[891] (*tunna*- = "hit" or "broken",

---

least mentions *valaya, kapāla* and *taruṇa* (cf. my footnotes 879, 880 and 881 on these terms).

884 SR śl.90cd–91 is identical to ŚG 9,46abcdef. But the ŚG contains different readings: *śarīra-saṅkhyā* instead of *śarīre saṅkhyā* ; *jalajāni, taraṇāni* and *navakāni,* instead of *jalajāni, taruṇāni* and *narakāni.*

885 D reads *śatānīti* instead of *śatāny atra.*

886 Optative. On the translation of Sanskrit optative, cf. Introduction to the *English translation.*

887 PARALLELS

SU śārīra., 5,26: *[sandhayas[...] /24/] saṅkhyātas tu daśottare dve śate.*

AS śārīra., 5,95.

AH śārīra, 3,16d–17ab: *sandhīnāṃ ca śata-dvayam /16d/ daśottaraṃ sahasre dve nijagādātri-nandanaḥ /17ab/.*

CA śārīra., 7,17: *[...] dve-sandhi-śate [...].*

AgniP 370,33d: *sandhīnāṃ dve śate daśa /33d/.*

ViṣṇudhP 2,114,83cd: *tathā caivātra sandhīnāṃ dve śate tu śatādhike /83cd/* (The reading *śatādhike* is obviously a mistake for *daśādhike*).

ANALYSIS

The SU, AS, AH, AgniP and ViṣṇudhP mention 210 junctures (*sandhi*) of bones. The statements of the AgniP and ViṣṇudhP are parallel to each other.

The AH is a unique exception, mentioning two different numbers, 210 and 2000. The AH attributes the number 2000 to the school of Atri (*sahasre dve nijagādātri-nandanaḥ*). But 2000 seems to be a mistake for 200, because the CA, which also belongs to Atri's school, mentions 200 junctures (*sandhi*) (cf. CA śārīra., 7,14).

Besides, the SU śārīra., 5,28 states that there are junctures (*sandhi*) not only of the bones, but also of the *peśī*-s, cords (*snāyu*) and vessels (*sirā*); their amounts are uncountable.

The YS does not mention the junctures (*sandhi*) at all.

888 For *tvasthisandhīnāṃ*, the Adyar edition notes an alleged variant of the Ānandāśrama edition (1896), *nāmabhedena.* But peculiarly the Ānandāśrama edition actually does not contain it.

889 Optative *syuḥ.*

890 *Gha.* reads *pracurāḥ* "large" instead of *pratarāḥ*. But this seems to be wrong, because the parallel in SU śārīra., 5,27 contains *pratara.* ZYSK 1986 reports that the commentary on SU śārīra., 5,27 explains it as *bhelaka* "raft-shaped", perhaps referring to the cross joints of a raft.

sīvanī = "seam/sewing"), *ulūkhala*-s ("mortar")[892], *sāmudga*-s[893] (a joint with cup-like socket[894]; shoulder- and hip-joint), *maṇḍala*-s[895] ("circular"), *śaṅkhāvarta*-s[896] (shaped like the involutions of a conch shell/"whorl of the conch"), *vāyasatuṇḍaka*-s[897] ("beak of a crow"). (śl.93)

Thus (*iti*), the junctures of bones are indicated [to be] eightfold by the chiefs of sages.[898] (śl.94ab)

---

891 The Ānandāśrama edition (1896) reads *sevanāḥ* instead of *sīvanyaḥ*.
Cf. Hastyāyurveda 3,9, v159 (p.449): *tūnasīvanavat sandhir vijñeyas tu kapālataḥ hanu-sṛkka-kapāleṣu [...].*

892 ZYSK 1986 cites the commentary on SU śārīra., 5,26: "the mortar proper for separating the chaff."
Cf. Hastyāyurveda 3,9, v157 (p.449): *ulūkhalākhyaḥ sandhis tu kalā-bhāge ca dantinaḥ / vitāne śravaṇe caiva bimboḥ kumbhāntare tathā /157/.*

893 Cf. Hastyāyurveda 3,9, v156 (p.448): *samudga-sandhiḥ sarveṣu yathāvad iti niścayaḥ / lāṅgūla-vaṃśe lāṅgūle jaghana-try-asthi-deśayoḥ /156/.*

894 The *sāmudga* is derived from *samudga* "a box or casket which has its cover". ZYSK 1986, p702, cites the commentary on SU śārīra., 5,27, "shaped like a hemispherical bowl".

895 Cf. Hastyāyurveda 3,9, v160cd–161 (p.449): *netre vartmany apāṅge ca karīṣāśrāva (= srāva) eva ca /160/ hṛdaye cākṣikūṭe ca kaṇṭhe klomni gude tathā / jñeyo maṇḍala-sandhis tu bhāgeṣv eteṣu dantinaḥ /161/.*

896 Cf. Hastyāyurveda 3,9, v162 (p.449): *śambuke pratimāne ca vāhitre danta-veṣṭayoḥ / śaṅkhāvarte bhavet sandhiḥ śṛṅgāṭaḥ śrotasi sthitaḥ.* The second half of this verse is obviously corrupted. Compare it with the parallel, SU śārīra., 5,27 (*śrotra-śṛṅgāṭakeṣu śaṅkhāvartāḥ*).

897 Cf. Hastyāyurveda 3,9, v159cd–160ab (p.449): *hanu-sṛkka-kapāleṣu sugadeṣv eva dantinaḥ /159/ sandhir vāyasatuṇḍaḥ syād bhāgeṣv eteṣu dantinaḥ.*

898 SR śl.92cd–94ab is identical to ŚG 9,47–48. But the ŚG contains *syātāṃ tatra* instead of *syātām atra* (SR śl.92d); *rauravāḥ prasarāḥ skanda-secanāḥ* instead of *korakāḥ pratarās tunnāḥ sīvanyaḥ* (SR śl.93ab); *samudgā maṇḍakāḥ* instead of *sāmudgā maṇḍalāḥ* (SR śl.93c); *samuddiṣṭāḥ śarīreṣv* instead of *samuddiṣṭā munīndrair* (SR śl.94ab).
The list of the bone-junctures accords with that in SU śārīra., 5,27. I quote ZYSK's translation of the SU, "These joints [are] eight-fold: *kora* (movable), *ulūkhala* (mortal-shaped), *sāmudga* (cup-shaped), *pratara* (grossed?), *tunnasevanī* (sutures), *vāyasatuṇḍa* (crow-beak-shaped), *maṇḍala* (circular), and *śaṅkhāvarta* (shaped like the involutions of a conch shell). Of them, the movable joints [are] in the fingers, the wrists, the ankles, the knees and the elbows; the mortar-shaped joints [are] in the armpits, the groins and the roots of the teeth; the cup-shaped joints [are] in the seats of the shoulders (glenoid cavity), the anux (coccyx), the vulva (pubis) and the hips; the crossed (?) joints [are] in the nape of the neck and vertebral column; the sutures [are] in the flat

There is[899] deemed [to be] a pair of thousands (= 2000) of the junctures of *peśī*-s[900] (masses_of_flesh/muscles), cords (*snāyu*) and vessels (*sirā*).[901] (śl.94cd)

There are[902] nine hundreds of cords (*snāyu*)[903]: cords (*snāyu*) are

---

bones of the hips and the head; but the crow-beak-shaped joints [are] in the tubes of the throat, the heart, the eyes and the *kloman*; the joints shaped like the involutions of a conch-shell [are] in the ears and the sinuses." (ZYSK 1986, p.702).

A parallel is found in AS śārīra., 5,99, too.

The AH, CA, YS, AgniP and ViṣṇudhP do not contain this list. But the term *danta-ulūkhala* is mentioned in CA śārīra., 7,6; this might perhaps accord with the SU's statement that *ulūkhala*-s are situated in the roots of the teeth etc.

The Hastyāyurveda mentions *ulūkhala, samudga, maṇḍala, śaṅkhāvarta* and *vāyasa-tuṇḍa*. In the description of the shapes of the bones and bone-junctures, the Hastyāyurveda accords with the SU.

899 Optative.

900 On the term *peśī*, cf. DAS 2003A, pp.562–563.

901 PARALLELS

SU śārīra., 5,28cd: *peśī-snāyu-sirāṇāṃ tu sandhi-saṅkhyā na vidyate.*

AS śārīra., 5,100: *snāyu-peśī-sirāśritais tu saha sahasra-dvayam.* It is explained that these kinds of junctures are not mentioned individually, because surgical science is not concerned with them (*na punaḥ śastra-praṇidhāne teṣu vācyam iti*).

AH śārīra., 3,16d–17ab: *sandhīnāṃ ca śata-dvayam /16d/ daśottaraṃ sahasre dve nijagādātri-nandanaḥ /17ab/.*

ANALYSIS

The SR mentions the number 2000 for the junctures of *peśī* etc., while the SU considers them to be uncountable. The AS mentions 2000.

The AH mentions two different numbers, 210 and 2000, for the junctures (*sandhi*). These might, however, be the amount only for the bone-junctures. The AH accredits the number 2000 to the school of Ātreya, but the CA does not mention this type of junctures.

The YS, AgniP and ViṣṇudhP do not mention the junctures (*sandhi*) of cords (*snāyu*) etc.

902 Optative.

903 SU śārīra., 5,29: *nava snāyu-śatāni.*

AS śārīra., 5,101 makes an expression very similar to that of the SU.

AH śārīra., 3,17a: *snāvā nava-śatī.*

CA śārīra., 7,14: *nava snāyu-śatāni.*

YS 3,101: *ekonatṛṃśal-lakṣāṇi tathā nava śatāni ca / ṣaṭ pañcāśac ca jānīta śirā dhamani-sañjñitāḥ.*

AgniP 270,34d: *snāyor nava-śatāni ca.*

deemed [to be] fourfold[904]: *Pratānavatī*-s ("having tendrils"), *suṣirā*-s ("hollow" or "perforated"), *kaṇḍarā*-s[905], further *pṛthulā*-s ("wide"). (śl.95)

As a ship bound with many bindings becomes (opt. *bhavet*) able to bear much burden in water, so is the body bound with hundreds of cords (*snāyu*)[906] [able to bear much burden]. (śl.96)[907]

VisṇudhP 2,115,85: *[antarādhau tv (= try-) aśītiś ca kathita, bhṛgunandana /84cd/] na ca (= nava-) snāyu-śatāny atra dve tu triṃśādhike mate / antarādhau tu kathitā hy ūrdhvagāś caiva saptatiḥ /85/.*

ANALYSIS

All the texts mention 900 cords (*snāyu*). The statements of the YS, AgniP and VisṇudhP are closely related to each other.

904 PARALLELS

SU śārīra., 5,30: *snāyuś caturdhā vidyāt [...] / pratānavatyo vṛttāś ca pṛthvyaś ca śuṣirās tathā.* (5,31 mentions *kaṇḍarā* as another name for *vṛttā*.)

AS śārīra., 5,105.

ANALYSIS:

The SU mentions four kinds of cords (*snāyu*): *pratanavatī, vṛttā, pṛthvī* and *śuṣirā*. But it also mentions *kaṇḍarā* as another name of *vṛttā*. The AS does the same. The AH, CA, YS, AgniP and VisṇudhP do not mention them.

905 This kind of cord (*snāyu*) is called *vṛtta* ("round" or "globular") in SU śārīra., 5,30. But, at the same time, SU śārīra., 5,31 mentions another name, *kaṇḍarā*, too. It states, *vṛttās tu kaṇḍarāḥ sarvā vijñeyāḥ kuśalair iha* "the round ligaments (*vṛtta*) are here by the specialists (i.e., those skilled in *śalya-tantra*) to be known as all of the [sixteen] cords (of the body)" (translated by ZYSK).

According to this statement, it is suggested that Śārṅgadeva's ancestors might have been physicians specialised in surgery (*śalya-tantra*).

The Hastyāyurveda 3,9, v.94ab contains a term, *kaṇḍurā*, which is an adjective describing the shape of a cord (*snāyu*) (*snāyur valkāvanaddhā ca ghanā pṛthuvī ca kaṇḍurā*).

906 Śl.96d: *snayu-śata-baddha-tanus.* Instead, the Ānandāśrama edition (1896) reads *snāyu-śata-baddhā tanus.* This is not noted by the Adyar edition.

According to SU śārīra., 5,33–34, the comparison of the human body to a ship illustrates the importance of the cords (*snāyu*). 5,35–36 states that the damage of the cords is more serious to the body-structure, than that of the bones, muscles, vessels, or joints; the knowledge of the cords is necessary to remove a hidden dart from the human body.

907 PARALLELS

SU śārīra., 5,33: *naur yathā phalakāstīrṇā bandhanair bahubhir yutā / bhāra-kṣamā bhaved apsu nṛyuktā susamāhitā // evam eva śarīre 'smin yāvantaḥ sandhayaḥ smṛtāḥ / snāyubhir bahubhir baddhās tena bhārta-sahā narāḥ //.*

AS śārīra., 5,105: *snāva-bandhanair hi sandhiṣu subaddha tanur naur iva samyag īhate.*

The learned ones mention five hundred of *peśī*-s (masses_of_flesh/muscles) situated in the body.[908]

There are twenty extra of women[909] (= Women have twenty

ANALYSIS

The AS and SR contain the term *tanu* "body", but the SU contains *nara* "human being". The AH, CA and the other texts do not contain a parallel.

The simile of the human body = a ship is also found in the Caryāgīti, the song no.13 (*ti-śaraṇa ṇābī*) cf. SEN 1977 and CARYĀGĪTIKOŚA (BAGCHI 1956), pp.45–46. This simile occurs in the song no.49, too.

DAS 1992, p.421, describes a similar simile in the songs of the Bauls, though the body is here not directly compared to a ship: "Ar'kum, describing the upward journey of the cosmic principle to the head as a laden boat's struggle against the current, […] Now usually certain qualities, properties, or agents are associated with the boat in such images in the songs – plexus or the like usually being associated with the river or land along or through which the boat plies."

908 PARARELLS

SU śārīra., 5,37: *pañca peśī-śatāni bhavanti.*

AS śārīra., 5,108.

AH śārīra., 3,17.

CA śārīra., 7,14 mentions 400 *peśī*-s.

YS 3,100d: *pañca peśī-śatāni ca.*

AgniP 270,36a: *pañca peśī-śatāny.*

ViṣṇudhP 2,115,86c: *pañca-peśī-śatāny atra catvāriṃśat tathordhvagāḥ /86cd/.*

ANALYSIS

All the texts mention the 500 *peśī*-s, except for the CA which mentions the number 400.

909 PARARLLELS

SU śārīra., 5,39: *strīṇāṃ tu viṃśaty adhikā / daśa tāsāṃ stanayor ekaikasmin pañca pañceti, yauvane tāsām parivṛddhiḥ, apatya-pathe catasraḥ tāsām prasṛte 'bhyantarato dve mukhāśrite bāhye ca vṛte dve garbhacchidrasaṃśritās tisraḥ śukrārtava-praveśinyas tisraḥ eva / pitta-pakvāśayor madhye garbhāśayā yatra garbhas tiṣṭhati /.*

AS śārīra., 5,112: *strīṇāṃ tu viṃśatir adhikā / tatra daśa stanayoḥ tāsāṃ yauvane parivṛddhir bhavati /112/ tāsām abhyantarāśrite dve / mukhāśrite vṛtte dve / tisro garbha-mārgāśriyāḥ / yasyāṃ garbhas tiṣṭhati.*

AH śārīra., 3,17d: *pañca puṃsām peśī-śatāni tu /17d/ adhikā viṃśatiḥ strīṇām yonī-stana-samāśritāḥ /18ab/.*

AgniP 270,36: *pañca peśī-śatāny eva catvāriṃśat tathordhva-gāḥ / catuḥ-śataṃ tu śākhāsu antarādhau ca ṣaṣṭikāḥ /36/ strīṇāṃ caikādhikā vai syād viṃśatiś catur-uttarā / stanayor daśa yonau ca trayodaśa tathāśaye /37/ garbhasya ca catasraḥ syuḥ […] /38a/.*

ViṣṇudhP 2,115,86cd: *pañca-peśī-śatāny atra catvāriṃśat tathordhvagāḥ /86cd/ catvāriṃśac chatāny atra tathā śākhāsu paṇḍitaiḥ / antarādhau tathā ṣaṣṭiḥ*

extra); in that connection (= among these), ten are[910] of/at the two breasts. (śl.97)

They grow full in youth. And, in that connection (= among these), there are, however[911], ten in the genital tract (*yoni*)[912]. Two are extended inside, two external, three going to the passage of the embryo (*garbha-mārga*). (śl.98)

The genital tract is that whose form is [like] the navel[-like cavity] of the conch, [and] that which consists of three whorls. Here, in the third whorl, is the bed of the embryo (= womb), between the receptacles (*āśaya*)[913] of bile and ripe (= digested) [food][914], resembling the fish called *rohita* ("red")[915]. (śl.99)[916]

---

peśyas tu kathitā budhaiḥ /87/ strīṇāṃ caivādhikā jñeyā triṃśatiś catur-uttarā / stanayor daśa vijñeyā yonau, rāma, tathā daśa /88/ garbhāśaye tathā jñeyāś catasro garbha-cintakaiḥ /89ab/.

Hastyāyurveda 3,9, v.87–88 (p.451): *kareṇvām adhikās triṃśat tāsāṃ vakṣyāmi lakṣaṇam / stanayor daśa vijñeyā daśa vidyād bhagāśritāḥ /87/ dve vṛtte mukham āśritya prasṛte 'tha bhagottare / garbhacchidrāntare tisras tisro garbhopajanmani /88/.*

ANALYSIS

The SR accords with the SU.

The AS presents the same theory as the SU, and makes expressions similar to the SU. Still, it contains the term *garbha-mārga* which occurs also in the SR, but not in the SU. This might suggest that the AS and SR are based on another version of the SU than that of today.

The AH deals with this theory very briefly. The CA and YS mention nothing about this topic. The AgniP and ViṣṇudhP, which are parallel to each other, present a theory similar to that of the SU, but the numbers given there are sometimes different from those in the SU. The reading *catasro* (AgniP 38a) might be a mistake for *ca tisro.*

910 Optative.

911 D reads *ca* "and" instead of *tu*.

912 The translation of *yoni* as "vagina" seems too narrow because of SR śl.99, cf. DAS 2003A, pp.572–574.

913 For *āśaya*, cf. JOLLY 1901, p.43.

914 The expression *pitta-pakvāśayāntare* occurs in SR śl.82d, too.

915 SHRINGY (1999, vol.I, p.79) proposes reading *matsyāsya* "of the mouth of the fish" instead of *matsyasya* "of the fish", according to the parallel in the SU which states that the uterus resembles the mouth of the *rohita* fish.
The *rohita* fish is mentioned in SU sūtra., 46, too, cf. DAS 2003A, p.572.

916 PARALLELS

In that, [there] are three *peśikā*-s[917] ("small masses of flesh"), (śl.100ab)
deemed [to be] coverings[918], characterised by the entrance of semen

SU śārīra., 5,43–44: *śaṅkha-nābhy-ākṛtir yonis tryāvartā sā prakīrtitāḥ / tasyās tṛtīye tv āvarte garbha-śayyā pratiṣṭhitā /43/ yathā rohita-matsyasya mukham bhavati rūpataḥ / tat-saṃsthānāṃ tathā-rūpāṃ vidur budhāḥ /44/.*
SU śārīra., 5,43, "The female genital tract (*yoni*) has the form of the navel of a conch-shell; it is proclaimed as having three turns (folds). The uterus is indeed in its third turn (fold)." (Translated by ZYSK 1986.)
SU śārīra., 5,39: *śukrārtava-praveśinyas tisra eva.*
AS śārīra., 5,113: *śaṅkha-nābhy-ākṛtis try-āvartā / tasyās tṛtīya āvarte pitta-pakvāśayāntare rohita-matsya-mukhākārā garbha-śayyā tasyāṃ śukrārtava-praveśinyas tisraḥ peśyaḥ /113/.*
ANALYSIS
The SR's theory accords with that of the SU and AS. The AS contains the expression *pitta-pakvāśayāntare* which does not occur in the SU but in the SR. Besides this, the three *peśī*-s are mentioned by the AS, in the same order as in the SR. In the SU they are not mentioned in that order; the *peśī*-s in SU śārīra., 5,39, are mentioned separately from SU śārīra., 5,44. That means that these passages of the AS may perhaps be based on another version of the SU than the version which we know today. The AH, CA, YS, AgniP and ViṣṇudhP do not contain parallels.

917 According to the parallel, AS śārīra., 5,113, we should read plural *peśikāḥ* instead of singular *peśikā* in SR śl.100b. This statement accords with the classical medical texts, in considering the uterus to resemble the *rohita* fish, and mentioning the three *peśī*-s (cf. TIVĀRĪ 1986, part I, pp.21–22. SU śārīra., 5,39: *śukrārtava-praveśinyas tisra eva*).

918 The term *pracchādakā* is mentioned in SU śārīra., 5,40:(*śukrārtava-praveśinyas tisra eva [...] /39/*) *tasām bahala-pelava-sthulāṇu-pṛthu-vṛtta-hrasva-dīrgha-sthira-mṛdu-ślakṣṇa-karkaśa-bhāvāḥ sandhy-asthi-sirā-snāyu-pracchādakā yathā-praveśaṃ svabhāvata bhavanti /40/.* It is stated that the *peśī*-s take various forms, and that they are the coverings (*pracchādakā*) of the joints, bones, vessels (*sirā*) and cords (*snāyu*).
AS śārīra., 5,113–114 is parallel to SU śārīra., 5,39–40 and SR śl.100ab. I quote the AS's passages: *tasyās tṛtīya āvarte pitta-pakvāśayāntare rohita-matsya-mukhākārā garbha-śayyā tasyāṃ śukrārtava-praveśinyas tisraḥ peśyaḥ /113/ tābhir hi śarīre tanu-bahala-sthūlāṇu-pṛthu-vṛtta-hrasva-dīrgha-sthira-mṛdu-ślakṣṇa-karkaśābhiḥ sandhy-asthi-sirā-snāvāni pracchāditāni /114/.*
Comparing SR śl.100cd with its parallels in the SU and AS, I suspect that the SR made an omission from its source-text. In the SR it seems that a long passage from *śukrārtava-praveśinyas tisraḥ* to *pracchādikāḥ*, which is contained in the SU and AS, was omitted. If my suspicion that this omission was made is correct, the translation of SR śl.100cd, to which I have

and procreatory-menstrual fluid. (śl.100cd)

There are, however, twenty-nine[919] lakhs, together with half (*sārdha*) [a lakh] of vessels (*sirā*)[920] and ducts (*dhamanikā*), further, nine hundred joined with fifty-six (= 2,950,956). (śl.101)[921]

---

supplemented the, supposedly, missing passage, should be, "Three *peśi*-s are characterised by the entrances of the semen and the menstrual-procreatory fluid. [The *peśī*-s] are deemed [to be] the coverings [of the joints, bones, vessels (*sirā*) and cords (*snāyu*)]".

919 B reads *nava dviśate iti* instead of *navaviṃśatiḥ*. But this seems unsuitable, when it is compared with the parallels in the medical texts (see below), and it also violates the metre.

920 ROŞU 1978, p.206 (especially note 1) gives references for the various kinds of vessels like *srotas, sirā, dhamanī, nāḍī*, etc.

921 PARALLELS

The SU contains no parallel.

AS śārīra., 5,115: *sirā-dhamanī-mukhānāṃ tv aṇuśo vibhajyamānānām ekona-tṛṃśac-chata-sahasrāṇi nava ca śatāni ṣaṭ-pañcāśāni bhavanti.*

CA śārīra., 7,14: *ekona-tṛṃśat sahasrāṇi nava ca śatāni ṣaṭpañcāśatkāni sirā-dhamanīnām aṇuśaḥ pravibhajyamanānām mukhāgra-parimāṇaṃ.*

YS 3,101: *ekona-tṛṃśal-lakṣāṇi tathā nava śatāni ca / ṣaṭpañcāśad vijāniyāc chirā-dhamani-sañjñitāḥ.*

AgniP 270,38bcd: *[...] śirāṇāṃ ca śarīriṇām / triṃśac-chata-sahasrāṇi tathā 'nyāni navaiva tu /38/ ṣaṭpañcāśat-sahasrāṇi rasaṃ dehe vahanti tāḥ /39ab/.*

ViṣṇudhP 2,114,89cd: *triṃśac-chata-sahasrāṇi tathānyāni daśaiva tu /89cd/ ṣaṭpañcāśat-sahasrāṇi śirāṇāṃ kathitāni tu / tā vahanti rasaṃ dehe kedāram iva kulyakāḥ /.*

ANALYSIS

The AS and YS mention the amount 2,900,956! Therefore, *sārdhāni* in SR śl.101c could be a mistake. It might be a corrupted form of a word which could be *sārdham* "together with", or something similar. It could also be a mistake for *sarthaḥ* "flock, collection" or some word which sounds similar. If we read *sārthaḥ* instead of *sārdhāni*, the whole sentence would mean: "The flocks of vessels (*sirā*) and ducts (*dhamanī*) are 2,900,956". But unfortunately, all these deviate from the metre.

Though the CA's wording is very close to that of the AS, it mentions the number 29956. This might be a mistake caused by taking *sahasrāṇi* for *śata-sahasrāṇi*. JOLLY 1901, p.44 reports that the amount 3956000 or 2900956 of vessels (*sirā*) and ducts (*dhamanī*) is mentioned in the CA, YS and the text entitled Vaṅgasena, but he does not give any reference.

On this topic, the AgniP and ViṣṇudhP are parallel to each other and seem to be derivatives of the YS. Instead of *lakṣa* in the YS, they contain *śata-sahasra* which also occurs in the AS. The SR contains *lakṣa* like the YS.

There are ten root/main vessels (*mūla-sirā*)[922] carrying the vital fluid (*ojas*), whose resort/base is the heart.[923] (śl.102ab)[924]

Surprisingly no parallel is found in the passages following SU śārīra., 5,45. The seventh and ninth chapter (adhyāya) in the śārīra sthāna of the SU deal with the vessels (*sirā*) and ducts (*dhamanī*), but these two chapters do not seem to contain any parallels.
In summary, the AS is the most closely related to the SR.

922 The term *mūla-sirā* might mean "the root of a vessel".

923 The Ānandāśrama edition (1896) reads *hṛdayāśritāḥ* "[being] resorted to the heart" instead of *hṛdayāśrayāḥ*. This variant is not noted by the Adyar edition. According to the footnote of the Ānandāśrama edition, the reading *hṛdayāśrayāḥ* is contained in the manuscripts *gha.* and *ṅa.*

924 PARALLELS
SU śārīra., 7,3: *sapta-sirā-śatāni bhavanti, [...] druma-patra-sevanīnām iva tāsām pratānāḥ.*
AS śārīra., 6,2: *daśa mūla-sirā hṛdaya-pratibaddhāḥ sarvāṅga-pratyaṅgeṣv ojo nayanti tat-pratibaddhā hi śarīra-ceṣṭāḥ. 6,3: tās tu dvy-aṅgulam aṅgulam ardhāṅgulaṃ yavaṃ yavārdhaṃ ca gatvā druma-patra-sevanī-pratāna-vad bhidyamānāḥ sapta-śatāni bhavanti.*
AH śārīra., 3,18cd: *daśa mūla-sirā hṛt-sthās tāḥ sarvaṃ sarvato vapuḥ /18cd/ rasātmakaṃ vahanty ojas tan-nibaddhaṃ hi ceṣṭitam / sthula-mulāḥ su-sūkṣmāgrāḥ patra-rekhā-pratāna-vat /19/ bhidyante tās tataḥ sapta-śatāny āsām bhavanti tu /20ab/.*
ANALYSIS
The SR accords with the AS and AH. In general, the AS is closer to the SR, than the AH is. But some expressions of the AH, such as *ojas vah-, bhidyante* etc., better resemble those of the SR. The AS and AH might be based on another version of the SU than the one we have today.
SU śārīra., 7,3–5 mentions 700 vessels (*sirā*) and compares their bifurcation to the veins of a leaf. But according to SU, the vessels (*sirā*) are not rooted in the heart, but in the navel (SU śārīra., 5,3: *tāsāṃ nābhir mūlaṃ*), unlike the SR's statement. The navel is compared to a wheel (*cakra*) in SU śārīra., 7,5. A parallel to SU śārīra., 7,3–5, is found in AS śārīra., 5,32–33. But this passage in the AS does not deal with the vessels (*sirā*), but the ducts (*dhamanī*) (AS śārīra., 5,32–33: *tābhiś ca nābhiś cakra-nābhir ārakair ivāvṛtā /32/ tasyāṃ nābhyāṃ viśeṣeṇa prāṇā vyavasthitāḥ / yataś ca sāntaragny-adhiṣṭhānam /33/*). According to the AS, the ducts (*dhamanī*) are rooted in the navel, in contrast to the vessels (*sirā*) which are rooted in the heart. The SR's statement accords with that of the AS (cf. SR śl.106).
SU śārīra., 7,6, mentions forty main vessels (*mūla-sirā*); ten are allotted to wind, bile, phlegm and blood, each. But in this passage, the SU does not mention the ones carrying *ojas*, while the SR, AS and AH do.
The CA, YS, AgniP and ViṣṇudhP contain no parallels.

And when, having gone two *aṅgula*-s[925], half an *aṅgula*, a *yava*, further half a *yava*, they are spread[926] like the veins (*sīvanī*)[927] of the leaf of a tree[928],
then they get split into seven hundred by (= in) total. (śl.102cd, 103)
Among them (= vessels/*sirā*-s), two each (*dve dve*) are situated in the tongue, [being] the cause of speech and the knowledge of taste (*rasa*) [929] [respectively],

925 *Aṅgula* "finger's breadth". A *yava* (literally "barley") is one sixth or one eighth of an *aṅgula*.

926 *Pratata* "spread". The term *pra√tan* may denote the bifurcation of plants, cf. *pratāna* "tendril". SU śārīra., 7,3 contains a simile which resembles that of SR śl.103ab, (<u>*druma-patra-sevanīnām*</u> *iva tāsām (= sirānām)* <u>*pratānāḥ*</u>).
The comparison of the human body to a tree is found in the following texts.
Caryāgīti: *kā'ā tarubara pañca biḍāla*, "The body is a beautiful tree; its branches are five" (Song no. 1 in SEN, NILRATAN 1977; Caryāgītikoṣa, ed. by BAGCHI 1956, p.1).
Bṛhadāraṇyakopaniṣad 3,9,28 and followings: *yathā vṛkṣo vanaspatiḥ, tathaiva puruṣo 'mṛṣā / tasya lomāni parvāṇi, tvag asyotpāṭikā bahiḥ //9,28,1// tvaca evāsya rudhiram prasyandi, tvaca utpalaḥ / tasmāt, tad ātṛṇṇāt praiti, raso vṛkṣād ivāhatat //2// māṃsāny asya śakarāṇi, kināṭaṃ snāva, tat sthiram / asthīny antarato dārūṇi, majjā majjopamā kṛtā //3// yad vṛkṣo vṛkṇo rohati mūlān navataraḥ punaḥ / martyaḥ svin mṛtyunā vṛkṇaḥ kasmān mūlāt prarohati //4// retasa iti mā vocata, jīvatas tat prajāyate / dhānāruha iva vai vṛkṣo 'ñjasā pretya-sambhavaḥ //5//.*
The body components like hair, skin, flesh, cords/ligaments (*snāva*), bones, marrow and semen are compared to the components of a tree. The chain of these components, hair etc., is the prototype of the metabolic chain of the seven elements (*dhātu*) in the classical medical theory, cf. JAMISON 1986. Also cf. SR śl.77b which compares the *kalā*-s to *kāṣṭha-sāra* "the pith of wood".
SU śārīra., 5,21 compares the bony cores to the piths of wood. "As plants stand erect by [their] inner piths; so also the bodies of embodied creatures are held fixed by the bony cores: it is the firm (or constant part)" (translated by ZYSK 1986, p.701).

927 The Ānandāśrama edition (1896) reads *sevanyaḥ* instead of *sīvanyaḥ*.

928 "As is the leaf of the Ashvattha [...] tree (*Ficus religiosa*) there are minute fibres, so is the body permeated by Nāḍīs [...]." (Śāṇḍilyopaniṣad chapter I, as quoted by AVALON 1924, p.112.)
The SU also makes the same allusion for the vessels stretching from the navel. SU śārīra. 7,3: <u>*druma-patra-sevanīnām*</u> *iva tāsām (= sirānām) pratānāḥ, tāsāṃ nābhir mūlaṃ, tataś ca prasaranty ūrdhvam adhas tiryak ca*. (Also see my footnote 1116 on SR śl.150.)

929 SU śārīra., 7,22 mentions two vessels (*sirā*) carrying taste (*rasa-vahā*) and two carrying speech (*vāg-vahā*) under the tongue; besides, there are two carrying

two each carrying smell, in the nose, two causing winking (*meṣa*) and blinking (*unmeṣa*), in/of the eyes. (śl.104)

Among them (= vessels/*sirā*-s), two carrying[930] sound (*śabda*), in the ears, are mentioned by Śārṅgin[931]. (śl.105ab)[932]

---

sound (*śabda-vāhinī*) in the ears. In its parallel in AS śārīra., 6,19, the vessel (*sirā*) carrying taste is called *rasa-vedinī*, and the one carrying speech, *vāk-pravartinī*.

930 The Ānandāśrama edition (1896) reads *śabda-grāhiṇyau* "seizing/receiving sound" instead of *śabda-vāhinyau*.
D reads *grāhiṇyau śrotrayoḥ śabdaṃ*.

931 SHRINGY 1999, p.79 translates "Dhanvantari". He identifies *śārṅgin* which literally means "someone who has a horn/bow (*śārṅga*)" or "archer" with Dhanvantari. He seems to associate the term *dhanvantari* with the words, *dhanu, dhanva* and *dhanvan* "bow". However, the term *dhanvantari* in itself does not necessarily mean "archer". I doubt if SHRINGY's translation is right. Dhanvantari is a famous personality considered the god of healing, associated with the science of medicine. Suśruta identifies Dhanvantari with his teacher Divodāsa. Cf. MEULENBELD 2000 (Vol.IIA), pp.358–361. MEULENBELD does not mention *śārṅgin* as a further name of Dhanvantari.
The term *śārṅga* might be associated with another medical author, Śārṅgadhara, which also mean "a holder of a bow". For Śārṅgadhara, cf. MEULENBELD ibid., pp.205–206. MEULENBELD here mentions P.V. Sharma's hypothesis that Śārṅgadhara might be the same person as Śārṅgadeva, the author of the SR, but he himself doubts the validity of this hypothesis.
However, in his medical work, the Śārṅgadharasaṃhitā (see under *Original texts* in the bibliography), Śārṅgadhara does not deal with the two vessels carrying sound in the ears, although he mentions two openings (*randhra*) in the ears (1,5,40). The twenty-four ducts (*dhamanī*) carrying the nutrient fluid (*rasa*) are also mentioned in this work (1,5,39ab), but this information is not restricted to this author. Also the SU, AS and AH deal with this topic (see the parallels in footnote 934). Therefore, I must conclude that the support to identify Śārṅgin as this author is very weak.
Actually, the Śārṅgadharasaṃhitā contains many discrepancies with the SR. E.g., it begins the metabolic chain of the seven *dhātu*-s with the nutrient fluid (*rasa*), cf. 1,5,11; 1,6,11.
However, the Śārṅgadharasaṃhitā (1,5,4) adopts the appellation *kaṇḍarā* for a certain kind of cord/sinew (*snāyu*), the same term as SR śl.95. The SU (śārīra., 5,30) considers it to be a technical term of the chirurgical experts (cf. my footnote 905 on SR śl.95, *kaṇḍarā*).
Intriguingly, this work implies the breath control practice of Haṭhayoga in 1,5,48cd–50ab: *nābhi-sthaḥ prāṇa-pavanaḥ spṛṣṭvā hṛt-kamalāntaram /48cd/ kaṇṭhād bahir viniryāti pātuṃ viṣṇupadāmṛtam / pītvā cāmbara-pīyūṣam punar āyāti vegataḥ /49/ prīṇayan deham akhilaṃ jīvayañ jaṭharānalam /50ab/.*

Twenty-four ducts (*dhamanī*) carrying the juice_[of_food]/ nutrient_fluid (*rasa*)[933] are proclaimed. (śl.105cd)[934]

---

932 PARALLELS and ANALYSIS:
The vessels (*sirā*) carrying various substances are mentioned by SU śārīra., 8,22, AS śārīra., 6,19–22 and AH śārīra., 3,28–32ab. The theories of SU, AS and AH are in concordance. But the SU lacks some pieces of information which the AS, AH and SR contain. For example, the SU does not mention the function of the two vessels (*sirā*) in the nose which is "making [one] perceive smell" (*gandha-vedinī*), while the AS and AH do mention it. The SU, again, does not mention the function that causes "winking and blinking", while the AS mentions *unmeṣa-nimeṣa*, and the AH mentions *meṣonmeṣa* which is the same expression as in the SR.

933 This *rasa* does not mean taste, but *āhāra-rasa*, as mentioned by Ḍalhaṇa on SU śārīra., 9,8–9.

934 PARALLELS
SU śārīra., 9,3: *catur viṃśatir dhamanyo nābhi-prabhavā abhihitāḥ.*
SU śārīra., 7,3: *yābhir idaṃ śarīram ārāma iva jala-hāriṇībhiḥ kedāra iva ca kulyābhir upasnihyate 'nugṛhyate ca.*
SU śārīra., 7,4cd: *nābhyām sarvā nibaddhās tāḥ pratanvanti samantataḥ /4cd/ nābhi-sthāḥ prāṇinām prāṇāḥ prāṇān nābhir vyupāśritā / sirābhir vṛtā nābhiś cakra-nābhir ivārakaiḥ /5/.*
AS śārīra., 6,30: *dhamanyas tu caturviṃśatiḥ /30/ tābhiḥ kāyo 'yam ārāma iva jala-hāriṇībhiḥ kedāra iva kulyābhir upasnihyati /31/ tābhiś ca nābhiś cakra-nābhir ārakair ivāvṛtā /32/ tasyāṃ nābhyāṃ viśeṣeṇa prāṇā vyavasthitāḥ / yataś ca sāntaragny-adhiṣṭhānam /33/.*
AH śārīra., 3,39: *dhamanyo nābhi-sambaddhā viṃśatiś catur-uttarā tābhiḥ parivṛtā nābhiś cakra-nābhir ivārakaiḥ /39/.* The AH does not mention the simile of channels and fields (*kulya-kedāra*).
CA śārīra., 7,14, only mentions the amount of two hundred ducts (*dhamanī*), without explaining them separately.
The YS has no parallel.
AgniP 370,39: *[śirāḥ] rasaṃ dehe vahanti tāḥ / kedāra iva kulyāś ca kleda-lepādikaṃ ca yat /39/.*
ViṣṇudhP 115, 90cd: *[śirāḥ] tā vahanti rasaṃ dehe kedāram iva kulyakāḥ /90cd/ abhyaṅgādi tathā sarvaṃ sveda-lepādikaṃ ca yat /91ab/.* (*Sveda* seems to be a mistake for *kleda*.)
ANALYSIS
The SU, AS and AH also mention twenty-four ducts (*dhamanī*) originating in the navel, like the SR does. All the consulted texts, except for the AH, CA and YS, mention the simile of irrigated fields (*kulya-kedāra-nyāya*). The AH (śārīra., 1,56), however, mentions this smile (*kedāra-nyāya*) in a different context which is solely the nourishing of the foetus through the umbilical cord, cf. DAS 2003A, pp.270–271, note 923.

The body grows through them like irrigated fields (*kedāra*)[935] through channels.

These are fixed at the navel like the spokes at the nave of a wheel (*cakra*). (śl.106)

Ten are stretched out upward, ten downward, four obliquely. (śl.107ab)[936]

Those going upward, [having] reached the heart, are spread out three-fold[937] separately (= respectively).[938] (śl.107cd)[939]

Two each (*dve dve*) [of the thirty branches respectively] release wind, (*vāta*), bile (*pitta*), phlegm (*kapha*), blood [and] the nutrient fluid (*rasa*). (śl.108ab)[940]

As above, the SR, SU, AS and AH share the comparison of the navel to a wheel (*cakra*).

The expression "carrying *rasa*" (*rasaṃ* √*vah*), which is contained in SR śl.105c, occurs only in the AgniP and ViṣṇudhP.

935 The Ānandāśrama edition (1896) reads sg. *kedāras* instead of pl. *kedārās*. The parallels in the consulted texts always contain the sg. (see above). Also cf. DAS 2003A, pp.270–271, note 923.

936 PARALLELS

SU śārīra., 9,4: *tāsāṃ tu khalu nābhi-prabhavānāṃ dhamanīnām ūrdhvagā daśa, daśa cādho-gāminyaḥ, catasras tiryaggāḥ.*

AS śārīra., 6,34: *tāsāṃ khalu dhamanīnāṃ daśordhvam prasṛtā daśādhaś catasras tiryak.*

AH śārīra., 6,40ab: *tābhiś [= dhamanībhiḥ] cordhvam adhas tiryag deho 'yam anugṛhyate.*

ANALYSIS

The theories of the SU, AS and AH are the same as that of the SR. The CA, YS, AgniP and ViṣṇudhP do not have parallels.

937 I.e., ramify into thirty branches.

938 The Ānandāśrama edition (1896) reads *hṛdayaprāptāḥ* instead of *hṛdayam prāptāḥ*.

939 PARALLELS

SU śārīra., 9,5: *[…] tās tu hṛdayam abhiprapannās tridhā jāyante.*

AS śārīra., 6,35: *tāsām ūrdhvagā hṛdayam abhiprapannāḥ pratyekaṃ tridhā jāyante.*

ANALYSIS

Only the SU and AS deal with this theory. The AS contains *pratyekam*, an equivalent to the expression *pṛthak* in the SR.

940 PARALLELS

SU śārīra., 9,5: *tāsāṃ tu vāta-pitta-kapha-śoṇita-rasān dve dve vahatas tā daśa.*

AS śārīra., 6,35: *tatas triṃśato madhye dve dve vāta-pitta-kapha-rakta-rasān vahataḥ.*

In_that_regard/There (*tatra*), two each (*dve dve*) [respectively] perceive sound, shape[941], taste[942] [and] smell. (śl.108cd)[943]

And two each [respectively] make talking, shouting, sleep, waking [and] crying. (śl.109ab)[944]

Two[945] ooze semen in a man, but[946] milk[947] in a woman.

ANALYSIS
Only the SU and AS mention these. The AS shares the expression *rakta* with the SR, while the SU contains a synonym, *śoṇita.*

941 D reads *sparśaṃ* "touch" instead of *rūpaṃ*. This reading is irrelevant, because the sense of touch is perceived through the pores (*roma-kūpa*) of the skin (cf. my footnote 961 on SR śl.113 *rasa*).

942 These are the extensions of the ducts (*dhamanī*) connected with the heart (cf. SR śl.107). The connection of the heart with the sense organs situated in the head is more explicitly explained by the Hastyāyurveda (3,9, v.28: *sirā rasavahā dṛṣṭā jihvāyāṃ hṛdayāśritāḥ / daśaiva tāḥ susūkṣmās tu yābhir vedayate rasān*). There, the vessel (*sirā*) in the tongue which carries taste is said to be connected with the heart.
The vessels (*sirā*) carrying sensations (mentioned by SR śl.104–105a), originating in the sense organs which are situated in the head, are connected with the navel passing through the heart.

943 PARALLELS
SU śārīra., 9,5: *śabda-rūpa-rasa-gandhān aṣṭābhir gṛhṇīte.*
AS śārīra., 6,35: *aṣṭābhiḥ śabda-rūpa-rasa-gandhā gṛhyante.*
ANALYSIS Only the SU and AS mention them.

944 PARALLELS
SU śārīra., 9,5: *dvābhyām bhāṣate, dvābhyāṃ ghoṣam karoti, dvābhyāṃ svapiti, dvābhyām pratibudhyate, dve cāśru-vāhinyau.*
AS śārīra., 6,35: *dvābhyāṃ dvābhyām bhāṣate, ghoṣaṃ karoti, svapiti, pratibudhyate ca / dve cāśru vahataḥ.*
ANALYSIS
Only the SU and AS mention this topic. The vocabulary of the two texts is almost the same as that of SR, except for *pratibudh* and *aśru-vah-* in contrast to *budh* and *rud-* of the SR.

945 According to AV 6,138 v.4 (a curse aiming at the impotence of an adversary), the tubes carrying semen are dual: *yé te nā́ḍyau devákṛte yayós tíṣṭhati vṛ́ṣṇyam / té te bhinadmi śámyayā 'muṣyā́ ádhi muṣkáyoḥ /4/*, "Your two tubes made by god, in which there is virility, I shall break those two of yours over the testicles with a peg." WHITNEY 1987, vol.I, p.309, translates *amuṣyā adhi muṣkayoḥ* differently as "on yon woman's loins".

946 D reads *dve* "two" instead of *tu* "but".

947 The process of milk production in the breasts is mentioned in SU śārīra., 10,14: *dhamanīnāṃ hṛdisthānāṃ vivṛtatvād anantaram / catūrātrāt trirātrād vā strīṇāṃ stanyam pravartate.*

(śl.109cd)[948]

The ones gone (= going) downward too, are separately situated three-fold[949], in the receptacle of the ripened/digested [food] (*pakvāśaya*).

The first ten there (= among them) make wind (*vāta*) etc. proceed, as before[950]. (śl.110)[951]

Two ducts (*dhamanī*) carry eaten food from the site of water (*ambu-samāśraya*)[952],

948 PARALLELS
SU śārīra., 9,5: *dve stanyaṃ striyā vahataḥ stana-saṃśrite, te eva śukraṃ narasya stanābhyām abhivahataḥ*. Cf. DAS 2003A, pp.93–94.
AS śārīra., 6,35: *dve ca stanāśṛte nāryāḥ stanyaṃ narasya śukram.*
ANALYSIS
Only the SU and AS mention this theory. According to SU śārīra., 9,5, men's semen is carried from the chest/breasts (*narasya stana*). On the other hand, SU śārīra., 4,23 states that the semen originates in the whole body (*kṛtsna-dehāśritaṃ śukraṃ*).
For SR śl.108cd–109, compare it with Hastyāyurveda 3,9, v.117–121 (p.446).

949 I.e., giving thirty in all.

950 In the same manner as described in śl.105ab: two each carry *vāta, pitta, kapha, rakta* and *rasa*.

951 PARALLELS
SU śārīra., 9,7: *[…] āma-pakvāśayāntare ca tridhā jāyante.*
AS śārīra., 6,36: *adhogamāḥ pakvāśayasthā evaṃ tridhā jāyante / tatra daśādyāḥ pūrvavat.*
ANALYSIS
The SU and AS accord with the SR. The AS's wording is closer to that of the SR's, than the SU's is.

952 The Ānandāśrama edition (1896) reads *vahato 'tra samāśrayāt* instead of *vahato 'mbu-samāśrayāt*. "With regard to this (*atra*), two ducts carry eaten food from the site/receptacle (*samāśraya*)". In this case, *samāśraya* would denote the *pakvāśaya*.
Indeed, SU śārīra., 9,7 states that the two vessels carrying food are situated in the intestine (*antrāśrite*), i.e. *pakvāśaya*. The Adyar edition's reading *ambu-samāśraya* "the site of water" is not in accordance with this. If we compare it with the parallel in the SU, *antrasamāśrite*, we should perhaps suspect the Ānandāśrama's reading *vahato 'tra samāśrayāt* to be a corruption of *vahato 'ntrasamāśrayāt*. The fact that manuscript D reads *vahantontaḥ samāśrayāt* seems to support this assumption.
In contrast to the SU, the parallel in the AS (śārīra., 6,36) states that the two ducts carrying food are connected to the mother (*mātrāśrite*). That means, the AS deals with the way the foetus is nourished through the umbilical cord. With

two each [respectively] water (*toya*), urine [and] strength/power (= semen[953]) (*bala*);
these, however, [carry] the menstrual-procreatory fluid (*ārtava*) of (= in) women. (śl.111, 112a)[954]

Two discharge the streams (*srotas*)[955], two, joined to the large

---

this, it might be assumed that the SR's reading *ambu-samāśrayāt* was originally *ambā-samāśrayāt* "through resorting to the mother". We thus have various possibilities we could consider.

953 The term *bala* is also contained in SR śl.118.

954 PARALLELS
SU śārīra., 9,7: *dve anna-vāhinyāv antrāśrite, toya-vahe dve, mūtra-bastim abhiprapanne mūtravahe dve, śukra-vahe dve śukra-prādurbhāvāya, dve visargāya, te eva raktam abhivahato visṛjataś ca nārīṇām ārtava-sañjñaṃ, dve varco-nirasanyau sthūlāntra-pratibaddhe, aṣṭāv anyās tiryag-gāminīnāṃ dhamanīnāṃ svedam arpayanti [...].*
AS śārīra., 6,36: *dve vahato 'nnam mātrāśrayeṇa / dve toyam / dve mūtram / dve śukraṃ vahato dve ca muñcataḥ / te eva nārīṇām ārtavam varco-nirāsinyau sthūlāntrapratibaddhe evaṃ dvādaśāśeṣās tv aṣṭau dhamanyās tiraścīnāḥ svedam abhitarpayanti.*
ANALYSIS
The SR accords with the AS.
The SU deals with this topic, too, but this text mentions four ducts (*dhamanī*) for semen in total. Namely, two for the appearance of semen and two for its discharge. On the other hand, the SU mentions only two for the menstrual-procreatory discharge, namely one for its carrying and one for its discharging.

955 The term *srotas* literally means "stream" or "flow". Here, it is not clear, which substance this term actually denotes. The two parallels to SR śl.112, i.e. SU śārīra., 9,7 and AS śārīra., 6,36 (which are given in the previous footnote), do not contain this term. The term *srotas* here in SR śl.112a might denote the streams of the fluids mentioned in the previous verse SR śl.111ab, i.e. water, urine, semen, the menstrual-procreatory discharge and perhaps excrement, too. Namely, there are two ducts (*dhamanī*) each for carrying (SR śl.111b, *vahato*) these fluids, and two each for discharging (SR śl.112a, *vimuñcato*) them. My assumption is supported by the parallel AS śārīra., 6,36, which contains *dve śukraṃ vahato dve ca muñcataḥ*. The parallel SU śārīra., 9,7 also contains *dve śukra-prādurbhāvāya, dve visargāya.*
However, in medical texts, *srotas* usually denotes "aperture" or "channel", cf. DAS 2003A, pp.585–590.
The SU adopts this term to denote something else than a vessel (*śirā*) or duct (*dhamanī*). The term *srotas* is defined by SU śārīra., 9,13, *mūlāt khād antaraṃ dehe prasṛtaṃ tv abhivāhi yat / srotas tad iti vijñeyaṃ sirā-dhamani-varjitam //*, "That which, spreading inside of the body, [starting] from the opening/vessel (*kha*) [of] the root, is the conveyer (*abhivāhin*) [of substances], is to be known

intestines[956], excrement. (śl.112ab)[957]

---

as *srotas*, except for vessels (*sirā*) and ducts (*dhamanī*)." According to SU śārīra., 9,12, *srotas*-s seem to denote branches of ducts (*dhamanī*). SU śārīra., 9,12 lists two *srotas*-s, each for carrying the vital wind (*prāṇa*), food (*anna*), water, the nutrient fluid (*rasa*), blood, flesh, fat (*medas*), urine, excrement, semen and the menstrual-procreatory discharge.
The AS contains a similar theory, but this theory is not completely identical to that of the SU. The AS (śārīra., 6,38–39) states that, in men, the *srotas*-s are nine, namely, the ears, eyes, nostrils, mouth, anus and urinary canal (*mūtra-patha*); in women, three in addition, namely two in the breasts, and one for the menstrual-procreatory discharge. AS śārīra., 6,40f mentions thirty *srotas*-s called *antaḥsrotas*-s "inner streams" in addition to the nine mentioned above. They are the ones carrying the vital wind (*prāṇa*), water, food, the nutrient fluid (*rasa*), blood, flesh, fat, bone, marrow (*majjan*), semen, urine, excrement and sweat. The same theory is briefly dealt with by AH śārīra., 3,40cd–42, too.
On the other hand, AS śārīra., 6,67 also informs us of a different theory. According to this theory, *srotas* denotes an opening or a canal/vessel in general, including *sirā* and *dhamanī*.
The fifth chapter of the CA's vimānasthāna deals with *srotas*. The theory of CA vimāna., 5,8 accords with that of AS śārīra., 6,40f. Besides, CA vimāna., 5,9 is parallel to AS śārīra., 6,67, considering *srotas* to be a general appellation for a canal/vessel. (For further discussion on this the term *srotas*, cf. DAS op.cit.)
According to DAS' (op.cit.) detailed discussion, *srotas*-s usually do not denote "streams/flows (of fluids)", but "apertures" or "channels" as we have seen above. So I am not sure if my translation is correct.
Another possibility might be to interpret *srotāṃsi* in SR śl.112a as meaning "channels", that means, to interpret it as the apposition of the latent subject of the whole sequence of the passages beginning with SR śl.105a, i.e. *dhamanyo* "the ducts". If it is so, we would have to translate SR śl.112ab as "two [ducts (*dhamanī*)], [which are called] channels, joined to the large intestines, discharge excrement". In this case, however, the plural *srotāṃsi* becomes problematic, as *vimuñcato dve* is dual.
Whatever it may be, the interpretation of the term *srotas* as standing for *dhamanī* in SR śl.112a does not contradict the statements of AS śārīra., 6,67 and CA vimāna., 5,9. But the reason, why the SR calls the vessel in question "*srotas*", might be that it is a branch of the large intestines (cf. SR śl.112b). If it is so, this statement would accord with SU śārīra., 9,12 (see above). Indeed, SU śārīra., 9,12 states that the *srotas*-s carrying excrement are rooted in the receptacle of digested/cooked food (*pakvāśaya*) and the anus (*guda*). CA vimāna., 5,8 also makes a similar statement that they are rooted in the *pakvāśaya* and *sthūla-guda*.

956 The term *antra* occurs in the hymn on anatomy in AV (9,8, v.17).

957 PARALLELS
SU śārīra., 9,7: *dve varco-nirāsinyau sthūlāntra-pratibaddhe.*

Eight deliver sweat. (śl.112c)[958]

The oblique ones are deemed [to be] manifold/in_many_[branches]. (śl.112d)[959]

They have (lit: Of these, there are) openings/mouths at the hair-pores[960], for the release of sweat. And [these] make/let *rasa*-s[961]

AS śārīra., 6,36: *dve varco-nirāsinyau sthūlāntra-pratibaddhe.*
ANALYSIS
Only the SU and AS, which are completely identical to each other in wording, accord with the SR.

958 PARALLELS
SU śārīra., 9,7: *aṣṭāv anyās tiryag-gāminīnāṃ dhamanīnāṃ svedam arpayanti.*
AS śārīra., 6,36: *śeṣās tv aṣṭau dhamanyas tiraścīnāḥ svedam abhitarpayanti.*
ANALYSIS
According to the SU and AS, these eight ducts (*dhamanī*) deliver (*arpayanti*) sweat to the oblique ducts (*dhamanī*).

959 PARALLELS
SU śārīra., 9,9: *tiryag-gānāṃ tu catasṛṇāṃ dhamanīnām ekaikā śatadhā sahasradhā cottarottaraṃ vibhajyante, tās tv asaṅkhyeyāḥ.*
AS śārīra., 6,37: *tiryag-gāminyaś catasro bhidyamānāḥ subahudhā bhavantīty uktam prāk.*
ANALYSIS
According to the SU, the four main oblique ducts (*dhamanī*) gradually bifurcate into hundreds, then thousands; therefore, they are uncountable. The AS, like the SR, simply states that the four bifurcate into many (*bahudhā*). According to the commentator Indu, this statement is already made in the fifth chapter (on *aṅga-vibhāga*) of the AS's śārīrasthāna. This seems to point at AS śārīra., 5,115 which is parallel to SR śl.101. AS śārīra., 5,115 mentions 2,900,290 openings (literally, "mouths") of the vessels (*sirā*) and ducts (*dhamanī*).

960 The CA mentions two ways for the foetus' nutrition, i.e., through the hair-pores and the umbilical cord, cf. CA śārīra., 6,23 (*tadanantaraṃ hi asya kaścil lomakūpāyanair upasnehaḥ kaścin nābhināḍyaiḥ*). Cf. DAS 2003A, p.466ff. For various manners of the foetus' nutrition, cf. COMBA 1981, footnote 40.
The Tandulaveyāliya (sūtra 2) mentions, in the seventh month, 700 vessels (*sirā*), 500 *pesī*-s, 9 ducts (*dhamanī*) or tubes (*nāḍī*), 9,900,000 hair-pores (which is equal to the amount of the hairs excluding the head-hairs and beard-hairs) and 35,000,000 as the total amount of the hairs including the head-hairs and beard-hairs (compare these amounts with their equivalents in the SR). The Tandulaveyāliya states that various kinds of vessels and hair-pores participate in the foetus' nutrition.

961 PARALLELS
SU śārīra., 9,9: *tāsām mukhāni romakūpa-pratibaddhāni, yaiḥ svedam abhivahanti, rasaṃ cābhitarpayanty antar bahiś ca, tair eva cābhyaṅga-pariṣekāvagāhālepana-vīryāṇy antaḥśarīram abhipratipadyante tvaci vipakvāni,*

arisen from smearing/unguents (*abhyaṅga*)[962], ointments etc. enter. (śl.113)

---

tair eva ca sparśaṃ sukhāsukhaṃ vā gṛhṇīte, tās tv etāś catasro dhamanyaḥ sarvāṅgagatāḥ savibhāgā vyākhyātā /9/ yathā svabhāvataḥ khāni mṛṇāleṣu biseṣu ca / dhamanīnāṃ tathā khāni raso yair upacīyate /10/.

AS śārīra., 5,117: *tāsāṃ hi mukhāni tat pratibaddhāni tābhir āpyāyyante / taiś ca tāḥ svedam abhivahanti / tathā 'bhyaṅga-lepādi-vīryaṃ tvaci vipakvam antar-nayati, sparśaṃ ca gṛhṇanti.*

The CA does not contain parallels.

ANALYSIS

The SR's usage of the term *rasa* here is unusual, as explained in the following. Parallels are found in SU śārīra., 9,9 and AS śārīra., 5,117 (see above). The statements of AS śārīra., 5,117 resemble that of SU śārīra., 9,9, but the term *rasa* does not occur in the AS.

In SU śārīra., 9,9, the term *rasa* obviously means "nutrient fluid or juice", unlike in SR śl.113. The commentator Ḍalhaṇa also adapts this meaning, cf. Ḍalhaṇa: *rasaṃ rasa-dhātum. [...] yair mukhaiḥ samyak-pariṇatāhāra-rasa-vāhibhiḥ. [...] bahiś ca santarpayanti "tvacam" iti śeṣaḥ.* Further, Ḍalhaṇa comments on *abhyaṅga* etc., saying that the ducts (*dhamanī*) deliver the *vīrya* (literally, "energy") of smearing/unguent (*abhyaṅga*) to the inside of the body. What the term *vīrya* actually denotes here is not clear. Ḍalhaṇa states that the *vīrya* of smearing/unguent is ripened/cooked (*vipakva*) by the fire/heat of the skin, which provides the skin with brilliance (*bhājakenāgninā*). The same ducts (*dhamanī*) also transfer the sensation of touch.

But the meaning of *rasa* in SR śl.113 is ambiguous. In SR śl.113, the *rasa*, being produced through smearing/unguent etc., cannot be *āhāra-rasa*. This *rasa* seems to be equivalent something to the *vīrya* of smearing/unguent mentioned by SU śārīra., 9,9. The AS also mentions the *vīrya* of smearing/unguents, ointments etc., (*abhyaṅga-lepādi-vīrya*) which is completely ripened (*paripakva*).

ViṣṇudhP 115,90cd–91ab, which deals with the same topic as in SR śl.105cd–106ab, includes smearing/unguents (*abhyaṅgādi kleda-lepādikam*) into the category of *rasa*.

On the other hand, it might be suspected that something is omitted from the source-text of the SR. My theory is as follows:

The source-text had once content similar to SU śārīra., 9,9. The two passages which originally stood separate, were combined awkwardly. Namely, the passage on the perspiratory canals' carrying *āhāra-rasa* and the passage on the perspiratory canals' making the *vīrya* of smearing/unguents enter (the pores) were combined.

I suspect a similar kind of omission in SR śl.100cd, too (cf. my footnote 918 on SR śl.100cd, "*coverings*"), though I cannot prove it.

962 The expression *abhyaṅgalepa* might also mean "ointment for massage".

[They] know one hundred and seven[963] vital points (*marman*)[964], [i.e.] places of life (*jīva*). (śl.114ab)[965]

[There are] three and a half[966] crores (=35,000,000) of body-hairs (*roman*). The beard [hairs] and the head hairs (*keśa*) are three lakhs (=300,000). (śl.114cd)[967]

The manuscript *ṅa.* reads *cābhyaṅge* instead of *cābhyaṅga*. This would mean, "And [these] make the *rasa*-s arisen from the ointments at [the time of] massage (*abhyaṅge*) enter [the pores]".

963 Literally, "a hundred increased by seven".

964 SU śārīra., 6,15 explains the vital points (*marman*) as "the meeting points of muscle, vessel, ligament, bone and joint; they are the seats of the vital life principle (*prāṇa*) and injury to them can and often does cause the death" (ZYSK 1986, p.699, footnote 17).
According to DAS 2003A, p.568, some *srotas*-s or ducts are included in *marman*-s. Intriguingly the SR mentions *marman*-s in a context dealing with the *srotas*-s for perspiration.

965 One hundred and seven vital points (*marman*) are mentioned by SU śārīra., 6,3, AS śārīra., 7,2, AH śārīra., 4,1a and YS 3,102c.
In contrast to these texts, CA śārīra., 7,9 mentions only six vital points (*marman*) which are considered to be the sites of the vital wind (*prāṇāyatana*); they are the head (*mūrdhan/śiras*), throat (*kaṇṭha*), heart (*hṛdaya*), navel (*nābhi*), anus (*guda*) and bladder (*vasti*). But CA sūtra., 29,3 mentions only three vital points (*marman*).
A parallel to CA śārīra., 7,9 is found in AS śārīra., 5,81, which mentions ten sites of the vital wind (*prāṇāyatana*). There, the first seven of them are called great vital points (*mahā-marman*); six of them are the same ones as the six vital points (*marman*) mentioned by CA śārīra., 7,9, and the one is the "bond of the tongue" (*jihvā-bandhana*). The AH (śārīra., 3,13) also mentions ten "sites of life" (*jīvita-dhāmāni*) which are the same points as mentioned by the AS. But the AH does not call these ten points "*marman*". YS 3,93 mentions ten sites of the vital wind (*prāṇasyāyatanāni*), but this text does not call them *marman*, either. AgniP 370,21 and ViṣṇudhP 115,69cd–70, which are parallel to each other, mention ten sites of the vital wind (*prāṇasyāyatana*). This statement of the two Purāṇa-s, mentioning the tongue and urinary bladder (*basti*), but not mentioning *ojas* nor shoulders, deviates from that of the YS.

966 D reads *sārdhaṃ* instead of *sārdha*.

967 SR śl.114cd is identical to ŚG 9,49ab.
ANALYSIS
AS śārīra., 5,117–116 identifies the total amount of the head-hairs, beard-hairs and body-hairs (*keśa-śmaśru-loman*) with that of the openings of the vessels (*sirā*) and ducts (*dhamanī*), namely, 2,900,956. SR śl.101 mentions 2,900,956 as the amount of the vessels (*sirā*) and ducts (*dhamanī*). In contrast to that, SR

546,750,000 (lit. fifty-four crores, with sixty-seven and a half of lakhs) of body hairs (*roman*) with channels (*srotas*), vessels (*sirā*)[968], beard [hairs] and head hairs (*keśa*) are, however, enumerated. (śl.115, 116a)

Now, the quantity of the combination (*saṃhitā-māna*)[969] (= the total quantity) of water etc. is mentioned: Water is to be known as being [to the measure of] ten hand-cavities (*añjali*)[970]. There are[971]

śl.114 gives 35,000,000 + 300,000 for the hairs. Thus, the SR does not accord with the AS in this regard.

The YS gives three lakhs as the amount of the head- and beard-hairs, then 54 *koṭi*-s plus 76 and a half lakhs, i.e. 54,675,000, as the amount of the body-hairs (YS 3,102: *trayo lakṣās tu vijñeyāḥ śmaśru-keśāḥ śarīriṇām* /102ab/ *[...]* /102/ *romṇāṃ koṭyas tu pañcāśac catasraḥ koṭya eva ca sapta-ṣaṣṭis tathā lakṣāḥ sārdhāḥ svedāyanaiḥ saha* /103/). As already remarked in my footnote 921 on SR śl.101, the YS also mentions the amount 2,900,956 of the vessels (*sirā*) and ducts (*dhamanī*). In summary, all three amounts mentioned by the YS are respectively identical to those of the SR.

AgniP 370,40ab and ViṣṇudhP 115,91, which are parallel to one another, mention 72 *koṭi*-s (72,000,000) as the amount of the body-hairs. (The AgniP's reading, *vyomnāni*, may be a mistake for *romṇāni*, though *vyoman* may also denote a "pore" or "body cavity".) 72 *koṭi* does not accord with the statement of the SR. The amount 72,000,000 might have been influenced by the Yoga theory considering the amount of the tubes (*nāḍī*) to be 7,200 (cf. my footnote 754 on SR śl.65).

The GaruḍaP considers the amount of the body-hairs to be 3,500,000, which is one-tenth of the amount mentioned by the SR (GaruḍaP (KIRFEL 1954) v.48ab: *romṇāṃ koṭyas tathā tisro 'py ardha-koṭi-samanvitāḥ* /48ab/). In the next passage, this Purāṇa gives seven lakhs as the amount of the head-hairs. The Tandulaveyāliya (sūtra 2) mentions, in the seventh month, 9,900,000 as the amount of the pores (*roma-kūva*), excluding the pores of the beard-hairs and head-hairs. Further, the text relates that the total amount of the pores is 35,000,000, including the pores of the beard-hairs and head-hairs. The latter amount accords with the SR.

In summary, the YS best accords with the SR. The GaruḍaP and the Tandulaveyāliya are partly in concordance with the SR.

968 *Srotaḥ-sirā-śmaśru-keśaiḥ* is obviously one compound.

969 The Ānandāśrama edition (1896) does not read *saṃhitā* and *mānaṃ* as a compound, but as two separate words. Thus, *saṃhitā* would denote the combination/ assemblage of body-hairs (*romṇāṃ saṃhitā*), while *mānaṃ*, the amount of water etc. (*mānaṃ jalāder*).

970 One *añjali* equals the amount that can be held in the cavity of the hand. According to CA śārīra., 7,15, it corresponds to the amount which can be measured by the cavity of one's own hand.

nine hand-cavities of the nutrient fluid (*rasa*)[972], (śl.116bcd)[973]
eight of blood, seven of faeces, but[974] six of phlegm, (śl.117ab)[975]
five of bile, four hand-cavities of urine, three of tissue-fat (*vasā*)[976], but two of fat (*medas*). [There is] deemed [to be] one hand-cavity of marrow (*majjan*)[977]. (śl.117cd–118ab)[978]

---

971 Optative.

972 AS śārīra., 5,121 considers *rasa* to be the nutrient fluid (*āhāra-rasa*).

973 SR śl.116cd is identical to ŚG 9,43cd.
AS śārīra., 5,121–123 and AH śārīra., 3,80–82 contain the same theory as the SR, stating that the mentioned amounts (of water etc.) are accumulated by a single hand-cavity (*añjali*), one by one. The first of them is the amount of marrow which is one hand-cavity; that of semen, brain (*mastiṣka*) and *ojas* is each a half hand-cavity. The substance which the SR calls "the essence of phlegm" is called *ojas* by the AS and AH.
CA śārīra., 7,15 also contains the same theory. Here, *rasa* is explained as the first element (*dhātu*) which is the result of the digestion of food (*pūrva āhāra-pariṇāma-dhātu*). In the CA, the substance which the SR calls "the essence of phlegm" (*śleṣma-sāra*) is called the *ojas* pertaining to phlegm (*ślaiṣmakam ojas*).
SU cikitsā., 15,37 states that the amounts of *doṣa, dhātu* and *mala* cannot be estimated because human bodies vary and are instable. This statement is quoted by AS śārīra., 6,129, as the opinion of Dhanvantari's school.
YS 3,105–107ab contains the same theory as the SR.
AgniP 370,40cd–42 and ViṣṇudhP 2,115,92–94ab, which are parallel to one another, contain the same theory as the SR. The two texts call the substance which is called *śleṣma-sāra* by the SR "*ojas*". But, the two texts do not mention the substance which is called *śiro-majjan* by the SR. That means, in these passages, the two Purāṇa-s deviate from the SR and YS.
GaruḍaP (KIRFEL 1956) v.49cd–52 mentions the amounts of various substances, but these substances are different from those mentioned in the SR. The substances listed by this Purāṇa are flesh, blood, fat (*medas*), skin, bone, marrow (*majjan*), *mahā-rakta* (lit. "large blood" seemingly menstrual-procreatory discharge), semen, blood, phlegm and excrement.

974 The Anandāśrama edition (1896) reads *śleṣmaṇaś ca* instead of *śleṣmaṇas tu*.

975 SR śl.117ab is identical to ŚG 9,44ab, which contains, however, a variant *śleṣmaṇaś ca* instead of *śleṣmaṇas tu*. The reading of the ŚG is the same as that of the Ānandāśrama edition (1896) of the SR.

976 DAS suggests the translation "tissue-fat" instead of Meulenbeld: "muscle-fat", cf. DAS 2003A, p.580, under *vasā*.

977 Grammatically, it should be *majjñaḥ* (genitive), not *majjaḥ*. Instead, the Ānandāśrama edition (1896) and the identical verse in the ŚG (9,45b) both read *majjā* (nominative), which seems to be correct (see below).

The marrow of the head (*śiro-majjan* = brain) is half a hand-cavity [in quantity], [also] the essence of phlegm (*śleṣma-sāra*), further strength/power (= semen) (*bala*). (śl.118)[979]

Thus (*iti*) a summary (*saṅkṣepa*, lit. "conciseness") of the secondary appendages.[980] The detail[s] with regard to this, however, may be accurately (*tattvataḥ*) perused by the wise ones in the Adhyātmaviveka[981] ("Investigation of the self") composed by us (= the author). (śl.119)

Comm. K on SR śl.86cd–119

[The part] whose beginning is "**Ears, eyes**" (śl.86a. *śravane nayane*), [and] whose end is "**may be perused by the wise in the Adhyātmaviveka composed by us**" is [that] whose meaning is clear.

Comm. S on SR śl.86cd–119

He relates the orifices carrying impurities outside: "**Ears**" (śl.86a. *śravane*). (śl.86–87)

He relates sixteen nets (*jāla*), six bunches (*kūrca*), seven tendons/

---

978 SR śl.117cd–118ab is identical to ŚG 9,44cd–45ab. But the ŚG reads *majjā tv añjali-sammitaḥ* instead of *majja eko 'ñjalir mataḥ*.
As a matter of fact, the Ānandāśrama edition (1896), too, reads *majjā tv añjali-sammitā*. If we take *majjā tv añjali-sammitaḥ*, it would mean, "but the marrow is measured by one hand-cavity".

979 SR śl.118cd is identical to ŚG 9,45cd. The ŚG, however, deviates, reading *ardhāñjalis tataḥ śukraṃ tad eva balam ucyate,* instead of *ardhāñjaliḥ śiromajjā śleṣmasāro balaṃ tathā*. It would mean, "thereafter (*tataḥ*), the semen is a half hand-cavity. It indeed is called strength".

980 This remark seems to conclude the part of the text beginning with śl.75ab (*atha vakṣyante pratyaṅgāny akhilāny api*).

981 SHRINGY 1999 (preface xiv.) states that this work by Śārṅgadeva is not available.
AVALON 1924, pp.140–141, however, discussing the fruits in the petals of the *cakra*-s, refers to a text entitled Adhyātmaviveka. According to AVALON, passages of this text are quoted in the Dīpikā on v.7 of the Haṃsopaniṣad. To my regret, AVALON does not further inform us of this text. But, AVALON often quotes the statements on the fruits of the petals from this Adhyātmaviveka. It is remarkable that these statements accord with those of the SR. I am discussing this topic in more detail in the respective footnotes on SR śl.124 (note 1014), śl.125 (note 1023), śl.128 (note 1029) and śl.137 (note 1071).

cords of flesh (*rajju*) and five sutures/seems (*sīvanī*) through "**bones**" (śl.88a. *asthi-*) etc. (Comm. on SR śl.88–89)

He relates the groups of bones with a difference of opinions/ doctrines: "**Fourteen**" (śl.90a. *caturdaśa*).

He counts the bones: "**Of bones**" (śl.90c. *asthnām*). (Comm. on SR śl.90)

He mentions the bones as fivefold through the particularity of form: "***Valaya*-s**" (śl.91a. *valayāni*). (Comm. on SR śl.91)

With a difference of opinions/doctrines, he makes known the number of the bones differently: "**Three**" (śl.92a. *trīṇi*).

He enumerates the bone-junctures: "**Two hundred**" (śl.92c. *dve śate*). (Comm. on SR śl.92)

He relates the eightfoldness of the bone-junctures through the particularity of form: "***Koraka*-s**" (śl.93a. *korakāḥ*). (Comm. on SR śl.93–94)

Enumerating the junctures[982] of *peśī*-s, cords (*snāyu*) and vessels (*sirā*), he relates the varieties of cords (*snāyu*): "***Peśī*[-s]**" (śl.94c). (Comm. on SR śl.94–95)

He relates the use of the cords (*snāyu*): "**With bindings**" (śl.96a. *bandhanair*). (śl.96)

He enumerates the *peśī*-s with the difference between man and woman: "**Five**" (śl.97a, *pañca*). (Comm. on SR śl.97–100)

He enumerates the vessels (*sirā*) and ducts (*dhamanikā*): "**Vessels**" (śl.101a, *sirā-*). (Comm. on SR śl.101–102)

Showing the particular appearance (*saṃsthāna*) of the sutures/ seems (*sīvanī*) by an example, he relates [their] number, place and classification: "**Two *aṅgula*-s**" (śl.102c, *dvy-aṅgulaṃ*). **Carrying smell**, [namely,] [being] the cause of the knowledge/perception of smell. (Comm. on SR śl.102–104)

Enumerating the ducts (*dhamanī*) carrying the nutrient fluid (*rasa*), he mentions their purpose: "**Ducts**" (śl.105c, *dhamanyo*). (Comm. on SR śl.105–106)

He relates their place and appearance: "**These**" (śl.106c, *etāḥ*). (Comm. on SR śl.106–107)

It is mentioned that (*iti*) [there are] **ten** stretched out **upward**, **ten**

982 See my footnote 901 on SR śl.94cd. According to this, the interpretation, "enumerating *peśī*-s, cords (*snāyu*), vessels (*sirā*) and junctures (*sandhi*)", is not correct.

stretched out **downward, four stretched out obliquely**. With regard to that, he relates the threefoldness of the ones stretched out upward through the difference of functions: "**those going upward**" (śl.107c, *ūrdhvagāḥ*). (Comm. on SR śl.107–109)

He relates the threefoldness of [those] going (lit. gone) downward: "**The ones going (lit. gone) downward**" (śl.110a, *adhogatāḥ*). He relates their function[s]: "**Make proceed**" (śl.110a, *pravartayanti*). **Strength/Power (*bala* = semen)** is vigour (*vīrya* = semen). (Comm. on SR śl.110–112)

He relates the particular[s] of [those] oblique: "**The oblique ones**" (śl.112d, *tiraścyaḥ*). (Comm. on SR śl.112–113)

He fully enumerates the vital points: "**Places of life**" (śl.114a. *jīva-sthānāni*)[983]. (Comm. on SR śl.114–115)

He relates the quantity of the combinations: "**Sixty-seven**" (śl.115d, *sapta-ṣaṣṭyā*). **A combination (*saṃhitā*)** is the union of the [body-]hairs etc.[984]

Introducing the quantity of [each] fluid (*drava*), he relates [it]: "**Of water etc.**" (śl.116b, *jalāder*). **Of juice (*rasa*)**[985], [namely] of the juice of food (*annarasa*).[986] **Of marrow**[987]: the element (*dhātu*) called marrow.

983 A and B read *oja* instead of *jīva*. It means "places of *ojas*". But, as the vital points (*marman*) are often called "places of life" (*jīvasthāna*) by the parallels in the other texts (cf. my footnote 964 on SR śl.114a, on *marman*), I consider this variant to be incorrect.

984 According to the commentary, it seems that a *saṃhitā* is the sum total of the hairs and other structures (*srotaḥ-sirā*, sweat gland and capillary vessel?) connected to them. But the meaning of the original text (*srotaḥ-sirā-śmaśru-keśa*) is obscure, as it includes the beard- and head-hair in the same compound. Grammatically, the past participle *ākhyātāḥ* (śl.115c), pl. feminine, agrees with *koṭayaḥ* (śl.115b), but not with *saṃhitā-mānam*, sg. neuter. Therefore a new sentence seems to begin with *saṃhitāmānam*. If it is so, *saṃhitā-mānam* means the total sum (of water etc.).

985 In the *mūla* text, I translated *rasa* as "the nutrient fluid".

986 For the terms, *rasa* and *annarasa*, cf. DAS 2003A, p.528 on *āhāra-rasa*.

987 This is problematic. In the original text, *majjaḥ* is obviously a grammatical mistake for *majjñaḥ*, because of the other genitives. But the commentator reads and explains it as a nominative *majjaḥ* (instead of the common *majjā*). On the other hand, he correctly reads *rasasya* as genitive. It would be interesting to know how *eko añjalir mataḥ* is understood, but, unfortunately, the commentary says nothing on this.

Through this, [namely] "**the detail[s], however, may be perused in the Adhyātmaviveka**", the maker (= author), thinking (*iti*): "That book made (= written) by me too exists", affirming his own pre-eminence, says: "that book too is to be looked into". (Comm. on SR śl.119)

SR śl.120–122ab

Between the anus and the genitals is a four-petalled *cakra*[988] called *Ādhāra*[989] (lit. "base"). (śl.120ab)[990]

988 The term *cakra* literally means a "circle" or "wheel". In the body, there are several *cakra*-s, or energy centres, arranged vertically along the spine. The *kuṇḍalinī*, the vital or sexual energy, which is at first situated in the lowermost centre, gradually ascends the spine through the *cakra*-s, until it attains the uppermost centre, the *Sahasrāra*. The *Sahasrāra* which is the place of bliss is situated over the top of the head. That means, the crude vital energy develops into a refined form which is utilised for the attainment of bliss. Each *cakra* corresponds to one of the stages of this development. Namely, the higher *cakra* the energy (*kuṇḍalinī*) reaches, the more it gets refined. For the term *cakra*, cf. AVALON 1924, p.105; pp.117–118.
HEILIJGERS-SEELEN 1990, p.57, states that the *cakra*-s "serve as the seat of some particular energy or power forming part of the sonic and/or phenomenal creation". The course from the *Ādhāra Cakra* through the *Ājñā Cakra* up to the *Sahasrāra* is concerned with the manifestation of sound. This seems to be the reason that musicological texts deal with the *cakra*-s.
The so-called six-*cakra* system contains six *cakra*-s plus the *Sahasrāra*. So, it is actually contains seven centres. But the *Sahasrāra* centre seems to take a unique position in this system, therefore it is very often not referred to by the term "*cakra*". This system is usually considered the most representative by modern studies (cf. AVALON 1924; DAS 1992). But there are various traditions which differ from it; these do not necessarily contain seven centres (DAS 1992, p.397). The information concerning the six-*cakra* system is mainly based on rather late Sanskrit texts, such as the Ṣaṭcakranirūpaṇa and the Śivasaṃhitā, cf. HEILIJGERS-SEELEN 1990.
The theory of *cakra*-s presented in SR śl.120–145ab contains nine *cakra*-s plus the *Sahasrāra*. That means, the SR's system adds three *cakra*-s to the above-mentioned six *cakra*-s which are today regarded as representative.
The term *cakra* is sometimes translated with "plexus" (cf. DAS 1992, p.396). Some *cakra*-s certainly seem to correspond to the nerve plexuses, e.g. the *cakra*-s in the navel and chest, but this does not necessarily have to be so; the uppermost *cakra* over the head, called *Sahasrāra*, has no corresponding physical object.
For the origin of the notion of energy centres, cf. ELIADE 1936, p.123.

In it, the fruit[991] on the north-eastern petal etc. is[992] [respectively]: the supreme (*parama*) [joy (*ānanda*)], similarly, the natural/inborn (*sahaja*) joy, [the joy] connected with a hero (*vīra-pūrvaka*)[993], and the joy of Yoga.[994] (śl.120cd, 121ab)

The *kuṇḍalinī*, [namely] the power (*śakti*) of Brahman, is/exists (*asti*) in the lotus (= *cakra*) of *Ādhāra*. (śl.121)

Led to straightness as far as the aperture of Brahman (*brahma-*

---

989 Cf. A. AVALON's (1924, p.118) explanation of the *Mūlādhāra Cakra*. Also see the description of the *Mūlādhāra Cakra* in v.4 of the Ṣaṭcakranirūpaṇa (AVALON 1924).
Cakrakaumudī 1,6ab: *guda-liṅgāntare padmaṃ catur-dala-samanvitam / [...].*
Dhyānabindūpaniṣad (in the Yogopaniṣad) v.43cd.ff (p.197): *ādhāram prathamaṃ cakraṃ svādhiṣṭhānaṃ dvitīyakam /43/ yonisthānaṃ tayor madhye kāma-rūpaṃ nigadyate / ādhārākhye guda-sthāne paṅkajaṃ yac caturdalam /44/ tan-madhye procyate yoniḥ kāmākhyā siddha-vanditā /45ab/.*

990 A parallel is found in the Cakrakaumudī 1,6ab (cf. the preceding footnote).

991 AVALON 1924 calls the fruits in the petals "*vṛtti*-s". AVALON ibid., p.42, states, "It is stated that particular Vrittis are assigned to a particular lotus, because of a connection between such Vritti and the operation of the Shaktis of the Tattva at the centre to which it is assigned. That they exist at any particular Chakra is said to be shown by their dissapearance when Kundalī ascends through the Chakra. Thus the bad Vritti of the lower Chakras pass away in the Yogī who raises Kundalī above them."

992 Optative *syāt.*

993 The manuscript D reads *parapūrvaka* instead of *vīrapūrvaka*. It would mean "[joy] connected with the_best/other[s]".
The expression *-pūrvaka* is merely adopted to avoid repeating the term *ānanda*. For example, *vīra-pūrvaka* stands for *vīra-ānanda* "the joy of a hero".

994 The order of these fruits given by AVALON 1924, p.118, is different: *paramānanda, sahajānanda, yogānanda* and *vīrānanda*. AVALON states that these four *vṛtti*-s or fruits are not mentioned by the Ṣaṭcakra-nirūpaṇa. Instead, he refers to Tarkālaṅkāra's commentary to the Mahānirvāṇa Tantra. AVALON 1913 (p.lvii and followings), too, describing the six *cakra*-s and the *Sahasrāra*, mentions the fruits (*vṛtti*) of each *cakra*. There, he lists *yogānanda, paramānanda, sahajānanda* and *vīrānanda* in this order which is converse of the order in the SR.
Explaining the system of the Sahajiyā cult, DASGUPTA 1976 (p.99) mentions four kinds of bliss whose names are the same as or similar to the fruits mentioned in SR śl.120cd–121a, namely, *ānanda, paramānanda, viramānanda* and *sahajānanda*. The *viramānanda* seems to stand for the *vīrānanda* of the SR. These four kinds of bliss are, however, one by one distributed to the four *cakra*-s of the Sahajiyā theory, unlike in the SR.

randhra), this (= *śakti*) [is/becomes] the bestower of ambrosia (*amṛta*). (śl.122ab)

Comm. K on SR śl.120–122ab

"**Between the anus and the genitals**" (śl.120a, *guda-liṅgāntare*): The fruits (*phala*), the supreme joy etc., are on the group of four petals, [namely] the north-eastern [petal] etc., of the *Ādhāra Cakra* situated between the anus and the genitals, through/according_to the staying/ position of the individual self (*jīva-sthiti*)[995], at the time of birth.[996] **In the lotus of *Ādhāra*,** [namely] in that very *Ādhāra Cakra*.[997] "**The *kuṇḍalinī***" ("spiral/snake") is the power of Brahman (*brahmaśakti*), having the form of a snake in accordance with the [true] sense [of "snake"] (*anvarthatayā*), [namely] the power of ignorance, bringing forth the shape connected with agency etc. (*kartṛtvādy-upahitākāra-kāriṇī*), of Brahman, of the non-attached, indifferent supreme self.[998] That very [power of Brahman], having a [substantial] form, is called

995 The term *jīva-sthiti*, lit. "the standing of the individual self", might mean something more concrete than "the condition of the individual self". It might mean "the position of the individual self [in one of the petals of a *cakra*]", namely "the petal which the individual self chooses to ride on at the time of birth". As a matter of fact, SR śl.140–145ab (cf. *sthito jīvo* in śl.140c) states that one's musical success in life depends on which *cakra* or petal which the individual self chooses to ride on. And this is decided as early as at the moment of birth (*janma-kāle*), according to the commentary K on SR śl.120, (but the Ānandāśrama edition (1896) lacks *janma-kāle*).

996 The reading of the Ānandāśrama edition (1896) differs from that of the Adyar edition. It reads *madhye tasyādhāra-* instead of *madhye sthitasyādhāra-*; *-dale* instead of *-dala*; It lacks *janma-kāle*; *paṭalāny* instead of *phalāni*. The editor of the Adyar edition does not note these variants.
With the variant *madhye tasyādhāra-*, it would mean "of that *Ādhāra Cakra* between" instead of "of the *Ādhāra Cakra* situated between".

997 In the Ānandāśrama editon (1896), *ādhāra-paṅkaje* belongs to the foregoing sentence. The Ānandāśrama edition reads *ādhāra-paṅkaje* in place of *ādhāra-cakre*, and lacks the *daṇḍa* after it.

998 As noted in the Adyar edition (p.60, variant 6, *1 ed.*), the Ānandāśrama edition (1896) reads *-upahitākāriṇy-avidyā-* in place of *-upahitākāra-kāriṇy-avidyā-*. This reading seems to be more natural, meaning "[…] the power of ignorance, having the shape connected […]".

"*kuṇḍalinī*"[999] through/because_of [its] having a crooked shape.

As far as the aperture of Brahman: the tip of the *Suṣumnā* [tube], named the aperture of Brahman, going to the middle of the thousand-petalled *cakra* in the head — as far as that.

"Led to straightness" (śl.121ab, *ṛjutāṃ nītā*): [this means,] when the *kuṇḍalinī* is made to attain (*prāpyate*, passive of causative) straightness by blocking (*nirudhya*) the movement of the wind in the *Iḍā* and *Piṅgalā* through the method taught by the preceptor, [namely] the process/sequence of *yama*, *niyama* etc.[1000], [and] by removing (*apasārya*), through that wind together with fire, the hood of the *kuṇḍalinī* of the knot of the aperture of the root of the *Suṣumnā* at the knot of Brahman (*brahma-granthi*),[1001] and by having abandoned its own crookedness [after] the piercing[1002] of the knots of Brahman/Brahmā[1003], Viṣṇu and Rudra[1004] successively occurs (*sati*) through that [wind] having been made to enter [these knots].[1005]

999 ROŞU 1978, p.59, informs us of a metaphor, in which the Haṭhayogic practice controlling the *kuṇḍalinī* is compared to the magic of a serpent charmer.

1000 *Yama* "forbearance" and *niyama* "religious observances" belong to the "five exterior (*bahiraṅga*) methods" of the kind of Yoga called *aṣṭāṅgayoga*. The five methods are *yama, niyama, āsana, prāṇāyāma* and *pratyāhāra*. Cf. WOODROFFE 1990, pp.129–130.

1001 The Ānandāśrama edition (1896) reads differently: *suṣumnā-mūla-randhrācchādanīm (-chādinīm) phaṇām* instead of *suṣumnā-mūla-randhra-granthīnāṃ kuṇḍalinī-phaṇām*. It would be translated, "the hood which covers the aperture of the root of the *Suṣumnā* at the knot of Brahman". I prefer this reading.
The Adyar edition's reading *-granthīnām* seems to be a mistake during transcription, caused by *brahma-viṣṇu-rudra-granthīnāṃ* in the line below.

1002 *Bhede*. The Ānandāśrama edition (1896) reads *vibhede* instead of *bhede*. This makes no difference to the meaning, but *vibhede* seems to be in better concordance with *vibhidya* in the eighth line.

1003 Here this commentator seems to consider *brahma-* of *brahmagranthi* as the male god, one of the trinity.

1004 *Rudra* here refers to Śiva.

1005 The process described here seems to be the following: the *kuṇḍalinī* which has the shape of a cobra lies in a coil at the bottom of the *Suṣumnā* tube. Its expanded hood hinders the cobra from entering the aperture of this tube. Through the breath control of Haṭhayoga, this hindrance of the hood is got rid of, and the cobra is now able to go up through the *Suṣumnā* tube, stretches its body upwards, so that it attains straightness.

"**The bestower of ambrosia**" (śl.122d, *amṛta-pradā*). It is called so because (*iti*) at that time the *kuṇḍalinī*, piercing (*vibhidya*) the thousand-petalled *cakra*, [i.e.] the receptacle/base of nectar (*sudhādhāra*)[1006], with the tip of its own tail (*vāla-agra*) gone out [from/through] the aperture of Brahman because of its (= the *kuṇḍalinī*'s) own straightness, makes the ambrosia flow from that [*cakra*]. [1007]

Comm. S on SR śl.120–122ab

Thus mentioning the limbs and the secondary appendages, he commences to mention the *cakra*-s: "**Between the anus and the genitals**" (śl.120a, *guda-liṅga-antare*)[1008].

The four-petalled *cakra* named "***Ādhāra***", whose form is a lotus, lies **between,** [namely] in the middle of, **the anus and the genitals.**

He relates the fruit[s] of its four petals: "**Supreme**" (śl.120a, *parama*): The supreme joy (*parama-ānanda*) is on the north-eastern petal, the natural/inborn joy (*sahaja-ānanda*) on the south-eastern petal, the joy of a hero (*vīra-ānanda*) on the south-eastern petal, the

---

The Dhyānabindūpaniṣad (contained in the Yogopaniṣad), v.65cd.ff. (p.201), describes the process of awakening of the *kuṇḍalinī* which sleeps in the *Mūlādhāra*: *mukhenācchādya tad-dvāram prasuptā parameśvarī / prabuddhā vahni-yogena manasā marutā saha* /66/ *sūcivad guṇam ādāya vrajaty ūrdhvaṃ suṣumnayā* /67ab/. The commentary explains that, covering the door of the *Suṣumnā* tube by closing one's mouth (which is a Haṭhayoga technique called *kumbhaka*), a Yogin wakes the *kuṇḍalinī* up; he leads it upward through the *Suṣumnā*, together with the fire of the *Mūlādhāra*, the mind and the wind.

Also cf. SR 1,3, śl.3–6, which states that sound (*nāda*) is produced by the union of wind and fire.

1006 The Ānandāśrama edition (1896) reads *sudhādharaṃ* instead of *sudhādhāraṃ*. It means "which contains nectar".

1007 This description is not very explicit to me. The preceding sentence makes the impression that the cobra of the *kuṇḍalinī* enters the aperture of the *Suṣumnā* tube with its head first. This sentence here, however, suggests that the cobra inserts the tip of its tail first, and stretches itself upside down; so that the tip of the tail is the first part of the cobra's body to attain and pierce the aperture of Brahman.

The Ānandāśrama edition (1896) reads *vibhidyate* with a *daṇḍa* after it. It would be translated, "at that time, the *kuṇḍalinī* pierces the thousand-petalled *cakra*, […] own straightness. It makes ambrosia flow from that [*cakra*]."

1008 *Guda-liṅhāntara* is obviously a mistake for *guda-liṅgāntara*.

joy of Yoga (*yoga-ānanda*) on the north-western petal.

From that, it is understood that (*iti*) the four petals of the lotus of *Ādhāra* are situated just in the intermediate direction[s][1009] [of the compass].

"**Is/exists**" (śl.121c, *asti*): **The power of Brahman** named ***kuṇḍalinī***, because of [its] being the cause of attaining Brahman, is found **in the lotus (= *cakra*)** of ***Ādhāra***. It, **this**, being **led to straightness as far as the aperture of Brahman**, [namely] up to the aperture of Brahman, [is/becomes] **the bestower (*prada*) of ambrosia**, [namely] bestower of liberation.

SR śl.122bc–124ab

The *Svādhiṣṭhāna*[1010] is the six-leaved (*ṣaṭ-patraṃ*)[1011] *cakra* at the root of the genitals (*liṅga*). And they mention these fruits of it successively on the first/eastern etc. petals: (śl.122cd–123ab)

1009 *Vidiś*, i.e. north-east etc. I read *vidik-sthitāny* as one compound. The manuscript A reads *vidiśi sthitāny*.

1010 Cf. Dhyānabindūpaniṣad (in the Yogopaniṣad) v.48ab: *svādhiṣṭhānaṃ tataś cakram meḍhram eva nigadyate*. According to v.47cd, this *cakra* is so called because *sva-śabdena bhavet prāṇaḥ svādhiṣṭhānam tad-āśrayam* "the vital wind arises [accompanied] with its own sound".
A parallel is also found in Cakrakaumudī 2,2ab: *liṅga-mūle smaret padmaṃ sindūrābhaṃ tu ṣaḍ-dalam*.

1011 The Ānandāśrama edition (1896) reads *ṣaḍ-dalaṃ* instead of *ṣaṭ-patraṃ*. Besides this reading, this edition also notes *ṣaṭ-patraṃ* as the variant in the other manuscripts *ka, ga, gha* and *ṅa*. But the commentary K reads *ṣaḍ-dalam,* both in the Ānandāśrama and the Adyar edition. On the other hand, the commentary S, which is not contained in the Ānandāśrama edition, explains the original *ṣaṭ-patraṃ* with *ṣaḍ-dalaṃ*.

[There] is[1012] respectfulness/humbleness (*praśraya*), cruelty, destruction of pride/arrogance (*garva*)[1013], further (*tataḥ param*), delusion (*mūrchā*), disrespect, disbelief (*aviśvāsa*).[1014] (śl.123cd–124a)

This is the abode of the power of desire (*kāma*). (śl.124b)

Comm. K on SR śl.122–123

"***Svādhiṣṭhāna***" (śl.122c).[1015] The second, **six**-petalled (*ṣaḍ-dalaṃ*) ***cakra*** called ***Svādhiṣṭhāna*** is **at the root of the genitals**, above the *Ādhāra Cakra*.

He mentions the fruits, respectfulness/humbleness etc., on its first/eastern etc. petals, respectively: "**And of it**" (śl.122d, *asya ca*).[1016] "**[There] is disrespect**" (śl.124a, *avajñā syāt*). This indication has justice (*dharma*)[1017] as the chief matter.

Comm. S on SR śl.122–123

He relates the *cakra* known as *Svādhiṣṭhāna*: "***Svādhiṣṭhāna***" (śl.122a). **Six-leaved (*ṣaṭ-patraṃ*)**, [namely] six-petalled (*ṣaḍ-dalaṃ*), whose form is a lotus.

He relates the fruits of the six petals of this [*cakra*]: "**And of it**" (śl.122d, *asya ca*). **Respectfulness/Humbleness (*praśraya*)**, [namely] modesty (*vinaya*) is the fruit of the petal situated in the first/eastern direction. The meaning is that (*iti*) the self situated on it becomes

1012 Optative *syād*.

1013 The manuscript D reads *garvo nāśo* instead of *garvanāśo*. It would mean "pride, destruction". But with this variant, the number of the fruits would be seven, which has one too many.

1014 AVALON 1924 (p.141) mentions a text titled Adhyātmaviveka, which lists *praśraya, aviśvāsa, avajñā, mūrchā, sarvanāśa* and *krūratā* (transcription has been modified by me) in this order. They are also mentioned in AVALON 1913 (introduction, lix.). They are identical to those mentioned by the SR, except for *sarva-nāśa* instead of *garva-nāśa* in the SR. The order of listing is reverse to that of the SR.

1015 The Ānandāśrama edition (1896) locates *svādhiṣṭham iti* after *phalāny āha*.

1016 The Ānandāśrama edition (1896) lacks *asya ca*.

1017 The Ānandāśrama edition (1896) reads *dharmiparo*. The Adyar edition mentions it (p.62, variant 1). It would mean "has [someone] who has *dharma* (*dharmin* "a just person") as the chief matter." The Ānandāśrama edition (1896) notes the variant of *ka., dharmiyoge*, too. With this variant, it would mean, "This indication is for the Yoga/employment of a just person."

characterised by respectfulness/humbleness. **Cruelty** is adherence to evil deeds. **Pride** is the conception/imagination (*bhāvana*) of the superiority of the self; **destruction** of it (= pride). **Delusion** is an excess of confusion (*moha*). **Disrespect** is contempt. **Disbelief** is the state of being with doubt with regard to everything or being without confidence. This *Svādhiṣṭhāna Cakra* is **the abode**, the receptacle, **of the power (*śakti*)** producing **desire**.

SR śl.124cd–126ab

The ten-petalled *cakra* called *Maṇipūraka*[1018] is at/in the navel.[1019] (śl.124cd)

In this [*cakra*], [there] are[1020], however, on the first/eastern etc. petals[1021], respectively: deep sleep (*suṣupti*), thirst, jealousy, further slander,
shame, fear, disgust (*ghṛṇā*), confusion, dullness (*kaṣāya*), then[1022] depression. (śl.125abcd-126a)[1023]

And it (= the *Maṇipūraka Cakra*) is the abode of the sun[1024]. (śl.126b)

The *Anāhata Cakra* ("the *cakra* of the unstruck [sound]") is at/in

1018 On the name of this *cakra*, the Dhyānabindūpaniṣad states that the vital wind ascends and descends in the tube (*nāḍī*), piercing this *cakra*, like a thread piercing a bead (v.48cd–49ab: *maṇivat tantunā yatra vāyunā pūritaṃ vapuḥ /48cd/ tan nābhi-maṇḍalaṃ cakram procyate maṇipūrakam /49ab/*). For this *cakra*, also see the third pariccheda of the Cakrakaumudī.

1019 The Ānandāśrama edition (1896) reads *sañjñakam* instead of *sañjñitam.*

1020 Optative sg. *syād.*

1021 The Ānandāśrama edition (1896) reads *patreṣu* instead of *patre tu.*

1022 The Ānandāśrama edition (1896) reads *kaṣāyo 'py aviṣāditā* "dullness, too, non-depression" instead of *kaṣāyo 'tha viṣāditā*. But the Adyar edition's reading seems to be correct, because the parallel in the Adhyātmaviveka also contains *viṣāda*.

1023 The Adhyātmaviveka quoted by AVALON 1924 (p.141) lists *lajjā, piśunatā, īrśyā, tṛṣṇā, suṣupti, viṣāda, kaṣāya, moha, ghṛṇā* and *bhaya*, in this order (also cf. AVALON 1913, introduction, lix.). They are identical to the fruits mentioned by the SR. But the order of listing is the reverse of that of the SR.

1024 The Cakrakaumudī 4,2a: *hṛt-padmam bhānu-dalaṃ.*
AVALON 1924 (p.122) states that this is the *cakra* of fire.
The Kubjikāmatatantra (12,37–40ab), too, compares this *cakra* to the sun (HEILIJGERS-SEELEN 1990, p.59).

the heart[1025]. (śl.126c)

They regard (*icchanti*) that [*cakra*] endowed with twelve petals as the place of worship of Śiva whose shape is *praṇava* (= the sacred sound *Om*). (śl.126d-127ab)

They have sung these fruits of the self situated on the petal of the east etc.:

Destruction of inconstancy[1026], also clear[1027] reasoning, the state of being repentant, (śl.127cd)
hope (*āśā*)[1028], splendour/renown (*prakāśa*), and consideration/anxiety (*cintā*), striving/desire (*samīhā*), further equanimity (*samatā*), deceit/vanity (*dambha*), defectiveness/weakness (*vaikalya*), discernment, further egotism/the_consciousness_of_"I" (*ahaṅkṛti*), respecttively.[1029] (śl.128–129ab)

Comm. K on SR śl.124cd–129ab

Of the eight *cakra*-s called *Maṇipūraka, Anāhata, Viśuddhi, Lalanā, Ājñā, Manas, Soma* and *Sahasrapatra* at the eight places above it (=

1025 In ṚV 5,36,3, the heart connected with many radiating vessels is compared to a wheel (*cakra*). ZYSK 1993 (p.208, note 97) reports that the CA (siddhi., 9,4) inherited this Vedic simile, comparing the heart to the spokes in a hub (*nābhyām arā iva*).

1026 The Ānandāśrama edition (1896) reads *laulyam praṇāśaḥ* "inconstancy, destruction" instead of *laulya-praṇāśaḥ*. With this variant, however, the number of the fruits which should be twelve would be thirteen.

1027 The Ānandāśrama edition (1896) reads *kapaṭaṃ* instead of *prakaṭo*. With this, it would mean, "fraud, also reasoning [...]". On the other hand, the commentary S, which is not contained in the Ānandāśrama edition, reads *prakaṭa*.

1028 The manuscript *gha.* reads *praṇāśa* "destruction" instead of *āśā*. This is obviously a mistake, caused by *praṇāśaḥ* in the above line (śl.127c).

1029 The Adhyātmaviveka mentioned by AVALON 1924 (p.141) lists *āśā, cintā, ceṣṭā, mamatā, dambha, vikalatā, ahaṅkāra, viveka, lolatā, kapaṭatā, vitarka* and *anutāpa* in this order (also cf. AVALON 1913, introduction, p. lx). The order of listing is the same as in the SR, though starting from a different point, namely from *āśā*. The terms are identical to those mentioned by the SR, except for *ceṣṭā* instead of *samīhā* in the SR; *samatā* instead of *mamatā* in the SR; *kapaṭatā* instead of *prakaṭatā* in the SR. Except for *ceṣṭā*, these deviations seem to be secondary. Like the Adhyātmaviveka which reads *kapaṭatā*, the Ānandāśrama edition of the SR also contains *kapaṭaṃ*. On the other hand, the Adhyātmaviveka does not contain a counterpart for *prakāśa* of the SR.

Svādhiṣṭhāna Cakra), [namely] the navel, the heart, the throat, the uvula, the middle of the brows, the forehead, the base/root (*mūla*) of the hair and the aperture of Brahman respectively, he mentions the fruits separately for each petal, by the order [which has] the beginning with the first/eastern, according to the number of the petals contained in (*gata*) those *cakra*-s: "**the ten-petalled at/in the navel**" etc. (śl.124c, *nābhau daśadalam*).[1030]

Comm. S on SR śl.124cd–129ab

He relates the *Maṇipūraka Cakra* : "**At/In the navel**" (śl.124c, *nābhau*).

He relates the fruits of this [*cakra*] on the first/eastern etc. petal: "**Deep-sleep**" (śl.125a, *suṣupti*). **Deep-sleep** is the condition of cessation of the outer faculties/organs (*bāhyendriya*) and the mind. **Thirst** is covetousness. **Jealousy** is non-endurance of others' [good] qualities. **Slander** is pointing out[1031] others' existent and non-existent faults. **Shame [and] fear** are explicit. **Pity (*ghṛṇā*)** is the wish to remove (*prahāṇa*) [1032] others' pain (*duḥkha*).[1033] **Confusion** is incomplete perception of objects even in an awake condition. **Dullness** is the state of being_the_abode_of_impurity/having_an_impure_mental_ disposition. **Depression** (*viṣāditā* lit. "the state of a person with

1030 The Ānandāśrama edition (1896) reads *-randhreṣv aṣṭasu* (with *sandhi*) instead of *-randhreṣu aṣṭasu* (without *sandhi*); *tat-tac-cakra-gaṇita-dala-saṅkhyayā* instead of *tac-cakra-gata-dala-saṅkhyayā*. The Adyar edition does not mention these variants.
The reading *gaṇita* here could be taken in the sense of "reckoned". Thus the variant would mean: "according to the number of the petals reckoned for this_and_that/each (*tat-tac*) *cakra*[s]".

1031 The manuscript A reads *-vacanam* instead of *-sūcanam.* It means "Speech/Speaking of other's existent [...]".

1032 The manuscript B reads *praharaṇe°* "to remove" instead of *prahāṇe°*.

1033 The term *ghṛṇā* might mean "disgust" and "pity". Of these two meanings, I prefer "disgust", because the other fruits listed here, shame, fear etc., are all negative ones. But the commentary S seems to consider it to mean "pity". On the other hand, we could perhaps interpret *duḥkha-prahāṇecchā* as a *dvandva* compound. In this case, the gloss of the commentary S would mean, "the wish of others' pain (*duḥkha*) and abandoning/neglecting (*prahāṇa*)".

depression") is pain with anxiety[1034]. **It**, the *Maṇipūraka Cakra*, is **the abode**, [namely] the place, **of the sun (*bhāṇu*)**, [namely] of the vital wind (*prāṇa*) called "sun" (*sūrya*). (Comm. S on SR śl.125 and 126)

He relates the *Anāhata Cakra*: "**At/In the heart**" (śl.126c, *hṛdaye*). **The place of worship of Śiva whose shape is *praṇava***, [namely] whose shape is *Om*. The meaning is that [one] should worship with [one's] mind the Supreme Lord situated in it (=the *Anāhata Cakra*), meditating [on him].

He relates the fruits through the self's settling down on the twelve petals, [i.e.] the first/eastern [petal] etc.: "**Inconstancy**" (śl.127c, *laulya*). The state of the inconstant (*lola*) is **inconstancy (*laulya*)**; its **destruction** is immobility. "**Clear**" **(*prakaṭa*)**: [1035] qualification[1036] of reasoning through clearness. **Reasoning (*vitarka*)** is the reflection of acceptability of either of (lit. between) two [options]. **The state of being repentant** is the state of having remorse. **Hope** [means] longing for getting [an object] wished for. **Splendour/Renown (*prakāśa*)** is lack of hiding the form. **Consideration** [means] reflecting solely. **Striving/Desire (*samīhā*)** is the wish for preventing the unwanted/harmful.[1037] **Equanimity** is the consideration of all with absence of superiority or inferiority. **Deceit/ vanity** [means] carrying out a deed with disbelief for the sake of the conciliation/propitiation of people. **Defectiveness/weakness** is unsteadiness. **Discernment** [means] doing a work after reflecting thoroughly/ properly. **Egotism/consciousness_of_"I"** is the exertion of desiring to do, even when the work is unable [to be done].

1034 The manuscript B reads *sañcitaṃ* instead of *sacintaṃ*. It would mean, "Depression is an accumulated pain".

1035 I put it in [" "] because of *iti*.

1036 The manuscripts A and B read *kriyā-viśeṣaṇam* instead of *viśeṣaṇam*. It would mean, "the qualification of the verb of reasoning through clearness".

1037 The manuscript B reads *'niṣṭaniṣṭhanirvṛttīcchā* instead of *aniṣṭa-nivṛttīcchā*. It would mean, "[…] the wish of ceasing (*nirvṛtti*) related to (*niṣṭha*) the unwanted/harmful".

At/In the throat is the place of Speech (*Bhāratī* = *Sarasvatī*, the goddess of Speech), the sixteen-petalled *Viśuddhi*.[1038] (śl.129cd)

There,

praṇava[1039], *udgītha*[1040], *huṃ-phaṭ*[1041], *vaṣaṭ*[1042], further *svadhā*[1043], *svāhā*[1044], salutation (*namas*), ambrosia, the seven [musical] notes (*svara*), [i.e.] *Ṣaḍja* etc., [and] poison[1045],
such (*iti*) are the sixteen fruits when the self is situated on the first/eastern etc. petal.[1046] (śl.130–131ab)

The twelve-leaved/petalled *cakra* called *Lalanā*[1047] is at/in the

1038 Cf. Cakrakaumudī 5,1cd: *kaṇṭhe padmaṃ ṣoḍaśāraṃ svaraiḥ ṣoḍaśabhir yutam.*

1039 *Praṇava* is the sacred syllable *om.*

1040 *Udgītha* is the chanting of the Sāmaveda.

1041 *Huṃ-phaṭ* is problematic, because it can be counted either as one or two fruits. Also see my footnote 1045 on "poison" in this verse, śl.130.

1042 *Vaṣaṭ*. The exclamation used in making an oblation to a deity.

1043 *Svadhā*. An exclamation uttered in offering an oblation to the deceased ancestors (*pitṛ*).

1044 *Svāhā*. An exclamation used in offering oblations to the gods.

1045 The manuscript D reads *-dayaḥ kramāt* "[[...] *ṣaḍja*] etc., respectively" instead of *-dayo viṣam*. In this case, we would have to count *huṃ* and *phaṭ* as two separate fruits. Otherwise, the number of the fruits would be only fifteen.

1046 This accords with the statement of AVALON 1924, p.141, "in the Vishuddha (*sic.*) the seven subtle "tones" Nishāda, Rishabha, Gāndhāra, Shaḍja, Madhyama, Dhaivata, Panchama; certain Bījas, Hūng, Phat, Vaushat, Vashat, Svadhā, Svāhā, Namah; in the eighth petal "venom", and in the sixteenth "nectar" ". He does not give the reference, but it is presumably quoted from the Adhyātmaviveka, which is often referred to by AVALON 1924, in discussing the fruits of the petals. AVALON 1913 (introduction, lxi.) also makes the same statement.

1047 The *Lalanā Cakra* is not included in the six-*cakra* (+ *Sahasrāra*) theory. AVALON 1924 (p.125) describes it, based on another text than the Ṣaṭcakranirūpaṇa.
According to AVALON (ibid., p.125), the fruits in the petals of the *Lalanā Cakra* are *śraddhā, santoṣa, aparādha, dama, māna, sneha, śuddhatā, arati, sambhrama* and *ūrmi*. This list is almost identical to that given in the SR, although it contains some deviations. It has *santoṣa* instead of *toṣa* in the SR; *dama* instead of *mada*; *śuddhatā* instead of *lubdhatā*. It lacks *śoka* and *kheda* of the SR. The list starts from *śraddhā* which is the tenth member of the list in the SR. Though AVALON does not give the reference, it seems to be information contained in the Adhyātmaviveka.

uvula (*ghaṇṭikā*). (śl.131cd)

Intoxication (*mada*)[1048], pride, further affection, grief, distress/ weariness (*kheda*[1049]), greediness,
lack of rejoicing/pleasure, and agitation, the wave [of infirmity], trust, satisfaction and being obstructive (*śraddhā-toṣoparodhitāḥ*) [1050] (śl.132)
are[1051] the fruits on the petal of the east etc., in the *Lalanā Cakra*. Thus (*iti*) [they say]. (śl.133ab)

Comm. K on SR śl.129cd–133ab

"***Udgītha***" is the second part of the *Sāman*, for [there are] five parts of the *Sāman*, namely (*iti*) *prastāva* ("introduction")[1052] as the first, *udgītha* ("singing aloud") as the second, *pratihāra* (lit. "beating back") as the third, *upadrava* (lit. "assault") as the fourth [and] *nidhana* (lit. "destruction/loss") as the fifth.

"***Praṇava***" is the making/pronouncing of *Om*, to be utilised at the beginning of the *Udgītha*. **At/In the uvula (*ghaṇṭikā*),** [means] at the root of the tongue. **The wave (*ūrmiḥ*)** is waves; craving for eating,

---

Peculiarly, AVALON 1913 (introduction, lxii.) includes *śoka* and *kheda* in the above-mentioned list. The reason why these two terms are not mentioned in AVALON 1924 is not clear. It might just be a mistake caused by his carelessness, because AVALON 1924 lists only ten fruits, though at the same time he states that the *Lalanā Cakra* has twelve petals.

1048 The Ānandāśrama edition (1896) reads *medo* "fat" which is obviously a mistake. The editor himself corrects it into *mado*.

1049 The Adyar edition (p.64, variant 3, 1 ed.) notes that the Ānandāśrama edition reads *svehaś ca* instead of *khedaś ca*. But the Ānandāśrama edition (1896) actually reads *svedaś ca* "and sweat". This does not fit the context.

1050 The Ānandāśrama edition (1896) reads *śraddhā toṣoparodhitā* (which means the same, otherwise "belief, being obstructive of satisfaction"), which is not noted by the Adyar edition.
The manuscript D reads *śraddhā dveṣo virodhitā*, "trust, hatred, antagonism/being_obstructive".

1051 Optative *syuḥ*.

1052 The Ānandāśrama edition (1896) reads *praṇavaḥ* "the sacred syllable *Om*" instead of *prastāva*. The Adyar edition does not note it. This variant seems to be a mistake caused by the *praṇava ity udgīthasyā 'dau* in the line below.

craving for drinking, grief, confusion, old age [and] death are the six waves[1053] (*ṣaḍ-ūrmayaḥ*).[1054]

Comm. S on SR śl.129cd–133ab

He relates the *cakra* called *Viśuddhi*: "**At/In the throat**" (śl.129a, *kaṇṭhe*). **The place of Speech (*Bhāratī*): the place**, the resort, **of Speech (*Bhāratī*)**, of Sarasvatī. The meaning is that (*iti*) the deity of speech (*vāc*), called/summoned there into existence (*bhāvita*), gives the might (*vaibhava*) of speech.

He relates the sixteen fruits of this *cakra* with regard to the self's situation on the first/eastern etc. petal: "**There**" (śl.130a, *tatra*). ***Praṇava*** is the making/pronouncing of *Om*. Through the investigation/arrangement of *praṇava* on the first/eastern leaf, the self attains the highest bliss (*śreyas*). ***Udgītha*** is a particular division of the *Sāman*. The meaning is that (*iti*) one should ponder over [the self] situated on the second petal, in the intentness/meditation (*upāsanā*) upon *udgītha*. "***Huṃ-phaṭ***" is an indeclinable word (*avyaya*) with regard_to/during offering an oblation.[1055] "***Vaṣaṭ***" [is the same,] too. "***Svadhā***" is an indeclinable word with regard_to/during offering an oblation directed towards (*uddeśena*) the fathers. "***Svāhā***" is [the same] directed towards a deity. "**Salutation**" **(*namas*)** is an indeclinable word with regard_to/during salutation (lit. "the making of *namas*"). The meaning is that (*iti*) these are to be pondered over on the respectable (*tattad*) petals. **Ambrosia** is nectar (*pīyūṣa*). The meaning is that (*iti*) the self situated on that petal becomes pleased as if bathed/overflowed with nectar. **The seven [musical] notes, [i.e.] *Ṣaḍja* etc.**: *Ṣaḍja, R̥ṣabha, Gāndhāra, Madhyama, Pañcama, Dhaivata* [and] *Niṣāda*. Their respective (*tāni tāni*) petals are the

1053 I.e. the six waves of human infirmity. The dictionary of Apte quotes the Bhāgavatapurāṇa 10,70,17: *śoka-mohau jarā-mr̥tyū kṣut-pipāse ṣaḍ-ūrmayaḥ / prāviśad yan niviṣṭānāṃ na santy aṅga ṣaḍ-ūrmayaḥ //.*
This term is mentioned for the first time in the Br̥hadāraṇyakopaniṣad 3,5 (cf. HALBFASS 2000, p.61).

1054 *Ūrmiḥ ūrmayo 'śanāyā- […] -maraṇāni ṣaḍ-ūrmayaḥ*. The Ānandāśrama edition (1896) lacks the foremost word, *ūrmiḥ*.

1055 The manuscript A reads *pradāneṣv ayam* instead of *pradāne avyayam*. It reads the same for the next gloss *svadheti* (variant 7). But there is no variant for *nama iti namaskāre avyayam*, below.

[respective] places of intentness/meditation (*upāsanā*). **Poison** is to be explained like ambrosia. The meaning is that (*iti*) the self situated on the petal of poison becomes pained.

He relates the *cakra* called *Lalanā*: "**Called *Lalanā***" (śl.131a, *lalanākhyam*). **The *ghaṇṭikā*** [means] the nape of the neck (*avaṭu*). He relates the fruits on the petals of this [*cakra*]: "**Intoxication**" (śl.132a, *mada*). **Intoxication (*mada*)** [means] being intoxicated (*mattatā*). **Pride** [means] not seeing/caring_for what is to be done and what is not to be done, due to the notion of [one's] superiority. **Affection (*sneha*)** is attachment/fondness (*snigdhatā*). **Grief (*śoka*)** is the pain whose cause is known. **Distress/Weariness (*kheda*)** is the pain whose cause is unknown. **Greediness** is the excess of longing. **Lack of rejoicing/pleasure** is anxiety (*udvega*) even with regard to the means of pleasure. **Agitation (*sambhrama*)** is flurry (*āvega*). **The waves** [of infirmity] are six, well-known in the six traditional scriptures (*āgama*) and theoretical works (*śāstra*) (or: six *āgama-śāstra*-s): "Craving to eat as well as craving to drink, grief and confusion, aging and dying: these are (*iti*) the six waves, established together in the vital_wind (*prāṇa*), consciousness (*buddhi*) [and] the body.[1056]"

Trust is the consciousness of the existence [of the other world]. **Satisfaction (*toṣa*)**[1057] is well-known. **Being obstructive** is impoliteness (*adākṣiṇya*)[1058].

SR śl.133cd–145ab

In the middle of the eyebrows is the three-petalled[1059] *cakra* called *Ājñā*.

1056 The *mūla* text (SR śl.132c) contains sg. *ūrmiḥ*, while the comm. S pl. *ūrmayaḥ* as the *pratīka*.
The manuscript A (variant 4) reads *prāṇa-buddhir deheṣu* instead of *prāṇa-buddhi-deheṣu*.

1057 The manuscript A reads *roṣaḥ* "rage", and the manuscript B reads *doṣaḥ* "defect", instead of *toṣaḥ*.

1058 The Adyar edition gives *uparodhitā dākṣiṇyam*. But I take this to be a printing mistake and modify it to *uparodhitā a̱dākṣiṇyam*, although the Adyar's reading could be correct, if *uparodhitā* means "protection/favor".

1059 The Ānandāśrama edition (1896) reads *dvidalaṃ* "two-petalled" instead of *tridalaṃ*. This does not fit the context at all.

[Its] fruits, however, (śl.133cd)
are deemed [to be] the manifestations[1060] of *sattva*, *rajas* and *tamas* respectively.[1061]

Even after that [*cakra*][1062] is/exists (*asti*) a_six-petalled *cakra*, [i.e.] ["]the mind["] (i.e., the six-petalled *Manas Cakra*).[1063]

Its fruits, however, are (śl.134)
thus (*iti*): sleep/dream (*svapna*) and enjoyment of tastes[1064], smell, getting (= perception) of shape, touching and perception of sound, on the first/eastern etc. petals. (śl.135)

It is proclaimed that (*iti*) even after that [*cakra*] [is] the sixteen-petalled *Soma Cakra*.

Sixteen *kalā*-s[1065] are established together on its sixteen[1066] petals. (śl.136)

The[se] fruits arise, of the individual self (*jīva*) situated on (lit:

1060 The Adyar edition notes the variant of C.E. which reads singular, *āvirbhāvaḥ* and *mataḥ*.

1061 The fruits of the petals of the *Ājñā Cakra* and *Manas Cakra* mentioned here accord with those listed in the Adhyātmaviveka which is referred to by AVALON 1924, p.141 (also see AVALON 1913, introduction, lxiii.). The fruits of the petals of the *Manas Cakra* are listed in the order reverse to that of the SR: *śabda-jñāna, sparśa-jñāna, rūpa-jñāna, āghrāṇopalabdhi, rasopabhoga* and *svapna*.
It is remarkable that the SR and the Adhyātmaviveka are parallel to each other, in the statement on the *Manas Cakra* which is not included in the popular six-*cakra* (+ *Sahasrāra*) system. This fact indicates that the *cakra* system of the Adhyātmaviveka is the same one as that of the SR.

1062 The Ānandāśrama edition (1896) reads *ato* instead of *tato*.

1063 The *Manas Cakra* and *Soma Cakra* are not included in the six-*cakra* (+*Sahasrāra*) system of the Ṣaṭcakra-nirūpaṇa. AVALON's (1924, pp.130–131) description of these two *cakra*-s is based on the Adhyātmaviveka. AVALON (ibid.) mentions it on p.158, too.

1064 The manuscripts *ka., ga.* and *ṅa.* read *rasopayoga* instead of *rasopabhoga*. But obviously *rasopayoga* "employment of tastes" does not fit the context.

1065 The term *kalā* "a small part of something", "a bit" or "a sixteenth part" may mean "a digit of the moon", because the digits of the moon are deemed sixteen. This term presumably occurs here because *soma* is associated with the moon.
According to AVALON 1913 (introduction, lxii.; lxiii.), the *Soma Cakra* and *Lalanā Cakra* are called *Kalā Cakra*-s in some other Tantras.

1066 The Ānandāśrama edition (1896) reads *ṣoḍaśeṣv* instead of *ṣoḍaśasv*. But it is grammatically incorrect, unless derived from *ṣoḍaśa*- "sixteenth", which makes no sense here.

going to) the first/eastern etc. petals:[1067] (śl.138)

Pity, tolerance, straightness, patience (*dhairya*), detachment, firmness/resolution (*dhṛti*) and cheerfulness[1068], laughter, the heaping_up/collection (*nicaya*) of thrill [of joy or horror] (lit. "thrill_of_ hairs/horripilation"), the tear of meditation[1069], further steadiness[1070], (śl.137)

earnestness, striving, transparency/pureness (*acchatva*)[1071], generosity and fixedness/concentration, respectively.[1072]

The thousand-petalled (*Sahasrapatra*) *cakra* holding ambrosia[1073],

1067 The Adyar edition notes the variant of C.E. (Calcutta edition, cf. *On the editions of SR*), *phalam bhidyanti* instead of *phalāni udyanti*. Though *phalam bhidyanti* is grammatically incorrect, it perhaps refers to the piercing of the fruits [with the tail of the *kuṇḍalinī*], when the self is situated on the eastern petal etc. (The verb √*bhid* "to pierce" is also contained in the commentary K on SR śl.120–129.)

1068 The Ānandāśrama edition (1896) reads *vairāgyadhṛtisammadāḥ* instead of *vairāgyaṃ dhṛti-sammadau*. The meaning is the same.

1069 For the thrill of the hairs and the tear of meditation, see Kubjikāmatatantra 11,94cd–98ab (HEILIJGERS-SEELEN 1990), "For the Viśuddhi are mentioned sixteen states which are perceptible (*pratyakṣa*), such as a flow of tears and the thrilling of the hair".

1070 The manuscript *ṅa.* reads *dhyāna-susthiratā* "Much steadiness of meditation".

1071 The manuscript C reads *gāmbhīrya-madam auddhatya* "earnestness and intoxication, arrogance"; the manuscript D reads *udyamojastvam* "striving and being/having *ojas*".

The Ānandāśrama edition (1896) reads *udyamocchratvam*, though the Adyar edition does not note it. The editor himself modifies it into *uddyamo 'cchatvam*.

1072 According to AVALON 1924, p.141, the Adhyātmaviveka lists *kṛpā, mṛdutā, dhairya, vairāgya, dhṛti, sampad, hāsya, romāñca, vinaya, dhyāna, susthiratā, gāmbhīrya, udyama, akṣobha, audārya* and *ekāgratā*. The order is the same as that in the SR. The terms mentioned here are almost the same as in the SR, except for *mṛdutā* instead of *kṣamārjava* in the SR; *sampad* instead of *sammada* in the SR; *vinaya* instead of *nicaya* in the SR; *dhyāna* and *susthiratā* instead of *dhyānāśru* and *sthiratā* in the SR; *akṣobha* instead of *acchatva* in the SR.

The Adhyātmaviveka's readings, *vinaya* for SR *nicaya* and *dhyānasusthiratā* for SR *dhyānāśru sthiratā* might be mistakes, as there is a parallel for the thrill of the hairs and the tear (see my footnote 1069 on SR śl.137).

1073 The manuscript *gha.* reads *sudhādhare* instead of *sudhādharam*. In this case, it would qualify *brahma-randhre*, i.e. "the thousand-petalled *cakra*, however, is at the aperture of Brahman holding ambrosia".

however, is at the aperture of Brahman[1074].

With the streams of the essence of ambrosia, it (*tad*) augments the body. (śl.139)

The individual self (*jīva*), situated at the first/eastern, at the eighth, further at the eleventh and at the twelfth petal of the *Anāhata [Cakra]*, attains the perfection of song etc. (śl.140)[1075]

Through the fourth, sixth and tenth petals, song etc. are (lit. "is") destroyed.

But the eight petals of the *Viśuddhi [Cakra]* beginning with the eighth, resorted to [by the individual self], give the complete perfection of song etc. The sixteenth is its (= song's) destroyer. (śl.142ab)

The tenth and eleventh petals in[1076] the *Lalanā [Cakra]*, however, are givers of perfection. (śl.142cd)[1077]

They (i.e. the wise ones) know the first, fourth and fifth petals as destroyers[1078]. (śl.143ab)

The individual self (*jīva*), situated at the aperture of Brahman, satisfied as if inundated with ambrosia, perfects[1079] the actions of song etc. with excellence. (śl.144ab)[1080]

1074 Brahman, the Supreme Principle. WHITE 1996 translates *brahmarandhra* as the "cleft of Brahman".

1075 The Ānandāśrama edition (1896) reads *'vasthito* "situated/fixed" instead of *ca sthito*. The meaning is the same.

1076 The manuscript D has a genitive (*lalanāyās*) instead of the locative (*lalanāyāṃ*).

1077 The Ānandāśrama edition (1896) reads *susiddhide* "givers of good perfection" instead of *tu siddhide*. Besides this, the Adyar edition notes also the variant of C.E. (Calcutta edition, cf. *On the editions of SR*) *siddhaye* "for the sake of perfection".

1078 The Adyar edition notes that the commentary S reads *nāśanī* instead of *nāśakāni* (for *nāśanaṃ* of the *mūla* text). But actually *nāśanī* does not occur in the commentary S at all. However, on p.68, the variant *nāśanauti* of the manuscripts A and B is given, which can be explained as a mistake for *nāśanīti* (for *nāśakānīti*).

1079 Optative *sādhayet*. For the function of the optative, cf. footnote 1 on the Introduction to the *English translation*.
The Ānandāśrama edition (1896) reads *kārayet* "makes" instead of *sādhayet* "perfects".

1080 The Ānandāśrama edition (1896) reads *yadā* instead of *yathā*. "When (*yadā*) the individual self, situated at the aperture of Brahman, is inundated with ambrosia, [it], satisfied (*tuṣṭo*), perfects the work of song etc. with excellence." The manuscripts *ka., kha., ga.* and *gha.* read *yathā*.

Situated on the remaining petals of these [*cakra*-s] and in the other *cakra*-s,
the individual self (*jīva*) never attains the complete perfection of song etc.[1081] (śl.145ab)

(SR śl.145cd belongs to the next part of the text.)

Comm. K on SR śl.133–145

The tear of meditation:[1082] **The tear of meditation** is the tear born from meditation.

Comm. S on SR śl.133–145

He relates the three-petalled *Ājñā Cakra*: "**In the middle of the eyebrows**" (śl.133c. *bhrūmadhye*). He relates its fruits: "**The manifestations**" (śl.134a. *āvirbhāvāḥ*). On the first petal is the manifestation of *sattva*, on the second petal is the manifestation of *rajas*, on the third petal is the manifestation of *tamas*.

He relates the six-petalled *cakra* named *Manas*: "**After that**" (śl.134c. *tataḥ*). The ***Manas Cakra*** is the *cakra* whose appellation is ["]Mind["] (*manas*). The appellation ["]Mind["] in that [case] is, however, because of the location/situation of the mind.[1083] He mentions the fruits: "**Its fruits**" (śl.134d. *tat-phalāni*). **Sleep/Dream** is the state of sleeping. **Enjoyment of tastes**[1084] is the enjoyment of the taste[s] of food.[1085] **Smell** [means] knowing smell. **Getting (= perception) of form** is seeing. **Touching** [means] the knowing

1081 C.E. (Calcutta edition, cf. *On the editions of SR*) reads *anyeṣvavasthitaḥ* instead of *anyeṣu ca sthitaḥ*. The meaning remains unchanged, with *avasthitaḥ* meaning "situated, fixed".

1082 The Adyar edition repeats *dhyānāśru* twice, while the Ānandāśrama edition (1896) lacks the first of these.

1083 Or else: "But (*tu*) the name of "mind" is because of the mind's settling there (*tatra*)".
The manuscript E reads *tasya* instead of *tatra*.

1084 The manuscripts A and B read *rasopayoga* instead of *rasopabhoga*. But these two manuscripts seem to read *rasopabhogaś ca* for the *mūla* text (SR śl.135a). In contrast to that, the manuscripts *ka., ga.* and *ṅa.* read *rasopayogaś ca* for the *mūla* text.

1085 The manuscript B reads *rasanaṃ annarasasyopa°* instead of *annarasasyopa°*, "tasting, enjoying of the taste of food."

whose means (*karaṇa*) is the faculty/organ (*indriya*) skin.[1086] **Perception of sound** is hearing.

He relates the *cakra* called *Soma*: "**After that [*cakra*]**" (śl.136a, *tataḥ*). **Even after that**, [namely] in the upper region. He relates the fruits of the individual self (*jīva*) through location/situation upon its (= the *Soma Cakra*'s) first/eastern etc. petals: "**Pity**" (śl.137a, *kṛpā*). **Pity** is regard (*apekṣā*) to showing favour to others. **Tolerance/Forgiveness** is the absence of anger even when a reason of anger exists. **Straightness** is the state of having a non-crooked intellect (*buddhi*). **Patience** is having a consciousness (*cetas*) which has not wavered/faltered. **Detachment** is the cognition (*buddhi*) of the inessentiality of the world. **Firmness/Resolution (*dhṛti*)** is holding (*dhāraṇa*).[1087] **Cheerfulness** is gladness (*harṣa*). [The expressions], beginning with **laughter**, ending with **earnestness**, are well-known. **Striving (*udyama*)** is effort (*udyoga*). **Transparency/Pureness** is having a non-turbid mind. **Generosity** is bravery in granting by nature. **Fixedness/Concentration** [means] being fixed on a single object (*ekatānatva*).

He relates another *cakra*: "***Cakra***" (śl.139a, *cakraṃ*). **Holding ambrosia (*sudhā*)**, [namely] the holder of ambrosia (*amṛta*).[1088] He relates its function: "**It**" (śl.140, *tad*). Relating the individual self's (*jīva*) fruits in settling on these_and_those/the_respective petals of these *cakra*-s, he says [on each petal's] appointed utility: "**The *Anāhata***" (śl.140, *anāhata*-).

The individual self (*jīva*), situated **at the first/eastern (*pūrve*)** — ***pūrve*** [means] situated on the eastern side or the first — **and eighth** etc. **petal, of the *Anāhata Cakra*, attains the perfection of**

1086 The manuscript A reads *indriye karaṇakaṃ*; B reads *indriye kāraṇakaṃ*. In the case of A, it would mean "the knowledge/knowing, whose means (*karaṇa*) is in the organ/faculty (*indriya*) skin." In the case of B, "the knowledge/knowing, whose cause (*kāraṇa*) is in the organ/faculty skin."

1087 The manuscripts A and B read *tasyā dhṛtiḥ* instead of *dhṛtiḥ*. It could mean, "maintenance/firmness (*dhṛti*) of it (= *saṃsāra-asāratā-buddhi*)". This variant might be possible if we adopt, for the *mūla* text, the variant of the Ānandāśrama edition (1896), the compound *vairāgya-dhṛti-*, instead of *vairāgyaṃ dhṛti-* in the Adyar edition.

1088 The text *sudhādharam mṛtadhārakam* is obviously a mistake for *sudhādharam amṛtadhārakam*.

song etc., [namely] obtains the perfection of song, dance[1089] [and] instrumental [music], or else longs for them.[1090] When the individual self (*jīva*) is situated at **the fourth, sixth and tenth petals, song etc. are destroyed.** "**Of the *Viśuddhi***" (śl.141c, *viśuddher*): When the individual self (*jīva*) is established on **the eight petals, beginning with the eighth,** of the *cakra* named ***Viśuddhi***, song etc. are perfected.[1091] When [it] abides on **the sixteenth** petal, they are destroyed. He relates the individual self's (*jīva*) perfection or imperfection of song etc., with regard to [its] abiding on a particular petal of the *cakra* called *Lalanā*: "**The tenth**" (śl.142a, *daśama*). **The two petals**, namely (*iti*) **the tenth and the eleventh**, become **givers of the perfection** of song etc. The knowers of the *Yogaśāstra*-s **know** that (*iti*) **the first** etc. are, however, **destroyers** of song etc. He relates the perfection of song etc. of the individual self (*jīva*) through its abiding at the aperture of Brahman: "**The aperture of Brahman**" (śl.143a, *brahmarandhra-*). Being situated at **the aperture of Brahman**, having reached the cessation (*nirvṛti*) as if **bathed** with ambrosia, **satisfied**[1092], [i.e.] without longing, **the individual self (*jīva*) perfects**[1093], [i.e.] brings about, song etc. Thus relating the perfecting and destroying petals of song etc.,[1094] he relates the common [petals]: "**Of these**" (śl.144c, *eṣām*). The meaning is that (*iti*): **Situated on the remaining petals**, [i.e.] on the ones other than the perfecting and destroying [ones], **of these** *cakra*-s called *Anāhata, Viśuddhi* and *Lalanā*, **and in the other *cakra*-s**, [i.e.] in [the *cakra*-s] other than

1089 The manuscript B reads *nṛttya* instead of *nṛtta*. But I prefer the reading *nṛtta* (see my footnote 361 on SR śl.1, *nṛtta* "dance").

1090 The commentary perhaps read *saṃsthito* for *ca sthito* of the *mūla* text.
The manuscripts A and B read *dale pūrvadale* instead of *dale pūrve*.
The manuscript A reads *sthito* instead of *saṃsthito*. The A reads *pratikarṣati* "to reject" instead of *abhilaṣati*, though it does not fit the context.

1091 The commentator paraphrases the verse by using the original expressions. The expression *gītādi siddhyate* is the gloss to the original, *dadyur gītādi-saṃsiddhiṃ*.

1092 The commentary contains *plutaḥ* and *santuṣṭo* instead of *samplutaḥ* and *tuṣṭo* of the *mūla* text.

1093 Optative *sādhayet*.

1094 The manuscript A reads *tad-dalāni ca* "the petals of them (= the *cakra*-s)" instead of *ca dalāni*.

Viśuddhi etc., [namely] in the *Maṇipūraka Cakra* etc., **the individual self (*jīva*)** is indifferent with regard to the perfection **of song etc.**

SR śl.145cd–150[1095]

Upward by two fingers from the base (*ādhāra*)[1096] [and] downward by two fingers from the penis (*mehana*), (śl.145cd)[1097]
is the *dehamadhya* (lit. "middle of the body"), one finger [in breadth], whose lustre is [that of] heated/melted gold[1098].
A slender flame of fire is situated there.[1099] (śl.146abc)
At nine fingers from that *cakra*[1100] is the lump/bulb (*kanda*)[1101] of

1095 The theory dealt with in SR śl.145cd–163ab is different from the theory of the *cakra*-s explained by the foregoing verses (up to śl.145ab). Passages parallel to SR śl.145cd–163ab are contained in the fourth chapter entitled Śarīravyavacchedavidyā of the Yogayājñavalkya (YY) (cf. *Situating the text §2.3., "Parallelism of SR and YY"*).

1096 The term *ādhāra* here denotes something else than the *Ādhāra Cakra* mentioned in SR śl.120. The Yogayājñavalkya (YY) mentions the areas of the body which are important for the *kuṇḍalī*. These areas overlap the *cakra*-s of Haṭhayoga, but the YY does not call them *cakra*-s, except for the navel *cakra*. Cf. *Situating the text §2.3.6., "Cakra in YY and its parallel in SR"* and GEENENS 2000, p.202.

1097 YY 4,14 might be a parallel: *gudāt tu dvy-aṅgulād ūrdhvam adho meḍhrāc ca dvy-aṅgulāt / deha-madhyaṃ tayor madhyaṃ [...] //*. "[...] le centre du corps / est à deux doigts au-dessus de l'anus / et à deux doigts au-dessous du sexe, / juste au milieu entre les deux." (GEENENS 2000's translation). The YY contains the term *guda* "anus" instead of *ādhāra* "base". GEENENS ibid., p.79, reports that YY 4,45 considers the area of the anus to be the base of the tubes which radiate from it.

1098 The term *jāmbūnada* literally means "gold from the *Jambū* river".

1099 YY 4,11bc and 13ab are parallel to this. YY 4,11cd, *dehamadhye śikhi-sthānaṃ tapta-jāmbūnada-prabham*. YY 4,13ab *tan-madhye tu śikhā tanvī śivā tiṣṭhati pāvakī*. GEENENS 2000 translates, "La flamme mince, ou le feu, est toujours là / au milieu, chez tous ceux-là." I presume that the edition he used contains a term meaning "toujours", supposedly *sadā*, instead of *śivā*. As a matter of fact, Brahmānanda's commentary on the Haṭhayogapradīpikā, 3,66 quotes YY 4,11cd–13ab, reading *sadā* instead of *śivā*.

1100 YY 4,16 which is parallel to this verse (see below) states "nine fingers from the middle of the body" (*deha-madhyān navāṅgulam*). According to this, "that *cakra*" of the SR is the *dehamadhya* ("the middle of the body").
SHRINGY 1999 presents two possibilities of interpretation: "that *cakra*" might be either *ādhāra* (SR śl.145c) or *dehamadhya*. He prefers the latter. The commentary S also mentions the two possibilities.

the body, four fingers in elevation/thickness and width. (śl.146d–147ab)[1102]

Its name is mentioned as "the knot of Brahman" (*brahma-granthi*) by the ancient ones. (śl.147cd)[1103]

But in its middle, the twelve-spoked *Nābhi* ("navel") *Cakra*[1104] is situated. (śl.148ab)[1105]

The evidence suggests, however, that "that *cakra*" can be nothing but the *Nābhi Cakra*, because only the navel is called *cakra* in the old theory of Yoga which the YY deals with (cf. *Situating the text §2.3.6.*).

1101 The Haṭhayogapradīpikā makes a different statement that the *kanda* is situated between the penis and navel, cf. AVALON 1924, p.151. The Ṣaṭcakranirūpaṇa śl.1 also locates the *kanda* in the same area. Its commentary explains that the *kanda* is the root of all the tubes (*nāḍī*); its root (*kandamūla*) is situated two fingers above the anus and two fingers below the penis; it has the shape of a bird's egg and has the width of four fingers (cf. AVALON 1924, *translation*, p.4 and p.7).
The same statement is also made by the Dhyānabindūpaniṣad (in the Yogopaniṣad) v.50cd (p.198): *ūrdhvam meḍhro adho nābheḥ kando yo 'sti khagāṇḍavat /50cd/ tatra nāḍyaḥ samutpannāḥ sahasrāṇi dvisaptatiḥ / teṣu nāḍī-sahasreṣu dvisaptatir udāhṛtā /51/.*

1102 YY 4,16 is parallel to SR śl.147ab: YY 4,16 *kanda-sthānam manuṣyāṇāṃ deha-madhyān navāṅgulam / caturaṅgulam utsedham āyāmaṃ ca tathāvidham //.* "Chez l'homme, l'emplacement du Balbe de vie / est à neuf doigts au-dessus du centre du corps. / Sa longueur est de quatre doigts, / et sa largeur de même proportion" (GEENENS 2000's translation).
When compared with the YY, the variant of C.E. (Calcutta edition?, cf. *On the Editions of SR*) noted by the Adyar edition, *°ndohyutsadhoyannābhyāṃ*, is obviously a mistake.

1103 C.E. reads *°ktam asya* instead of *°ktaṃ tasya*. The meaning remains unchanged.
The YY has no parallel to SR śl.147cd.
For the term *brahma-granthi*, cf. AVALON 1924, p.151. It is mentioned again in the next chapter (SR 1,3, śl.4a).

1104 The Navel *Cakra* (*nābhi-cakra*) is different from the *Maṇipūraka Cakra* mentioned in SR śl.124. The Navel *Cakra* belongs to the old theory of Yoga. This theory is different from the *cakra* theory dealt with in SR śl.120–145ab, as already noted (cf. *Situating the text §2.3.6.*).

1105 YY 4,18 is parallel to SR śl.148ab: *tan-madhyaṃ nābhir ity uktaṃ nābhau kanda-samudbhavaḥ / dvādaśāra-yutaṃ tac ca tena dehaḥ pratiṣṭhitaḥ //* (Trivandrum edition). GEENENS 2000's translation is, "On affirme que le nombril est juste au milieu du ventre. / Là se trouve l'origine des Roues. / L'une part peut être comparée à une Roue / qui a douze rayons. Sur elle le corps est fixé." The edition which GEENENS ibid. consulted seems to read the

There, this individual self (*jīva*) wanders about like a spider[1106] situated in a net of [spider] threads. (śl.148cd)[1107]

It ascends through the *Suṣumnā* to the aperture of Brahman, [and] descends. (śl.149ab)[1108]

The individual self is mounted on the vital wind (*prāṇa*) like an acrobat (*kohlāṭika*) [1109] on a rope.[1110] (śl.149cd)[1111]

---

text differently, as shown by the fact that the Trivandrum edition consulted here contains no expression corresponding to "l'origine des Roues". GEENENS (ibid. p.202 and 218) reports that the old theory of the YY mentions only the navel *cakra* (cf. *Situating the text §2.3.6.*).

1106 This simile of a spider occurs in Bṛhadāraṇyakopaniṣad 2,1,90. For the occurrences of this simile in Tantric texts, cf. AVALON 1924, p.153. He refers to the second verse of the Ṣaṭcakranirūpaṇa stating that the tube (*nāḍī*) called *Citriṇī* which is situated inside the *Suṣumnā* tube, is as fine as the thread of a spider (*lūtā-tantūpameyā*).
The simile of a spider is also mentioned by KANE 1990 (p.447), as contained in the YS and Śaṅkara's Śārīrakabhāṣya.
Intriguingly in Bṛhatkathāślokasaṅgraha 17,140, *tantu-cakra* is used as meaning "a mass of webs"; it describes a *vīṇā*, whose body is filled with webs inside, cf. ZIN 2004, p.329.

1107 YY 4,19 is parallel to SR śl.148cd: *cakre 'smin bhramate jīvaḥ puṇya-pāpa-pracoditaḥ / tantu-piñjara-madhyastho yathā bhramati lūtikaḥ //* (Trivandrum ed.). GEENENS translates "À partir de là se meut l'âme individuelle, / mobilisée en effet par le bien et par le mal. / Comme une araignée prête à courir sur le fil, / tapie au milieu de sa toile."

1108 The Adyar edition notes the variant of the Ānandāśrama edition (1896), *suṣumnāyā*. But in reality, the Ānandāśrama edition contains the same reading as in the Adyar edition. Besides, the Ānandāśrama edition notes that the manuscript *gha.* lacks SR śl.148cd.
The manuscript D reads *rajjvā* instead of *rajjvāṃ.*

1109 The acrobat (*kohlāṭika*) is mentioned in the SR's seventh chapter on dancing: "The *kohlāṭika* is accepted to be one who can carry a heavy burden, who is well versed in *bhramarikā* etc., who is clever in rope-walking, who is an expert in dancing with a dagger, and who is clever in using weapons." (SR śl.1330–1331, translated by RAJA & BURNIER 1976, p.199). In SR śl.149cd, it means a rope-walker.

1110 The manuscript D reads *rajjvā* instead of *rajjvāṃ.*

1111 To SR śl.149, no parallel is found in YY, though YY 4,23–24 deals with a similar topic.

Beginning from the lump/bulb (*kanda*) to the aperture of Brahman,[1112] tubes (*nāḍī*)[1113] around the *Suṣumnā*, situated forming a lump/bulb[1114], spread over (*tanvate*) the body (*tanu*), the lump/bulb,[1115] through branches.[1116] (śl.150)[1117]

Comm. K on SR śl.145cd–150[1118]

He mentions the place and the nature[1119] of the knot of Brahman, which is (*bhūta*) the root of the multitude/totality of tubes (*nāḍī*) maintaining the body: "**By two fingers from the base (*ādhāra*)**" (śl.145c, *ādhārād dvyaṅgulād*). **From the base (*ādhāra*)**, [namely]

1112 D reads *ākandād* instead of *kandād ā*. "As far as the bulb (*kanda*), from the aperture of Brahman".

1113 The term *nāḍī* originally means "the tubular stalk of any plant". It also means "flute" in ṚV 10,135,7; Kāṭhakasaṃhitā 23,4,5 and 34,5,6. In Kāṭhaka 12,10, it means "drinking-straw". In summary, the term *nāḍī* seems to denote a tubular vessel in which fluids flow.
For the term *nāḍī* in a medical context, cf. JOLLY, p.109; DAS 2003A, p.560, on *nāḍī*.
But the *nāḍī* in SR śl.145cd–163ab which is parallel to the YY should be understood in a Haṭhayogic context. For *nāḍī* in Haṭhayoga, cf. AVALON 1924, pp.111–117.

1114 *Prāṇa* is mentioned as being situated in the middle of the body in ŚB 7,1,2,13 (WEBER's ed., p.576, l.4), *prāṇo mádhyamātmā* "The breath is in the middle of the body". Also cf. ŚB 7,3,1,2 (WEBER's ed., p.587, l.11), *ayam ātman prāṇo madhyataḥ*.

1115 The commentary S contains *kroḍīkṛtya* instead of *kandīkṛtya* of the *mūla* text.

1116 SU śārīra., 7,23 compares the navel to the bulb (*kanda*) of a lotus: *vyāpnuvanty abhito dehaṃ nābhitaḥ prasṛtāḥ sirāḥ / pratānāḥ padminī-kandād bisādīnāṃ yathā jalam*.

1117 The YY does not have a parallel to SR śl.150, although YY 4,25 mentions the tubes (*nāḍī*) which are situated around the *Suṣumnā*, in the *dehamadhya* (*kanda-madhye sthitā nāḍī suṣumneti prakīrtitā / tiṣṭhanti paritaḥ sarvāś cakre 'smin nāḍi-sañjñakāḥ*).

1118 The editor of the Adyar edition falsely separates SR śl.145cd from śl.146ab, although they actually seem to be a pair. But both commentators correctly begin with explaining śl.145cd.

1119 The Ānandāśrama edition (1896, p.28) reads *vā* instead of *cāha*. The Adyar edition does not note it. "The place or (*vā*) nature of the knot of Brahman [...].".

from the *Ādhāra Cakra*, which is (*bhūta*) the limit.[1120] "***Dehamadhya***" **("the middle of the body")**. (śl.146a): **The *dehamadhya*** (= middle of the body), because of its being situated in the middle of the upper and lower parts of the body. "**From that *cakra***" (śl.146d. *cakrāt tasmāt*). From the *cakra* of the *dehamadhya* (= middle of the body), [which is] the limit.[1121] "**The knot of Brahman**" (śl.147c): **The knot of Brahman** because of being the abode of (lit. "being stepped upon by") Brahman. **In its middle**, [namely] in the middle of the knot of Brahman. **The twelve-spoked *Nābhi Cakra***[1122]: that whose spokes are twelve; thus [it] is said. The net of [spider] thread is an example of the *Nābhi Cakra* because of [its] having the shape of an assemblage of tubes (*nāḍī*) whose nature is extension [and] expansion. **Forming a lump/bulb**: making [something which is] not a lump/bulb into a lump/bulb. "[The suffix] *cvi* with regard to (= in the case of) becoming that (*tadbhāva*) which was not [before]" (Pāṇini 5,4,50, Kāśikā accepts *abhūtatadbhāve* as part of the *sūtra*.) And those tubes (*nāḍī*) **spread over (*tanvate*) the body (*tanu*)**, [i.e.] the form of the body (*deha*), [namely] **the lump/bulb**, [which is] made into a lump/bulb, **through** their/its **own branches.**[1123]

1120 This commentary misunderstands the proper meaning of the term *ādhāra*. It simply means "the base area of the body" in the context here, cf. my footnote 1096 on SR śl.145c *ādhāra*.
The Ānandāśrama edition (1896) reads *ādhārād ity* instead of *ādhārād*. The Adyar edition does not note it.

1121 The Ānandāśrama edition (1896) reads *deha-madhya-cakrād avadher* and *brahmādhiṣṭhitatvād* in one line, lacking *brahmagranthir* between them. This obviously does not fit the context. The Adyar edition does not note this.

1122 The Ānandāśrama edition (1896) contains *nābhir eva cakraṃ* after *nābhicakraṃ*. This is not contained in the Adyar edition. It means "*Nābhi Cakra* (Navel *Cakra*): the very navel is a *cakra*."

1123 *Svaśākhābhiḥ* could qualify either the tubes or the bulb.
Instead of *kandaṃ*, the Adyar edition notes the variant of the Ānandāśrama edition, *kandasya* (cf. Adyar ed., p.69, variant 2). But in reality, the Ānandāśrama edition does not contain this variant, but reads the same as the Adyar.

He relates the place of the flame of fire: "**From the *Ādhāra*/base**"[1124] (śl.145a, *ādhārād*). In the region **upward** by the measure of **two fingers from the *Ādhāra*** *Cakra* [and] in the region **downward** by the measure of **two fingers from the penis (*mehana*)**, the place of the genitals, is the gold-coloured ***dehamadhya***, **whose size is one finger**, [namely] the middle (*madhya*) region of the body (*deha*). **There, a slender**, [namely] thin, **flame of fire is situated. At nine fingers**, [namely] at the distance of nine fingers, **from that *cakra***, [namely] from the *dehamadhya* or from the *Ādhāra*.

He relates the knot of Brahman: "**Of the body**" (śl.147a, *dehasya*). **The lump/bulb of the body (*deha*)**, [namely] the body frame (*śarīra*), is the cause of extension. **Elevation/Thickness** is long increase. Situated at the navel, [and] whose measure is **four fingers. Its name** is "**the knot of Brahman**".

"**In its middle**" (śl.145a. *tanmadhye*). **In the middle of it**, [namely] of the knot of Brahman, is the **twelve**-petalled *cakra* called ***Nābhi* ("navel") *Cakra*. A spider** (*lūtṛ*) is a spider (*ūrṇanābha*, lit. "the one whose navel is [of] wool"). Like it, **the individual self wanders about there.**

"**Through the *Suṣumnā***" (śl.149a. *suṣumnayā*). **The *Suṣumnā*** is the tube (*nāḍī*) extended from the *Ādhāra* as far as the aperture of Brahman. Through it (= the *Suṣumnā*), it (= the individual self) **ascends** from its (= the *Ādhāra*'s)[1125] vicinity up to (*paryanta*) **the aperture of Brahman**, and **descends** from the aperture of Brahman. **Mounted on the vital wind (*prāṇa*)**, [namely] in close connection with the vital wind. **An acrobat (*kohalāṭika* = *kohlāṭika*)** is a dancer/actor of the difficult (*viṣama-nartaka*).

"***Suṣumnā***" (śl.150a). **Tubes (*nāḍī*) situated around the *Suṣumnā* (*suṣumnām paritaḥ*),** [namely] on all sides of the *Suṣumnā* (*sarvataḥ suṣumnāyāḥ*), **embracing (*kroḍīkṛtya*)**, [namely] pervading, **the bulb/lump**, the knot of Brahman, **spread/stretch (*tanvate*) the body**,

1124 This commentary falsely considers the term *ādhāra* to denote the *Ādhāra Cakra*. But actually this term simply means the base area of the body, cf. my footnote 1096 on SR śl.145c "*ādhāra*".

1125 The manuscripts A and B have the variant *tasyāḥ* instead of *tasya*. These two manuscripts seem to consider *tasyāḥ sakāśāt* to explain the instrumental case of *tayā*, "Through it [i.e.] in the vicinity of it".

the body frame (*śarīra*), [namely] make [the body] extend (*vistārayanti*).[1126]

SR śl.151–155

And they are [very] numerous (*bhūritara*). The principal ones among them are declared [to be] fourteen[1127]:
Suṣumnā, *Iḍā* and *Piṅgalā*, further *Kuhū*, *Sarasvatī*, (śl.151)
Gāndhārī[1128] and *Hastijihvā* and *Vāruṇī*[1129], *Yaśasvinī*,
Viśvodarā and *Śaṅkhinī*, moreover *Pūṣā*, *Payasvinī* (śl.152)
Alambusā[1130]. With regard to that, the first three are deemed the most

1126 The commentary S contains *kroḍīkṛtya* instead of *kandīkṛtya* of the *mūla* text. For the *mūla* text, I have translated *tanvate* as "spread over [the body]". But the commentary understands it as "extend/lengthen [the body]", paraphrasing it with *vistārayanti.*

1127 These fourteen tubes (*nāḍī*) are mentioned in the Haṭhayoga texts, cf. ZYSK 1993, p.209. They are also mentioned by AVALON 1924, pp.115–116, whose statement is based on the SR.
In contrast to the fourteen mentioned in the SR, ten are listed as the main tubes in the Dhyānabindūpaniṣad, Śāradātilaka, Cakrakaumudī, GaruḍaP etc. The list of the ten tubes lacks the *Sarasvatī*, *Vāruṇī*, *Viśvodarā* and *Payasvinī* mentioned by the SR. The references for these are as follows: Dhyānabindūpaniṣad (in the Yogopaniṣad) v.52f. (p.199): *pradhānāḥ prāṇavāhinyo bhūyas tatra daśa smṛtāḥ / iḍā ca piṅgalā caiva suṣumnā ca tṛtīyakā /52/ gāndhārī hastijihvā ca pūṣā caiva yaśasvinī / alambusā kuhūr atra śaṅkhinī daśamī smṛtā /53/.*
Śāradātilaka 1,41–43ab.
Cakrakaumudī 1,15–16.
GaruḍaP (KIRFEL 1956) śl.40–41: *iḍā ca piṅgalā caiva suṣumnā ca tṛtīyakā / gāndhārī gajajihvā ca pūṣā caiva yaśā tathā /40/ alambuśā kuhūś caiva śaṅkhinī daśamī smṛtā / piṇḍa-madhye sthitā hy etāḥ pradhānā daśa nāḍayaḥ /41/.*

1128 The Adyar edition reads *gāndhāri*, but this seems to be a mistake, as the parallels in the other texts (see above) call it *gāndhārī*. Besides, the nominative of *gāndhāri* should be *gāndhārir*. The Ānandāśrama edition (1896) contains *gandhārī*. Indeed YY 4,28, which is parallel to SR śl.152, contains *gāndhārī*, too. The manuscript D has the variant *gāndhārā*.

1129 The Ānandāśrama edition reads *vāraṇā* instead of *vāruṇī*, which is contained in the Adyar edition. YY 4,27 (Trivandrum ed., 1938) which is parallel to SR śl.152 also contains *vāraṇā*. The Adyar edition does not note it.

1130 The Ānandāśrama edition (1896) reads *alambuṣā* instead of *alambusā*. The Adyar edition does not note it. On the other hand, the Ānandāśrama edition notes *alambusā* as the variant of *kha., ga.* and *ṅa.* The YY (4,28) and the other Haṭhayoga texts (see above) contain *alambusā*.

important. (śl.153ab)[1131]

The *Suṣumnā*, the best of the three, related to Viṣṇu, going the way of liberation[1132], (śl.153cd)
is situated in the middle[1133] of the bulb (*kanda*).[1134] The *Iḍā* is on its (= the *Suṣumnā*'s) right, further, the *Piṅgalā* on [its] left.[1135] In the *Iḍā* and *Piṅgalā*, the moon and the sun respectively, (śl.154)[1136]

1131 SR śl.151–153ab are parallel to YY 4,26–28. Cf. YY 4,26, *nāḍīnām api sarvāsām mukhyā gārgi caturdaśa / iḍā ca piṅgalā caiva suṣumnā ca sarasvatī //* (Trivandrum ed., 1938). GEENENS translates "D'entre les invisibles Rivières, les principales / sont au nombre de quatorze. / Ce sont Iḍā et Piṅgalā, / Suṣumnā, Sarasvatī,". YY 4,27, *vāraṇā caiva pūṣā ca hastijihvā yaśasvinī / viśvodarā kuhūś caiva śaṅkhinī ca tapasvinī //* (Trivandrum ed., 1938). The Trivandrum edition seems to differ from the edition used by GEENENS who translates, "Vāruṇī, Pūṣā, / Hastijihvā, Yaśasvinī, / Viśvodarā, Kuhū, / Śaṅkhinī, Payasvinī". YY 4,28, *alambusā ca gāndhārī mukhyāś caitāś caturdaśa / āsām mukhyatamās tisras tisṛṣv ekottamā matā //* (Trivandrum ed., 1938). "Alambusā et Gandhārī. Parmi les invisibles Rivières, / les quatorze qui sont plus remarquables sont bien celles-là. / Trois d'entre elles sont particulièrement importantes, / [...].")

1132 The manuscript *gha* reads *mārgadā* instead of *mārgagā*. "[...] giving the way of liberation".

1133 The Ānandāśrama edition (1896) reads *kanda-madhya-sthitā* instead of *kanda-madhye sthitā*.

1134 SR śl.153–154a are parallel to YY 4,28cd–29: *āsām mukhyatamās tisras tisṛṣv ekottamā matā /28cd/ mukti-mārgeti sā proktā suṣumnā viśva-dhāriṇī / kandasya madhyame gārgi suṣumnā supratiṣṭhitā /29/.* GEENENS translates "En effet, Suṣumnā est cell qui soutient tout l'ensemble, / et elle est appelée "la voie de la liberation" / Suṣumnā, ô Gārgī, est bien située / au milieu du Balbe de vie".

YY 4,29a contains *mukti-mārgā* "the way of liberation" instead of *mukti-mārgagā* "going the way of liberation" in SR śl.153d.

1135 I.e. *savye 'tha dakṣiṇe*. The manuscripts *ṅa* and D respectively read *tu* and *ca* instead of *'tha*.

1136 SR śl.154 is parallel to YY 4,31cd–32. YY 4,31, *iḍā ca piṅgalā caiva tasyāḥ savye ca dakṣiṇe* (Trivandrum ed.). "[...] / À sa gauche et à sa droite / sont Iḍā et Piṅgalā." (tr. by GEENENS). YY 4,32, *iḍā tasyāḥ sthitā savye dakṣiṇe piṅgalā sthitā / iḍāyām piṅgalāyāṃ ca carataś candra-bhāskarau //*, "Iḍā est située à gauche / et Piṅgalā à droite. / Se déplacent sur Iḍā la lune / et soleil sur Piṅgalā." (tr. by GEENENS). YY 4,32ab is a repetition of YY 4,31cd, and YY 4,33ab is a repetition of YY 4,32cd.

A parallel is found in the Dhyānabindūpaniṣad (in the Yogopaniṣad) v.54cd–56ab: *satatam prāṇa-vāhinyaḥ soma-sūryāgni-devatāḥ /54cd/ iḍā-piṅgalā-*

the two origins of the going/movement of time, move.[1137] The *Suṣumnā* is the desiccator/absorber of time.[1138] (śl.155ab)[1139]

The *Sarasvatī* and *Kuhū*, however, lie on the two sides of the *Suṣumnā*. (śl.155cd)[1140]

Comm. K on SR śl.151–155

"Of the going/movement of time" (śl.155a, *kālagater*). Here, by the word ["]time["] the moments of heating[1141] and palpitation, being/which_are the cause of the [material] body (*piṇḍa*) of the individual self, are mentioned. This is to say (*iti yāvat*), those same (*ta eva*, i.e. total sum of the moments) are the lifespan.[1142] Of its (= time's) **going/movement**, [namely] of decay/passing_away[1143]. Thus (*iti*) is the meaning. **"The desiccator/absorber of time"** (śl.155b, *kālaśoṣiṇī*). Here, by the word "time", death is mentioned. Its **desiccator/absorber**, [namely] obstructer.

---

suṣumnās tisro nāḍyaḥ prakīrtitāḥ / iḍā vāme sthitā nāḍī piṅgalā dakṣiṇe sthitā /55/ suṣumnā madhyasthā prāṇa-mārgās trayaḥ smṛtāḥ /56ab/.

Śāradātilaka 1,40: *agnī-ṣomātmako deho bindur yad-ubhayātmakaḥ / dakṣiṇāṃśaḥ smṛtaḥ sūryo vāma-bhāgo niśākaraḥ.* 1,41cd–42ab mentions the ten tubes such as the *Iḍā, Suṣumnā, Piṅgalā* etc.

1137 The Ānandāśrama edition (1896) reads singular *gater hetuḥ* instead of *gater hetū*. The manuscript D reads *gatī hy etau*, "These are the two [mode of] going/movement of time".

1138 For the correspondence of the *Iḍā* and *Piṅgalā* with the moon and sun, cf. AVALON 1924, p.114. For the mention in the Baul tradition, cf. DAS 1992, p.403.

1139 The YY does not contain a parallel to SR śl.155ab. But YY 4,33–35ab deals with a topic similar to that of SR śl.154cd–155ab. YY 4,34c states that the *Iḍā* and *Piṅgalā* are associated with time. YY 4,35ab makes a statement inconsistent with that of the SR, saying that the *Suṣumnā* is closely related to time.

1140 SR śl.155cd–159a deals with the respective positions of the tubes (*nāḍī*).
SR śl.155cd is parallel to YY 4,35cd: *sarasvatī kuhūś caiva suṣumnā-pārśvayoḥ sthite.* "Et Sarasvatī et Kuhū sont situées / de part et d'autre de Suṣumnā." (tr. by GEENENS).

1141 The Ānandāśrama edition (1896) reads *svapana* "dream/sleep" instead of *tapana*.

1142 The Ānandāśrama edition (1896) reads *ta eva hetur iti* instead of *ta evāyur iti*. It would mean, "those same are the cause". But this does not make sense.

1143 The manuscript C reads *kṣaṇasyety* instead of *kṣayasye*. This is obviously a mistake caused by °*kṣaṇā* in the preceding line.

He relates that (*iti*) the principal ones among those tubes (*nāḍī*) are fourteen: "**And they**" (śl.151a, *tāś ca*) [etc.] He relates their names: "***Suṣumnā***" (śl.151c) [etc.] He mentions the principalness of the three tubes (*nāḍī*), [namely] of the *Suṣumnā, Iḍā* and *Piṅgalā*, among the fourteen, and the principalness of the *Suṣumnā* among them (= the three): "**With regard to that, the first**" (śl.153a, *tatrādyāḥ*) [etc.] He mentions the cause of the principalness of the *Suṣumnā*: "**Related to Viṣṇu**" (śl.153d, *vaiṣṇavī*) [etc.] **Related to Viṣṇu**, [namely] that whose deity is Viṣṇu. Or else, Vaiṣṇavī is the possessor of the shape of that which is the power (*śakti*) of illusion. **Going the way of liberation**, [namely] bestowing liberation.[1144] Or else, the way of liberation is the means of/for liberation, [namely] the self; relating to it (= the self). The meaning is thus (*iti*): Being (*bhūta*) the base/abode (*adhiṣṭhāna*) of the self. **Situated in the middle of the bulb**, [namely] of the knot of Brahman.[1145] The meaning is that (*iti*) the *Iḍā* and the *Piṅgalā* are situated on the left and the right of it (= the *Suṣumnā*). He mentions the particular name of the vital wind (*prāṇa*) moving in them both: "**In the *Iḍā* and *Piṅgalā***" (śl.154c, *iḍāpiṅgalayoḥ*) [etc.] It is said that (*iti*) the vital wind following the left tube (*nāḍī*) is **the moon**, but (*tu*) the one following the right tube (*nāḍī*) is **the sun**. With regard to (*iti*) [the question:] "How is there the vital wind's [state of] being the moon and sun?", hence he says: "**The origins of the going/movement of time**" (śl.155a, *kālagater hetū*). The meaning is that (*iti*), like the moon and sun are the causes of knowledge of time[1146], so are those two, too. "**The desiccator/absorber of time**". **The desiccator/absorber of time**, because (*iti*) [it] dries_up/absorbs time.

1144 The manuscript B reads *mārgadā* instead of *mārgagā*. The manuscript *gha* reads *mārgadā* for the *mūla* text (SR śl.153d), cf. the Adyar edition, p.70, variant 3 (cf. footnote 1132).
The comm. S's gloss *mokṣa-pradāyinī* would better fit *mukti-mārgadā*.

1145 The manuscript B reads *sthite* (f. dual) instead of *sthitā* (f. sg.).

1146 The manuscript B reads *kālakāraṇe*, lacking *jñāna*. "[…] the two causes of time […]."

SR śl.156–167

The *Gāndhārī* and *Hastijihvā* are situated in the back and front of the *Iḍā*, (śl.156ab)[1147]
the *Pūṣā* and *Yaśasvinī*, in the back and front of the *Piṅgalā*, respectively. (śl.156cd)[1148]

The *Viśvodarā* would be in the middle region of (= between) the *Kuhū* and *Hastijihvā*. (śl.157ab)[1149]

The *Vāruṇī*[1150] is considered to be standing in the middle of (= between) the *Kuhū* and *Yaśasvinī*. (śl.157cd)[1151]

The *Payasvinī* lies upon (= in) the middle of (= between) the *Pūṣā* and *Sarasvatī*. (śl.158ab)[1152]

The *Śaṅkhinī* abides in the middle of (= between) the *Gāndhārikā* and *Sarasvatī*. (śl.158cd)[1153]

1147 SR śl.155ab is presumably parallel to YY 4,36ab: *gāndhārī hasti-jihvā ca iḍāyāḥ pṛṣṭha-pārśvayoḥ* (Trivandrum ed.), "D'Iḍā provient Gandhārī et Hastijihvā, / sises de part et d'autre, dans le dos." (tr. by GEENENS). The YY contains *pṛṣṭha-pārśvayoḥ* instead of *pṛṣṭha-pūrvayoḥ* of the SR. GEENENS interprets *pṛṣṭha-pārśvayoḥ* as meaning "on both sides of the back". The expression *pṛṣṭha-pūrvayoḥ* is also contained in the next verse, SR śl.156cd, to which the YY has no parallel.
In this regard, the manuscript D contains an interesting variant *pṛṣṭha-vaṃśasthau* "situated in the backbone", although *-sthau* (as f. dual) is grammatically incorrect.

1148 To SR śl.156cd, there is no parallel in the YY. In SR śl.155cd–159a, which describes the positions of the tubes (*nāḍī*), this verse is the only one to which the YY has no parallel.

1149 SR śl.157ab is parallel to YY 4,36cd: *kuhoś ca hastijihvāyā madhye viśvodarā sthitā,* "Viśvodarā est au milieu / entre Kuhū et Hastijihvā." (tr. by GEENENS).

1150 The Ānandāśrama edition (1896) reads *vāraṇā* instead of *vāruṇī.*

1151 SR śl.157cd is parallel to YY 4,37ab: *yaśasvinyāḥ kuhor madhye vāruṇā ca pratiṣṭhitā* (Trivandrum ed.), "Vāruṇī est au milieu, / entre Kuhū et Yaśasvinī." (tr. by GEENENS). The YY's edition (Bombay 1954) used by GEENENS seems to contain *vāruṇī* instead of *vāruṇā*, which is contained in the Trivandrum edition.

1152 SR śl.158ab is parallel to YY 4,37cd: *pūṣāyāś ca sarasvatyāḥ sthitā madhye tapasvinī,* "Entre Pūṣā et Sarasvatī, / au milieu est Payasvinī." (tr. by GEENENS).

1153 SR śl.158cd is parallel to YY 4,38ab: *gāndhāryāḥ sarasvatyāḥ sthitā madhye ca śaṅkhinī*, "Au milieu est Śaṅkhinī, / entre Gāndhārī et Sarasvatī." (tr. by GEENENS).

The *Alambusā*[1154] is in the middle of the bulb. (śl.159a)[1155]
With regard to that (*tatra*), the *Iḍā* and the *Piṅgalā* [extend], respectively
as far as the left and the right nostril. The *Kuhū* [extends] forward/in_front as far as the urinary organ,[1156] (śl.159bcd)[1157]
the *Sarasvatī* upward/above as far as the tongue. (śl.160a)[1158]

The *Gāndhārī*[1159] is situated at the back/backward (śl.160b)
as far as the left eye. (śl.160c, *āvāmanetram*)[1160]

1154 The Ānandāśrama edition (1896) reads *alambuṣā.*

1155 SR śl.159a is parallel to YY 4,38cd: *alambusā ca viprendre kanda-madhyād avasthitā,* "Et au-dessus est Alambusā, au milieu / du Bulbe du vie, ô reine des brahmanes." (tr. by GEENENS).

1156 A parallel is found in Yogacūḍāmaṇy-upaniṣad (in Yogopaniṣad) v.18cd ff. (p.341): *iḍā vāme sthitā bhāge dakṣiṇe piṅgalā sthitā /18cd/ suṣumnā madhyadeśe tu gāndhārī vāmacakṣuṣi / dakṣiṇe hastajihvā ca pūṣā karṇe tu dakṣiṇe /19/ yaśasvinī vāmakarṇe cānane cāpy alambusā / kuhūś ca liṅgadeśe tu mūlasthāne tu śaṅkhinī /20/ evaṃ dvāraṃ samāśritya tiṣṭhante nāḍayaḥ kramāt /21ab/.*
Also cf. Cakrakaumudī 1,16cd–18ab.

1157 In the following verses, another topic is dealt with; SR śl.159bc–163ab lists the respective extensions of the tubes (*nāḍī*).
SR śl.159bc is parallel to YY 4,40cd and 43cd. YY 4,40cd, *piṅgalā cordhvagā yāmye nāsāntaṃ viddhi me priye*; YY 4,43cd, *iḍā ca savya-nāsāntaṃ savya-bhāge vyavasthitā.* SR śl.159d is parallel to YY 4,39ab, *pūrva-bhāge suṣumnāyās tv āmeḍhrāntaṃ kuhūḥ sthitā.*

1158 SR śl.160a is parallel to YY 4,42ab: *sarasvatī tathā cordhvam ājihvāyāḥ pratiṣṭhitā,* "Et encore: Sarasvatī est localisée à partir de la langue, / juste au-dessus. [...] /" (tr. by GEENENS).

1159 The Adyar edition contains a strange form *gāndhāryā* (śl.160b). I adopt the reading of the Ānandāśrama edition (1896), *gāndhārī.* The manuscript D reads *gāndhārā.* As a matter of fact, the parallel in the YY (4,43ab) contains *gāndhārā*, according to the Trivandrum edition.
Perhaps, the editor of the Adyar edition might have read *gāndhāryāḥ*. Then it would mean "the *Sarasvatī* upward as far as the tongue, situated behind the *Gandhārī*", but this is obviously a mistake, when we compare it with its parallel in the YY (see below).

1160 SR śl.160b is parallel to YY 4,43ab: *gāndhārā savya-netrāntā iḍāyāḥ pṛṣṭhataḥ sthitā,* "Dans le voisinage de l'œil gauche, / à partir d'Iḍā, dans le dos, est localisée Gandhārī." (tr. by GEENENS). The YY's edition (Bombay, 1954) used by GEENENS reads *gāndhārī* instead of *gāndhārā* of the Trivandrum edition.

But (*tu*) the *Hastijihvā* is situated as far as the great toe of the left foot.[1161] (śl.160cd–161a)[1162]

The *Vāruṇī*[1163] is, however, going in all [directions]. Now (*atha*), the *Yaśasvinī*
is situated in the right foot, as far as the great toe[1164]. (śl.161abc, *sarvagā [...]*)[1165]

The *Viśvodarā* is in the whole body. (śl.161d)[1166]

The *Śaṅkhinī* is as far as the left ear. (śl.162a)[1167]

When compared with this, *pṛṣṭhataḥ* "at the back/backward" in SR śl.160b should mean "at the back/backward of the *Iḍā*".

1161 The Ānandāśrama edition reads *āsādya pādāṅguṣṭhaṃ* instead of *āsavya-pādāṅguṣṭhaṃ*. But this reading is obviously a mistake, when we compare it with the parallel, YY 4,44ab.

1162 SR śl.160cd (*āsavyapādā-*) up to śl.161a is parallel to YY 4,44ab: *hastijihvā tathā savya-pādāṅguṣṭhāntam iṣyate,* "Quant à Hastijihvā, on pense bien / qu'elle est près de l'orteil gauche." (tr. by GEENENS).

1163 The Ānandāśrama edition (1896) reads *vāraṇā.*

1164 The term *aṅguṣṭha* might mean "thumb" (of the hand), too. But the parallel, YY 4,40ab, contains *pādāṅguṣṭhānta* "the great toe".

1165 For SR śl.160cd, the Ānandāśrama edition (1896) reads *aṅguṣṭhād dakṣiṇāṅghristhād dehe viśvodare 'khile*. The reading *viśvodare* is obviously a mistake, and should be read *viśvodarā* instead. Then it would mean: "[The *Yaśasvinī* extends] from/as_far_as the great toe of (lit. situated in) the right foot. The *Viśvodarā* is in the whole body." I prefer this reading, because it is in concordance with the other expressions such as SR śl.160c *āvāmanetram* etc.; it is always indicated whether it is on the right or left side, in the other passages. As a matter of fact, it accords with the statement of the parallel, YY 4,40ab, which contains *yāmyasya pādāṅguṣṭhāntam.*
SR śl.161abc (*sarvagā [...] °āṅghristhā*) is parallel to YY 4,39cd–40ab. YY 4,39cd (Trivandrum ed.), *adhaś cordhvaṃ ca vijñeyā vāraṇā (-nī?) sarvagāminī.* "Vāruṇī s'étend partout au-dessous / et au-dessus de l'Enlovée (*kuṇḍalī*)." (The Bombay edition used by GEENENS seems to read differently from Trivandrum ed.) YY 4,40ab (Trivandrum ed.), *yaśasvinī ca yāmyasya pādāṅguṣṭhāntam iṣyate.* The Bombay edition of the YY used by GEENENS again seems to read differently, as GEENENS' translation suggests: "On pense que Yaśasvinī est sur l'hémisphère droit, et dans le voisinage du gros orteil".

1166 SR śl.161d does not have a parallel in the YY. YY 4,44cd locates the *Viśvodarā* in the middle of the belly (YY 4,44cd: *viśvodarā tu nāḍī tundamadhye vyavasthitā*). The Śāradātilaka 1,51b (*sā sarvagā viśvarūpiṇī*) is parallel to the SR. Ibid. 1,52b, too, contains *sarva-dehānugā.*

But (*tu*) the *Pūṣā* is as far as the right eye.

The *Payasvinī*, however, is extended as far as the right ear. (śl.162bcd)[1168]

The *Alambusā* is fixed, resting_upon/situated_near the root of the anus. (śl.163ab)[1169]

In this body of such manner, however, covered with the accumulation of dirt, (śl.163cd)
the wise ones gain/accomplish enjoyment (*bhukti*) and liberation by [certain] means.[1170]

With regard to that, from the contemplation (*dhyāna*) with quality (*saguṇa*)[1171], [there] is[1172] enjoyment/experience. But liberation from that without quality (*nirguṇa*).[1173] (śl.164)

Contemplation, accomplishable only through a concentrated

1167 SR śl.162a is parallel to YY 4,42cd: *ā savya-karṇād viprendre śaṅkhinī cordhvagā matā /*, "[…] Ô excellente, on pense que Śaṅkhinī est localisée / à partir de l'oreille gauche / ou jusqu'à elle, juste au-dessus." (tr. by GEENENS).

1168 The Adyar edition notes that the C.E. (= Calcutta edition?) reads *āvāmanetrataḥ* ("as far as the left eye") instead of *āyāmyanetratah*. But this reading does not seem to be suitable, when compared with YY 4,41, which contains *yāmya-* (see below).
SR śl.162bcd is parallel to YY 4,41: *yāmye pūṣā ca netrāntam piṅgalāyāṃ tu pr̥ṣṭhataḥ / tapasvinī tathā gārgi yāmya-karṇā(ntam iṣya)te //*, "À droite, Pūṣā est dans la région de l'œil, / à partir de Piṅgalā, dans le dos. / De même pour Payasvinī, ô Gargī, on trouve bon / qu'elle soit à droite, près de l'oreille." (tr. by GEENENS).

1169 The Ānandāśrama edition reads *avalambya* "hanging from" instead of *avaṣṭabhya*.
SR śl.163ab is parallel to YY 4,45ab: *alambusā mahābhāge pāyu-mūlād adhogatā,* "Alambusā, ô grandement fortunée, est / au-dessus du fondement (*mūla*) ou de l'anus." (tr. by GEENENS).

1170 Printing mistake: *dhīmatnto* instead of *dhīmanto.*
For the relationship between ascetic and medical tradition, cf. ZYSK (1990 and 1991). But also see DAS' (2003B) critique.

1171 The Ānandāśrama edition (1896) reads *saguṇa-dhyānād* instead of *saguṇād dhyānād.*

1172 Optative *syāt.*

1173 *Saguṇa, nirguṇa*: Vedānta distinguishes between the exoteric and esoteric nature of Brahman. The lower, attribute-possessing Brahman is the object of worship, while the higher, attributeless Brahman, of knowledge. Since the former is closely connected to the latter, the worship of the former can function as an entrance to the latter. (Cf. DEUSSEN 1912, p.102ff.)

mind, is not something whose doing (= accomplishment) is easy for people.

Therefore, in this case, the sages engage_in/serve (*samupāsate*) the enjoyable/easy means, [namely] the auspicious unstruck sound (*nādam anāhatam*)[1174], through the way instructed by the master[s]. (śl.165–166ab)

Even it is not attractive to the mind of people, because of [its] being devoid of pleasantness. (śl.166cd)

We shall therefore explain the arising/origination of the struck sound (*āhata-nāda*) unfolding/pervading all song (*geya*, lit. "that is to be sung") delighting people/the_world [and] breaking [through] existence, by means of (*dvāratas*) the microtone[s] (*śruti*)[1175] etc., [and] further, the [unstruck sound's] being cause of the microtone[s] etc. (śl.167–168ab)

Comm. K on SR śl.156–158ab

"The Vāruṇī[1176]**, however, going in all [directions]"** (śl.161a, *sarvagā tu vāruṇī*). Here, the tube called (*iti*) ***Vāruṇī*****, however**, is **going in all [directions]**, [namely] pervading the whole body. "**Delighting people/the_world [and] breaking [through] existence**" are the two qualifiers of "**that [which] is to be song**".

Comm. S on SR śl.156–158ab

He mentions the position of the tubes (*nāḍī*), [namely] of the *Sarasvatī* etc: "**the *Sarasvatī***" (śl.155c). **The *Sarasvatī*** stands on the right side **of the *Suṣumnā*, the *Kuhū*** on the left side, **the *Gāndhārī*** in the region in the back **of the *Iḍā*, the *Hastijihvikā*** in the region in front, **the *Pūṣā*** in the region in the back of the *Piṅgalā*, **the *Yaśasvinī*** in the region in front, **the *Viśvodarā*** in the middle of (= between) ***Kuhū* and *Hastijihvā*, the *Vāruṇī* in the middle of (between) *Kuhū***

1174 The Haṭhayogapradīpikā v.84 mentions various kinds of unstruck sound (*anāhata-nāda*), like those of the drum, bell, lute, flute etc. The Haṭhayogic practice utilising *nāda* is dealt with in the Haṭhayogapradīpikā v.79–101. Verse 101 identifies *nāda* with *śakti*.

1175 For the musicological term *śruti*, i.e. microtone or micro-interval of the octave, cf. NIJENHUIS 1970, pp.88–94.

1176 The Ānandāśrama edition (1896) reads *vāraṇā*.

and **Yaśasvinī**. **The Payasvinī lies upon (= in) the middle of (= between) Pūṣā and Sarasvatī.** The [state of] being an accusative [of *madhyam*] is because of (*iti*) "Accusative in the case of [the verbs] *adhiśī*, *sthā* etc." (Pāṇini 1,4,46). **The Śaṅkhinī** stands **in the middle of (= between) Gāndhārikā and Sarasvatī, the Alambusā in the middle** of the bulb mentioned before.

He relates the extent of these tubes (*nāḍī*): "**With regard to that**" (śl.159b, *tatra*). **The Iḍā** is as far as **the left nostril, the Piṅgalā** as far as **the right nostril, the Kuhū** as far as **the urinary organ, the Sarasvatī** as far as **the upper tongue,**[1177] **the Gāndhārī** as far as the region of the back, **the Hastijihvā** beginning from **the left eye** as far as **the great toe** of the left foot, **the Vāruṇī** in **all** regions. **The Yaśasvinī** is situated **in the right foot**, beginning from **the big toe, the Viśvodarā in the whole body, the Śaṅkhinī** as far as **the left ear, the Pūṣā as far as the** right **eye. As far as (avadhi) the ear**, [namely] up to (*paryantam*) the ear. **The Payasvinī** is up to **the right** ear. **The Alambusā stands resting upon/situated near the root of the anus (pāyu)**, [namely] the root of the anal opening (*guda*).

He sums up the investigation/explanation of the body (*piṇḍa*): "**In [the body] of such manner**" (śl.163c, *evaṃvidhe*). It is said that (*iti*) **enjoyment and liberation** are accomplished/effected **in the body through [certain] means.**[1178]

He relates the means of those two: "**With regard to that**" (śl.164b, *tatra*). The mention at first of the engagement_in/service_of that with qualities is, however, to hint at its being the cause for (*prati*) the engagement_in/service_of that without qualities. Thus it is said in the Vedāntakalpataru:

> Those who are dull, unable (*anīśvara*) to directly visualise the supreme Brahman without characteristics, are harmonised/made_to_sympathise

1177 *Ūrdhva-jihvā-paryantam*. In contrast, the *mūla* text contains *ūrdhvam ājihvam* "upward, up to the tongue". The correct reading might be *ūrdhvaṃ jihvāparyantam* "upward as far as the tongue".

1178 Manuscript B reads *bhukti-muktī dehe upāyāt sadhye*. The meaning is the same, except for "are to be completed (*sādhye*)" instead of "are completed (*sidhyate*)".

(*anukampyante*)[1179] [with the object of devotion] through the observation of [Brahman] with characteristics.

Intending (*iti*), "Those unable with regard to (= to render) the engagement_in/service_of Brahman with qualities or without qualities, engage_in/serve sound["], he says: "**Contemplation**" (śl.165a, *dhyānam*). **Not something whose doing (= accomplishment) is easy**, unable to be made with ease.

Intending (*iti*), "even the unstruck sound is not captivating with regard to the mind of people", he says: "**Even it**" (śl.166c, *so 'pi*).

Summing up that put forward before, he introduces that to be explained in the latter [part]: "**Therefore**" (śl.167, *tasmāt*). By the word "**microtone[s] (*śruti*) etc.**", note[s] (*svara*)[1180], basic scale[s] (*grāma*)[1181], scale[s]/mode[s] (*mūrchanā*)[1182], note-serie[s] (*tāna*), permutational note-serie[s] (*kūṭa-tāna*)[1183] etc. are grasped (= included). The relation of [the words] is thus (*iti*): **By means of** them (= the microtones etc.), **unfolding/pervading**, [i.e.] spreading out, **all song (lit. "that is to be sung")**, [namely] that fit for the action of singing, melody type[s] (*jāti*)[1184] etc., we will mention **the arising/origination of the struck sound** and [its] **being the cause of the microtone[s] etc.** "**Delighting people/the_world [and] breaking [through] existence**", a pair of qualifiers of ["]**song (lit. that [which] is to be sung)**["], designates [its] being a bestower of enjoyment and

1179 *anukampyante* is the passive of causative of *anu√kamp* which belongs to class 1A.

1180 The term *svara* in a musicological context means "interval" or "note", cf. NIJENHUIS 1970, p.94.

1181 The muscological term *grāma* means "tone-system" or "basic scale", cf. NIJENHUIS 1970, p.10; WIDDESS 1995, p.401; DANIÉLOU 1996, p.39.

1182 The musicological term *mūrchanā* means "scale/mode", cf. NIJENHUIS 1970, pp.131–139. WIDDESS 1995, p.404, explains: "A scale of seven *svara*s in ascending and/or descending order. seven *mūrchanā*s can be derived from each *grāma* by starting on each *svara* in turn." DANIÉLOU 1996, p.45, "plagalleitern".

1183 *Tāna* and *kūṭa-tāna* are musicological terms. The former is called *śuddha-tāna*. For the definition of the two terms, cf. SHRINGY 1999, p.15; NIJENHUIS 1970, p.143 and p.152.

1184 *Jāti* is a musicological notion which is an old name or a forerunner of the modern *rāga*, cf. NIJENUHIS 1970, p.172 and pp.168–172. Also see WIDDESS 1995, p.36.

liberation. The meaning is thus (*iti*): Otherwise, [the state of] being the substitute for the engagement_in/service_of Brahman with qualities and without qualities would not fit the engagement_in/service_of sound.[1185] (śl.156–167)

Thus (*iti*) is the second section, the Arising/Origination of the Body (*Piṇḍotpatti*), in the first chapter [called] ["]On the Notes [etc."] (*Svaragata*).

1185 I.e. the engagement_in/service_of sound could not serve as a substitute for the engagement_in/service_of Brahman.

Bibliography

Abbreviations

AgniP	Agnipurāṇa
AH	Aṣṭāṅgahṛdayasaṃhitā
AS	Aṣṭāṅgasaṅgraha
AV	Atharvaveda
CA	Carakasaṃhitā
Hasty.	Hastyāyurveda
SR	Saṅgītaratnākara
SU	Suśrutasaṃhitā
ŚB	Śatapathabrāhmaṇa
ŚG	Śivagītā
YS	Yājñavalkyasmṛti
YY	Yogayājñavalkya

Original texts

Agnipurāṇa (abbr. AgniP)
ŚrīmadDvaipāyanamunipraṇītam Agnipurāṇam. Poona (Puṇya-ākhyapattana) 1900. Ānandāśramasaṃskṛtagranthāvaliḥ 41.
Aitareyāraṇyaka
— DEO, MUNISHWAR 1992: Aitareyāraṇyaka with the commentary of Sāyaṇa. Ed. by Munishwar Deo. Hoshiarpur (Vishveshvaranand Vedic Research Institute). Vishveshvaranand Indological Series 82.
— KEITH, ARTHUR BERRIEDALE 1995: The Aitareya Āraṇyaka. Ed. from the manuscripts in the India Office and the Library of the Royal Asiatic Society with introduction, translation, notes,

indexes and an appendix containing the portion hitherto unpublished of the Śāṅkhāyana Āraṇyaka. Oxford at the Clarendon Press. Reprint: Delhi (Eastern Book Linkers). (Revised edition).

Aṣṭāṅgahṛdaya (abbr.: AH)

— ŚrīVāgbhaṭaviracitam Aṣṭāṅgahṛdayam. Sarvāṅgasundarī vyākhyā vibhūṣitam. Vyākhyākāraḥ: Śrī Lālacandra Vaidya. Dillī: Motīlāla Bānārasīdāsa Pabliśarsa. 1999. Ed. by Lālacandra Vaidya. Motilal Banarsidas. First edition: Varanasi 1963. Reprint: Delhi 1999.

— DAS, R.P. & EMMERICK, R.E. 1998: Vāgbhaṭa's Aṣṭāṅgahṛdayasaṃhitā. The romanised text accompanied by line and word indexes, compiled and ed. by Rahul Peter Das & Ronald Eric Emmerick. Groningen (Egbert Forsten). Groningen Oriental Studies Vol.XIII.

Aṣṭāṅgasaṅgraha (abbr.: AS)

Aṣṭāṅgasaṅgraha of Wāgbhaṭṭa. Popularly known as Vṛddha Wāgbhaṭṭa with the commentary Śaśilekhā by Mahāmahopādhyāya Indu. Part II. Śārīrasthānam with various tables, pictorial illustrations and notes critically ed. with an introduction and index by Pt. Ramchandrashastri Kinjawadekar. Poona (Chitrashala Press) 1939.

Atharvaveda (AV)

WHITNEY, WILLIAM DWIGHT 1987: Atharva-Veda-Saṃhitā. (Text with English Translation, Mantra Index and Names of Ṛṣis and Devas.) Translated into English, William Dwight Whitney. Introduction, Suryakant Bali. Revised and Edited by Nag Sharan Singh. Vol. I and II. Delhi (Nag Publishers). First edition.

Bhāgavatapurāṇa (BhāgP)

ŚrīmadBhāgavatapurāṇam. ŚrīmadBhāgavata Śrīdharī Ṭīkā. Delhi (Chaukhamba Orientalia) 1988.

Bhāvaprakāśana

Bhāvaprakāśana of Śāradātanaya. Ed. by Yadugiri Yatiraja Swami of Melkot & K.S. Ramaswami Sastri Siromani. Oriental Institute Baroda. 1930. Gaekwad's Oriental Series No.XLV.

Brahmasūtra

Śrīmad-Dvaipāyanapraṇīta-Brahmasūtrāṇi. Ānandagirikṛtaṭīkā-saṃvalita-Śāṅkarabhāṣya-sametāni. Prastāvena pādaviṣayādhikaraṇa-viṣayānukramaṇyā pramāṇatayā gṛhītāny anya-granthastha-vākyānāṃ sthalanirdeśena ca sanāthīkṛtāni. (Prathamodhyāyaḥ, dvitīyādhyāyasya pādadvayaṃ ca) Nārāyaṇa-śāstribhiḥ saṃśodhitāni. Puṇyākhya-pattane (Ānandāśrama-saṃskṛta-granthāvaliḥ. granthāṅkaḥ 21) Khristābdāḥ 1900. Poona, 1900. Ānandāśrama series 21.

Bṛhaddeśī (see P.L. SHARMA 1992)

Cakrakaumudī by Badarīnātha

JHA, JEEVESHWAR (ed.) 1979: Chakra Kaumudi, a treatise on the six plexus in the human body, by Badarinatha. Ganganatha Jha. Allahabad (Rashtriya Sanskrit Sansthan). Kendriya Sanskrit Vidyapeetha Text Series No. 4.

Carakasaṃhitā

SHARMA, PRIYAVRAT (ed.) 1996: Caraka-saṃhitā. Agniveśa's treatise refined and annotated by Caraka and redacted by Dṛḍhabala. Text with English translation. Ed. and translated by Priyavrat Sharma. Vol.I (Sūtrasthāna to Indriyasthāna). Varanasi (Chaukhambha Orientalia). Jaikrishnadas Ayurveda Series 36.

Caryāgītikoṣa

BAGCHI, PRABODH CHANDRA & ŚĀSTRI, ŚĀNTI BHIKṢU 1956: Caryāgīti-koṣa of Buddhist Siddhas. Santiniketan (Visva-Bharati).

Garuḍapurāṇa

The Garuḍa Mahāpurāṇam. Delhi (Nag Publishers) 1984.

Hastyāyurveda

Pālakāpyamuniviracito Hastyāyurvedaḥ. Śivadattaśarmaṇā saṃśodhitaḥ. Ānandāśrama-saṃskṛta-granthāvaliḥ 26. Puṇyapattana 1894. (Ed. by Śivadattavarman. Poona 1894. Ānandāśrama Series 26).

Haṭhayogapradīpikā

— SINH, PANCHAM 1980: The Hatha Yoga Pradipika. Allahabad (Panini Office) 1914–15. Third edition: New Delhi (Munshiram Manoharlal) 1980.

— WALTER, HERMANN 1893 (German tr.): Svātmārāma's Haṭhayogapradīpikā (Die Leuchte des Haṭhayoga) aus dem Sanskrit übersetzt, und als Inaugural-Dissertation der Philosophischen Fakultät Sektion I, der Universität München vorgelegt von Hermann Walter. München.

— The Haṭhayogapradīpikā of Svātmārāma with the commentary *Jyotsnā* of Brahmānanda and English translation. Madras (The Adyar Library and Research Centre). The Adyar Library General Series Vol.4.

Manusmṛti

Manusmṛti: Sanskrit text with English Translation of M.N. Dutt, Index of Ślokas and Critikal Notes. Introduction by R.N. Sharma. Delhi (Chaukhamba Sanskrit Pratishthan). First edition 1998. The Vrajajivan Indological Series 6.

Mārkaṇḍeyapurāṇa

The Mārkaṇḍeya Mahāpurāṇam, ed. by Rajendranātha Śarmaṇ. Delhi (Nag Publishers) 1983.

Nāṭyaśāstra

Nāṭyaśāstra of Bharatamuni with the commentary Abhinavabhāratī by Abhinavaguptācārya. Vol.IV (Chapters 28–37) Ed. by M. Ramakrishna Kavi & J.S. Pade. Baroda (Oriental Institute) 1964.

Nirukta

ROTH, RUDOLF 1852: Jâska's Nirukta sammt den Nighaṇṭavas herausgegeben und erläutert von Rudolf Roth. Göttingen (Verlag der Dieterischen Buchhandlung).

Ṛgveda (abbr. ṚV)

— see GELDNER 1951.

— VAN NOOTEN, BAREND A. & HOLLAND, GARY B. (ed.) 1994: Rig Veda. A metrically restored text with an introduction and notes. Published by the department of Sanskrit and Indian studies, Harvard University. Distributed by Harvard University Press, Cambridge, Massachusetts and London.

Sākhī

KABĪR: Sākhī (Veṃkaṭeśvar 1972): Śrī saty-Kabīr-kṛt, Saty Kabīr kī Sākhī. ŚrīYug'lānand'jīdvārā saṃgṛhīt. Bambaī (Śrī Veṃkaṭeśvar Pres Prakāśan) 1972.

Saṅgītaratnākara (SR)

— Adyar ed.:

SASTRI, SUBRAHMANYA S. 1992: Saṃgītaratnākara of Śārṅgadeva. With the Kalānidhi of Kallinātha and the Sudhākara of Siṃhabhūpāla. Vol. I. Adhyāya 1. Ed. by Subrahmanya Sastri. Revised by S. Sarada. Madras (The Adyar Library and Research

Centre). First edition 1943. Second edition 1992. The Adyar Library Series Vol. 30.
— Adyar ed. Vol.III = SASTRI, SUBRAHMANYA S. 1986: Saṃgītaratnākara of Śārṅgadeva. With the Kalānidhi of Kallinātha and the Sudhākara of Siṃhabhūpāla. Vol.III. Adhyāya 5 and 6. Ed. by Subrahmanya Sastri. Revised by S. Sarada. Madras (The Adyar Library and Research Centre). First edition 1951. Second edition 1986. The Adyar Library Series Vol. 78.
— Ānandāśrama ed. (First ed.): ŚrīniḥśaṅkaŚārṅgadevapraṇītaḥ Saṅgītaratnākaraḥ Catura-Kallinātha-viracita-Kalānidhy-ākhyaṭīkāsaṃvalitaḥ. (Svara-prastārādi-pariśiṣṭa-ṣaṭka-sanāthīkṛtaś ca) Tasyāyam ādyādhyāyam ārabhya pañcamāntaḥ prathamo bhāgaḥ. Etat pustakaṃ "Maṅgesa Rāmakṛṣṇa Telaṅga" ity etaiḥ saṃśodhitam. Puṇyākhyapattane 1895. Ānandāśrama-saṃskrita-granthāvaliḥ 35.
— Ānandāśrama ed. Reprinted 1985 with the preface of G.H. Tarlekar.
— Tāraḷekara, Ga. Ha. 1975 (original text with Marathi translation): Saṃgītaratnākara. Bhāga 1, adhyāya 1 te 4. Śrīniḥśaṅka-Śārṅgadeva viracita 'saṃgītaratnākara' yā saṃskṛta graṃthāce 'Kalānidhi' Ṭīkesaha saṭīpa marāṭhī bhāṣāṃtara. Mumbaī (Mahārāṣṭra Rājya Sāhitya Saṃskṛti Māṇḍaḷa) 1975.
— see also SHRINGY 1999 (Vol.I) and 1989 (Vol.II).

Śāṅkarabhāṣya (Śaṅkara's Bhāṣya)
ŚrīmadDvaipāyanapraṇītaBrahmasūtrāṇi Ānandagiri-kṛta-ṭīkā-saṃvalita-Śāṅkarabhāṣya-sametāni. Padādhikaraṇa viṣaya-anukramaṇyā sūtrāṇām akārādi varṇānukramaṇyā ca sanāthīkṛtāni. Dvitīyādhyāyāntima pāda dvaya sametau tṛtīyacaturthādhyāyau. Ekasaṃbekarety upāhvaiḥ Ve. Śā. Rā. Rā. Nārāyaṇaśāstribhiḥ saṃśodhitāni. Dvitīyeyam aṅkanāvṛttiḥ. Puṇyākhyapaṭṭana śakābdāḥ 1825 (= A.D. 1903). Ānandāśramasaṃskritagranthāvaliḥ 21.

Śāradātilaka
The Śardaatilakam (sic) by Lakśmaṇadeśikendra (sic) with the Padārthādarśa commentary by Raghavabhaṭṭa (sic). Ed. with Introduction, etc, by M.M. Pandit Śri Mukunda Jha Bkashi (sic =Bakhshi). Benares (The Chowkhamba Sanskrit Series Office)

1934. Reprint 1991. The Kashi Sanskrit Series (Haridās Sanskrit Granthamālā) No. 107, Tantra Śāstra Section No.1.
Śārṅgadharasaṃhitā
The Śārṅgadhara-saṃhitā by Paṇḍita Śārṅgadharācārya Son of Paṇḍita Dāmodara, with the commentary Aḍhamalla's Dīpikā and Kaśīrāma's (sic. = Kāśīrāma) Gūḍhārha- (sic. = Gūḍhārtha-) Dīpikā. Ed. with footnotes by Paṇḍita Paraśurāma Śāstrī, Vidyāsāgar. Bombay (Nirṇaya Sāgar Press) 1920.
Śathapathabrāhmaṇa (ŚB)
— WEBER, ALBRECHT (ed.) 1964: The Śathapatha-Brāhmaṇa in the Mādhyandina-śākhā with extracts from the commentaries of Sāyaṇa, Harisvāmin and Dvivedaganga. Ed. by Albrecht Weber. Varanasi (The Chowkhamba Sanskrit Series Office). The Chowkhamba Sanskrit Series No.96.
— Śathapatha Brāhmaṇa of the White Yajurveda in the Mādhyandina Recension. Complete (1–14 Kāṇḍ) together with comprehensive Brāhmaṇa Index, Critical Introduction and Notes. Ed. by A. Chinnaswami Śāstri; Pattābhirāma Śastry; Rāmanātha Dīkṣita. Varanasi (Caukhambha Sanskrit Sansthan). Third edition 1998. Kashi Sanskrit Series 127.
Śivagītā (abbr. ŚG)
Śivagītā (Bombay 1987): (Hindi title) Paṇḍita-Jvālāprasādamiśra-kṛta-bhāṣāṭīkāsamalaṅkṛtā. Mudraka aura prakāśaka, Gaṅgāviṣṇu Śrī-Kṛṣṇadāsa. Adhyakṣa, Lakṣmī-veṅkaṭeśvara Sṭīma Presa. Bambaī. (Skt. text with the comm. by Jvālāprasādamiśra.) Publisher: Gaṅgāviṣṇu Śrī Kṛṣṇadāsa. Bombay (Śrī Veṅkaṭeśvar Press).
Suśrutasaṃhitā (abbr. SU)
Suśrutasaṃhitā of Suśruta. With the Nibandhasaṅgraha Commentary of Śrī Ḍalhaṇāchārya and the Nyāyacandrikā Pañjikā of Śrī Gayadāsāchārya on Nidānasthāna. Ed. from the Beginning to the 9th Adhyāya of Cikitsāsthāna by Vaidya Jādavji Trikamji Āchārya and the rest by Nārāyaṇ Rām Āchārya "Kāvyatīrtha". Introduction by P.V. Sharma. Varanasi/Delhi (Chaukhamba Orientalia) 1980. Jaikrishnadas Ayurveda Series 34.
Tandulaveyāliya
Pratnapūrvadharanirmitaṃ Tandulavaicārikaṃ, Śrīmad-Vijaya-vimala-gaṇi-dṛbdha-vṛttiyutaṃ, sāvacūrikaṃ ca Catuḥśaranaṃ.

Bombay 1922. Śreṣṭha-Devacandra-Lāla-bhāī-Jaina-pustakoddhāra-granthāṅkaḥ 59.
Viṣṇudharmottarapurāṇa (abbr. ViṣṇudhP)
The Viṣṇudharmottarapurāṇam. Delhi (Nag Publishers) 1985.
Yājñavalkyasmṛti (YS)
Yājnavalkya Smṛti with 'Viramitrodaya' Commentary of Mitra Mishra and 'Mitakshara' Commentary of Vijnaneshwara. Ed. by Narayana Shastri Khiste & Jagannatha Shastri Hoshinga. Varanasi (Chowkhamba Sanskrit Series Office) 1997. Chowkhamba Sanskrit Series 62.
Yogayājñavalkyam (YY)
— Trivandrum ed.: The Yogayājñavalkya. Ed. by K. Sāmbaśiva Śāstrī. Trivandrum (Printed by the Superintendent, Government Press) 1938. Trivandrum Sanskrit Series No.CXXXIV. Śrī Citrodayamañjarī No.XXIII.
— Bombay ed. (used by GEENENS 2000): Yoga Yājñavalkya, a Treatise on Yoga as Taught by Yogī Yājñavalkya. Ed. by Sri Prahlad Divanji, Monograph no. 3, reprinted from the *Journal of the Bombay Branch Asiatic Society* (Vol.XXVIII, and XXIX, Part 1), Bombay 1954. (I did not consult this edition.)
Yogopaniṣad
SASTRI, A. MAHADEVA (ed.) 1968: The Yoga Upaniṣad-s with the commentary of Śrī Upaniṣad-brahmayogin. Madras (The Adyar Library and Research Centre). First Published 1920. Reprinted 1968.

Secondary literature

ANGOT, MICHEL 1993–94: "La notion de *napuṃsaka* dans les texts médicaux, grammaticaux et rituels." *Bulletin d'Études Indiennes* (L'Université de Paris-III) *11–12.1993–94*, pp.15–38.
APTE, VAMAN SHIVARAM 1992: The Practical Sanskrit Dictionary. Kyoto (Rinsen Book Company). Third reprinting 1992. (Reprinted from the Revised & Enlarged Edition, Poona 1957).

AVALON, ARTHUR (see also WOODROFFE) 1913: Tantra of the Great Liberation (Mahānirvāṇa Tantra). London (Luzac & Company) 1913. Reprint: New York (Dover Publications) 1972.
— 1924: The Serpent Power, being the Shat-chakra-nirūpana and Pādukā-panchaka. Two works on Laya Yoga, translated from the Sanskrit, with Introduction and Commentary. Second revised edition. Madras (Ganesh & Co).
BANDHU, VISHVA (VISHVA BANDHU)[1186] 1976: A Vedic Word Concordance (Skt. Title: Vaidikapadānukramakoṣaḥ). Vol.I in Six Parts, Saṃhitās, Part I. Second edition, revised and enlarged by Bhim Dev. Hoshiarpur (Vishveshvaranand Vedic Research Institute). Śantakuṭī Vedic Series 1. V.V.R.I. Publication 638.
BANSAT-BOUDON, LYNE 1992: Poétique du théâtre indien. Lectures du *Nāṭyaśāstra*. Paris. Publications de l'École Française d'Extrême-Orient, Vol.169.
BECK, GUY L. 1993: Sonic Theology. Hinduism and Sacred Sound. Columbia (University of South Carolina).
BEYER, NORBERT 1999: Lautenbau in Südindien. M. Palaniappan Achari und Seine Arbeit. Berlin (Staatliche Museen zu Berlin, Preussischer Kulturbesitz). Veröffentlichungen des Museums für Volkerkunde, Neue Folge 69; Musikethnologie XI.
BHATTACHARYA, A. N. 1986: Essence of Vedānta. Sadānanda's Philosophy of Vedāntasāra. Delhi (Durga Publications).
BHATTACHARYA, FRANCE 1998: "Les Mémoires des Binodinī Dāsī et les Débuts du Théâtre Bengali Moderne". *Théâtres Indiens. Études Réunies par L. Bansat-Boudon. Collection Puruṣārtha 20. 1998*, Paris (Éditions de l'École des Hautes Études en Sciences Sociales), pp.219–237.
BLOOMFIELD, MAURICE 1906: A Vedic Concordance. Being an alphabetic index to every line of every stanza of the published Vedic literature and to the liturgical formulars thereof, that is an index to the Vedic mantras, together with an account of their variations in the different Vedic books. Cambridge/Massachusetts (Harvard University). Harvard Oriental Series Vol.10.

1186 I have referred to him as BANDHU 1976 for the sake of convenience, although "Bandhu" is actually not the author's family name.

BRONKHORST, JOHANNES 1993: The Two Sources of Indian Asceticism. Bern (Peter Lang). Schweizer Asiatische Studien, Monographien Bd.13.

CAILLAT, COLLETTE 1974 (1): "Sur les doctrines médicales dans le Tandulaveyāliya." *Indologica Taurinensia* (Torino) *2.1974*, pp.45–55.

COMBA, ANTONELLA 1981: "Un capitolo della Śivagītā sulla medicina Āyurvedica." Torino (Accademia delle Scienze). Memorie dell'Accademia della Scienze di Torino, Serie V, 5. 1981, II. Classe di Scienze Morali, Storiche e Filologiche, pp.173–223.

— 1984: "Some priorities in non-medical texts." Procedings of the international workshop on priorities in the study of Indian medicine held at the State University of Groningen 23–27 October 1983. Ed. by G. Jan Meulenbeld. Groningen (Institute of Indian Studies, University of Groningen) 1984. Rijksuniversiteit te Groningen, Publikaties van het Instituut voor Indische talen en culturen 4, pp.223–249.

— 1987: "Carakasaṃhitā, Śārīrasthāna I and Vaiśeṣika Philosophy." Studies on Indian Medical History. Papers presented at the International Workshop on the study of Indian Medicine held at the Welcome Institute for the History of Medicine 2–4 September 1985. Ed. by G. Jan Meulenbeld and Dominik Wujastyk. Groningen (Egbert Forsten) 1987. Groningen Oriental Studies 2; pp.43–61.

DANIÉLOU, ALAIN 1996: Einführung in die indische Musik. Aus dem Franz. von Wilfried Sczepan. Wilhelmshaven (Noetzel, Heinrichshofen-Bücher) 4. Auflage 1996. (1. Auflage 1975, Erweiterte Neuausgabe 1982). Taschenbücher zur Musikwissenschaft 36.

DAS, RAHUL PETER 1988: Das Wissen von der Lebensspanne der Bäume. Surapālas Vṛkṣāyurveda. Kritisch ediert, übersetzt und kommentiert von Rahul Peter Das. Mit einem Nachtrag von G. Jan Meulenbeld zu seinem Verzeichnis 'Sanskrit Names of Plants and their Botanical Equivalents'. Stuttgart (Franz Steiner Verlag). Alt- und Neu-Indische Studien, hrsg. vom Seminar für Kultur und Geschichte Indiens an der Universität Hamburg, 34.

— 1990: "Miscellanea de Operibus Āyurvedicis." *Journal of the European Āyurvedic Society 1.1990*. Reinbeck (Dr. Inge Wezler, Verlag für Orientalistische Fachpublikationen), pp.47–68.

— 1992: "Problematic Aspects of the Sexual Rituals of the Bauls of Bengal." *Journal of the American Oriental Society 112.1992*. Ann Arbor, pp.388–432.

— 2003A: The Origin of the Life of a Human Being. Conception and the Female according to Ancient Indian Medical and Sexological Literature. Delhi (Motilal Banarsidas) 2003. Indian Medical Tradition Vol.VI.

— 2003B: "[Review of] Kenneth Zysk, Asceticism and Healing in Ancient India. Medicine in the Buddhist Monastery." *Traditional South Asian Medicine, 7.2003*. Wiesbaden (Ludwig Reichert), pp.228–232.

— 2004: "Kaste". Gemeinsame kulturelle Codes in koexistierenden Religionsgemeinschaften. Leucorea-Kolloquium 2003. Hrsg. von Ute Pietruschka. Halle (Saale) (Martin-Luther-Universität Halle-Wittenberg, Institut für Orientalistik) 2005. Hallesche Beiträge zur Orientwissenschaft 38.2004.

DAS GUPTA, SHASHIBHUSHAN 1976: Obscure Religious Cults. Calcutta (KLM).

DEUSSEN, PAUL 1883: Das System des Vedānta nach den Brahmasūtra's des Bādarāyaṇa und dem Commentare des Śaṅkara über dieselben als ein Compendium der Dogmatik des Brahmanismus vom Standpunkte des Śaṅkara aus. Leipzig (F.A. Brockhaus).

— 1912: The System of the Vedānta. According to Bādarāyaṇa's Brahmasūtras and Śaṅkara's commentary thereon set forth as a compendium of the dogmatics of Brahmanism from the standpoint of Śaṅkara. Chicago (The Open Court Publishing Company) 1912. (This is an English translation by Charles Johnston of DEUSSEN 1883.)

DIMOCK, EDWARD C., JR. 1966: The Place of the Hidden Moon. Erotic Mysticism in the Vaiṣṇava-Sahajiyā Cult of Bengal. Chicago/London (The University of Chicago Press).

DOSSI, BEATRICE 1998: Samen, Seele, Blut. Die Zeugungstheorien des Alten Indiens. München (Akademischer Verlag). Ganesha 11.

DVIVEDĪ, HAZĀRĪPRASĀD 1990: Kabīr. Kabīr ke vyaktitva, sāhitya aur dārśanik vicāroṃ kī āloc'nā. Nayī Dillī (New Delhi)

(Rāj'kamal' Prakāśan) Pah'lā saṃskaraṇ (first print) 1971. Chaṭhā saṃskaraṇ (sixth print) 1990.
FADDEGON, B. 1969: The Vaiçeṣika-System, described with the help of the oldest texts. Wiesbaden (Martin Sändig). Verhandelingen der Koninklijke Akademie van Wetenschappen te Amsterdam. Afdeeling Letterkunde, Nieuwe Reeks. Deel xviii n.2.
FILLIOZAT, JEAN 1975: La doctrine classique de la médecine indienne; Ses origines et ses parallèles grecs. Paris (Impremerie nationale) 1949. Paris (Ecole Française d'Extrême-Orient), 1975.
FOSTER, SUSAN LEIGH 1997: "Dancing Bodies". Meaning in Motion. New Cultural Studies of Dance, ed. by Jane C. Desmond. Durham/London (Duke University Press), pp.235–257.
FUNATSU, KAZUYUKI 1991: "Sangīta ratonākara ni okeru ongaku no keijijougaku" [i.e. "Metaphysics of music in SR"]. Ga no shisō. Maeda Sengaku Hakase Kanreki Kinen Ronshū. Tōkyō (Shunjūsha). (English title: Ātmajñāna. Professor Sengaku Mayeda Felicitation Volume), pp. 83–96.
GARBE, RICHARD 1917: Die Sāṅkhya-Philosophie. Eine Darstellung des indischen Rationalismus. Leipzig (H. Haessel Verlag).
GEENENS, PHILIPPE 2000: Yogayājñavalkyam. Corps et âme, yoga selon Yājñavalkya. Paris (Gallimard).
GELDNER, KARL FRIEDRICH 1951: Der Rig-Veda. Aus dem Sanskrit ins Deutsche übersetzt und mit einem laufenden Kommentar versehen. Teil 1. Erster bis vierter Liederkreis. Cambridge/Massachusetts (Harvard University Press). Harvard Oriental Series, Volume 33.
GHOSH, MANMOHAN (ed.) 1991: Pāṇinīya Śikṣā. Delhi (V.K. Publishing House).
GOMPERTS, AMRIT 2000: "Indian Music, the Epics and Bards in Ancient Java." Vorträge des Internationalen Musikarchäologischen Kolloquiums des Deutschen Archäologen Instituts. Studien zur Musikarchäologie III: Archäologie für Klangerzeugung und Tonordnung; Musikarchäologie in der Ägäis und Anatolien. Ed. by Ellen Hickmann, Anne D. Kilmer, and Ricardo Eichmann. Rahden (Leidorf), pp.573–596.

GOUDRIAAN, TEUN 1990: “The Ātman as charioteer: treatment of a Vedic allegory in the Kulālikāmnāya.” Panels of the VIIth World Sanskrit Conference, vol. I. The Sanskrit Tradition and Tantrism. Leiden/New York/Köln (E.J. Brill), pp.43–55.

HABERMANN, DAVID L. 2001: Acting as a Way of Salvation. A Study of Rāgānugā Bhakti Sādhanā. Delhi (Motilal Banarsidas).

HACKER, PAUL 1978: “Zur Methode der geschichtlichen Erforschung der anonymen Sanskritliteratur des Hinduismus”. Paul Hacker, Kleine Schriften, hrsg. von Lambert Schmitthausen. Wiesbaden (Franz Steiner Verlag) 1978. Glasenapp-Stiftung Bd.15, pp.483–492. (Originally published in: Der XV. Deutsche Orientalistentag Göttingen 1961).

HALBFASS, WILHELM 2000: Karma und Wiedergeburt im indischen Denken. München (Heinrich Hugendubel, Diedrichs). Diedrichs Gelbe Reihe 161.

HARA, MINORU 1977: “Shōku” (in Japanese). Festschrift K. Tamaki (Tamaki Koshiro Hakase Kanreki Kinen Ronbunshu). Tokyo, pp.667–683.

— 1980: “A note on the Buddha’s birth story”. Indienisme et Bouddhisme. Mélanges offerts à Mgr Étienne Lamotte. Louvain-La-Neuve. Université Catholique de Louvain, Institut Orientaliste. Publications de l’Institut Orientaliste de Louvain 23, pp.143–157.

HARDER, HANS 2011: Sufism and Saint Veneration in Contemporary Bangladesh. The Maijbhandaris in Chittagong. London/New York (Routledge).

HEDAYATULLAH, MUHAMMAD 1989: Kabir. The Apostle of Hindu-Muslim Unity. Interaction of Hindu-Muslim Ideas in the Formation of the Bhakti Movement with Special Reference to Kabīr, the Bhakta. Reprinted, Delhi (Motilal Banarsidas) 1989. (First edition, Delhi, 1977).

HEILIJGERS-SEELEN, DORY 1990: “The doctrine of the Ṣaṭcakra according to the Kubjikāmata.” Panels of the VIIth World Sanskrit Conference, Vol. I, The Sanskrit Tradition and Tantrism, ed. by Teun Goudriaan. Leiden (E.J. Brill), pp.56–65.

HOERNLE, A.F. RUDOLF 1907: Studies in the Medicine of Ancient India. Part I. Osteology or the Bodies of the Human Body. Oxford (At the Clarendon Press).

HOUBEN, JAN E.M. 1995: The Saṃbandha-Samuddeśa (chapter on relation) and Bhartṛhari's Philosophy of Language. A Study of Bhartṛhari's Saṃbandha-samuddeśa in the context of the Vākyapadīya with a translation of Helarāja's commentary Prakīrṇa-prakāśa. Groningen (Egbert Forsten). Gonda Indological Studies, Vol.II.

HULIN, MICHEL 1999: "*KUṆḌALINĪ* – Zur 'Mythischen Physiologie' des tantrischen Yoga." Raum-zeitliche Vermittlung der Transzendenz. Zur „sakramentalen" Dimension religiöser Tradition. Herausgeben von Gerhard Oberhammer und Marcus Schmücker. Wien. Österreichische Akademie der Wissenschaften, philosopshisch-historische Klasse, Sitzungsberichte, 665. Bd., pp.191–205.

INGALLS, DANIEL HENRY HOLMES 1951: Materials for the Study of Navya-nyāya Logic. Cambridge/Massachusetts (Harvard Oriental Press). London (Geoffrey Cumberlege, Oxford University Press). Harvard Oriental Series, Vol.40.

JACOB, G.A. 1985: Upaniṣadvākyakośaḥ. A Concordance to the Principal Upaniṣads and Bhagavadgītā. Delhi (Motilal Banarsidas). Reprint 1985.

JAINI, PADMANABH S. 1991: Gender and Salvation. Jaina Debates on the Spiritual Liberation of Women. Berkeley/Los Angeles/ California (University of California Press).

JAMISON, STEPHANIE W. 1986: "Brāhmaṇa Syllable Counting, Vedic *Tvac* 'Skin', and the Sanskrit Expression for the Canonical Creature." *Indo-Iranian Journal 29.1986.* Dordrecht/Boston/ Lancaster/Tokyo, pp.161–181.

JOLLY, JULIUS 1901: Medicin. Strassburg (Verlag von Karl J. Trübner). Grundriss der Indo-Arischen Philologie und Altertumskunde, III. Band, 10 Heft.

Kabīr → see Sākhī among the *Original texts* in this bibliography.

KANE, PANDURANG VAMAN 1990: Histrory of Dharmaśāstra, vol.I. Second edition. Revised and enlarged, Part I. Reprinted. Poona (Bhandarkar Oriental Research Institute).

KAPANI, LAKSHMI 1992: La Notion de Saṃskāra. Vol.I, II. Paris (Diffusion de Boccard). Collège de France, Publications de l' Institut de Civilisation Indienne, Fascicule 59.

KATRE, SUMITRA M. 1989: Aṣṭādhyāyī of Pāṇini. Delhi (Motilal Banarsidas). First Indian edition, 1989.

KATZ, JONATHAN 1983: "Indian Musicological Literature and its Context." Inde et Litératures. Études Réunies par Marie-Claude Porcher. Centres d'Études de l'Inde et de l'Asie du Sud. Collection Puruṣārtha n° 7. Paris (Éditions de l'École des Hautes Études en Sciences Sociales), pp.57–75.

KEITH 1995 → see Aitareya-āraṇyaka among the *Original texts* in this bibliography.

KIEHNLE, CATHARINA 1997: Songs on Yoga. Jñāndev Studies I & II. Texts and Teachings of the Mahārāṣṭrian Naths. Stuttgart (Frank Steiner Verlag). Alt- und Neu-Indische Studien, hrsg. vom Institut für Kultur und Geschichte Indiens und Tibets an der Universität Hamburg, 48,1.

KIRFEL, WILLIBALD 1951: Die fünf Elemente. Insbesondere Wasser und Feuer. Ihre Bedeutung für den Ursprung altindischer und altmediterraner Heilkunde. Eine medizingeschichtliche Studie. Walldorf-Hessen (Verlag für Orientkunde Dr. H. Vorndran). Beiträge zur Sprach- und Kulturgeschichte des Orients 4.

— 1954: "Ein medizinisches Kapitel des Garuḍapurāṇas." Asiatica. Festschrift Friedlich Weller zum 65. Geburtstag. Hrsg. von Johannes Schubert und Ulrich Schneider. Leipzig (Otto Harassowitz). pp. 333–356.

KOCH, LARS-CHRISTIAN 1995: Zur Bedeutung der *Rasa*-Lehre für die zeitgenössische nordindische Kunstmusik. Mit einem Vergleich mit der Affektenlehre des 17. & 18. Jahrhunderts. Bonn (Holos Verlag).

KUPPUSWAMY, GOWRI & HARIHARAN, M. 1984: Royal Patronage to Indian Music. Delhi (Sundeep Prakashan).

LAMOTTE, ÉTIENNE 1970: Le traité de la grande vertu de sagesse de Nāgārjuna (Mahāprajñāpāramitāśāstra). Tome III (Chapitres XXXI-XLII). Louvain (Université de Louvain, Institut Orientaliste), pp.1311–1328, Chapitres XXXV ("Les neuf notions des horreurs").

LATH, MUKUND 1997: A Study of Dattilam. A Treatise on the Sacred Music of Ancient India. New Delhi (Impex India).

LEACH, LINDA YORK 1995: Mughal and Other Indian Paintings from the Chester Beatty Library. Vol. II. London (Scorpion Cavendish).

MAYRHOFER, MANFRED 1992, 1996: Etymologisches Wörterbuch des Altindoarischen. I., II. Band. Heidelberg (Carl Winter Universitätsverlag) 1992 (I. Band), 1996 (II. Band).

MEULENBELD, G. JAN 1991: "The constraints of theory in the evolution of nosological classifications: A study on the position of blood in Indian Medicine." Panels of the VIIth World Sanskrit Conference, Vol.VIII. Leiden (E.J. Brill), pp.91–106.

— 1999: A History of Indian Medical Literature. Vol.IA (Text) and IB (Annotation). Groningen (Egbert Forsten).

— 2000: A History of Indian Medical Literature. Vol.IIA. Groningen (Egbert Forsten).

MEYER, J.J. 1930: "Die menschlichen Körperteile in ihrer Bedeutung für Schicksal und Charakter. Ein Beitrag zur Kulturgeschichte und zur Frage von der Entstehungsart der Purāṇas." *Wiener Zeitschrift für die Kunde des Morgenlandes 36.1930,* pp.108–155.

— 1928: "Über den anatomisch-physiologischen Abschnitt in der Yājñavalkya- und Viṣṇusmṛti." *Wiener Zeitschrift für die Kunde des Morgenlandes, XXXV. Bd., 1. u. 2. Heft. 1928,* pp.49–58.

MICHEL, WALTER 1987: "Die Darstellung der Affekte auf der Jesuitenlehre." Theaterwesen und dramatische Literatur, Beiträge zur Geschichte des Thaters, hrsg. von Günter Holtus. Tübingen (Francke Verlag). Mainzer Forschungen zu Drama und Theater, Bd. 1, pp. 233–251.

MITTWEDE, M. 1993: "Der Schlaf bei Caraka." *Journal of European Āyurvedic Society 3.1993,* pp.137–143.

MÖLLER, HANS JÜRGEN 1974: "Psychotherapeutische Aspekte in der Musikanschauung der Jahrtausende." Neue Wege der Musiktherapie, hrsg. von Revers, W.J.; G. Harrer & W.C.M. Simon. Düsserdorf/Wien.

MÜLLER, REINHOLD F.G. 1941–1942: "Zu Vorstellungen altindischer Ärzte über Fortpflanzungs-Stoffe." *Quellen und Studien zur Geschichte der Naturwissenschaften und der Medizin 8.1941–1942.* Berlin, pp.458–480.

— 1952: "Manas und der Geist altindischer Medizin." Leipzig (Johann Ambrosius Barth Verlag). Nova Acta Leopoldina, Abhandlungen der Deutschen Akademie der Naturforscher (Leopoldina) zu Halle/Saale, Neue Folge Bd. 15, Nummer 108.
— 1955: "Altindische Embryologie." Leipzig (Johann Ambrosius Barth Verlag). Nova Acta Leopoldina, Abhandlungen der Deutschen Akademie der Naturforscher (Leopoldina) zu Halle/Saale, Neue Folge Bd. 17, Nummer 138.
— 1961: "Medizinisches Sanskrit-Wörterheft." Berlin (Akademie-Verlag). Sonderdruck aus der Zeitschrift: Mitteillungen des Instituts für Orientforschung, Bd. VIII, Heft I. Deutsche Akademie der Wissenschaften zu Berlin, Institut für Orientforschung.

NAMOUCHI, NICOLE 1995: Käufliche Liebe. Prostitution im Alten Indien. Frankfurt am Main/Berlin/New York/Paris/Wien (Peter Lang). Europäische Hochschulschriften, Reihe 27, Asiatische und Afrikanische Studien, Vol.48.

NARUSE, MASAHARU 1986: "Kuchufuyo" (i.e., "Haṭhayogic Levitation") *Korrani* (*Kalyāṇī*), *Indo to hito no bunka, Tantora tokushu 11.1986.* Tokyo.

NIJENHUIS, EMMIE WIERSMA-TE 1970: Dattilam. A compendium of ancient Indian music. Leiden (E.J. Brill). Orientalia Rheno-Traiectina.
— (Emmie Te Nijenhuis) 1977: Musicological Literature. Wiesbaden (Otto Harrasowitz). A History of Indian Literature, Vol.VI, Fasc.1.
— 1992: Saṅgītaśiromaṇi. A medieval Handbook of Indian Music. Ed. with introduction and translation by Emmie Te Nijenhuis. Leiden/New York/Köln (E.J. Brill). Brill's Indological Library, Vol.5.

O'FLAHERTY, WENDY DONIGER 1980: Karma and rebirth in classical Indian traditions. Univ. of California Press.

OLIVELLE, PATRICK 1992: Saṃnyāsa Upaniṣads. Hindu Scriptures on Ascetisism and Renunciation. Translated with Introduction and Notes. New York/Oxford (Oxford University Press).

PADOUX, ANDRÉ 1992: Vāc, The concept of the word in Selected Hindu Tantras. Delhi (Sri Satguru Publications, Indian Books Centre) 1992. (This is an English translation from French. The data of the English translation published in USA is: New York (State University) 1990).

PESCH, LUDWIG 1999: The Illustated Companion to South Indian Classical Music. Delhi etc. (Oxford University Press).

PIETRUSCHKA, UTE 2001: "Hermes und der Musikant: Zu einer verlorenen Passage aus den Nawadir al-falasifa des Hunain b. Ishaq". Hallesche Beiträge zur Orientwissenschaft, hrsg. von Walter Beltz, Markus Mode u. Jürgen Tubach, Heft 32.2001. Halle (Institut für Orientalistik, Martin-Luther-Universität).

POTTER, KARL H. 1977: Encyclopedia of Indian Philosophies. Indian Metaphysics and Epistemology: The Tradition of Nyāya-Vaiśeṣika up to Gaṅgeśa. Ed. by Karl H. Potter. Delhi etc. (Motilal Banarsidas). First Edition 1977.

PREISENDANZ, KARIN 1994: Studien zu Nyāyasūtra III.1 mit dem Nyāyatattvāloka Vācaspati Miśras II. Teil. 1 & 2. Stuttgart (Franz Steiner Verlag). Alt- und Neu-Indische Studien. Hrsg. vom Institut für Kultur und Geschichte Indiens und Tibets an der Universität Hamburg. 46,1 & 46,2.

RADHAKRISHNAN, S. 1953: The Principal Upaniṣads. Ed. with Introduction, text, translation and notes. Delhi (Oxford University Press).

RAJA, K. KUNJUNNI & BURNIER, RADHA 1976: The Saṃgītaratnākara of Śārṅgadeva. Vol.IV. Chapter on Dancing. Madras (The Adyar Library and Research Centre). (This is an English translation of the text).

RAU, WILHELM 1986: Naturbeobachtung und Handwerkskunst im vorislamischen Indien. Stuttgart (Franzsteiner Verlag Wiesbaden). Sitzungsberichte der Wissenschatlichen Gesellschaft an der Johann Wolfgang Goethe-Universität Frankfurt am Main 22,4.

— 2002: Bhartṛhari's Vākyapadīya. Versuch einer vollständigen deutschen Übersetzumg nach der kritischen Edition der Mūla-Kārikās hrsg. von Oskar von Hinüber. Mainz (Akademie der Wissenschaften und der Literatur). Stuttgart (Franz Steiner Verlag). Akademie der Wissenschaften und der Literatur, Abhandlungen der Geistes- und sozialwissenschaftlichen Klasse, Einzelveröffentlichung Nr.8.

RINPOCHE, S. 1986: A review of rare Buddhist texts (Hindi title: *durlabh bauddh granth śodh patrikā*). Vol.2. Varanasi (Vārāṇasī) (Vraj Vallabh Dwivedī, Kendrīya Ucca Tibbatī Śikṣā Sansthān).

ROCHER, LUDO 1986: The Purāṇas. Wiesbaden (Otto Harrassowitz). A History of Indian Literature, Vol.II, fasc.3.

ROŞU, ARION 1978: Les conceptions psychologiques dans les textes médicaux indiens. Paris (Collège de France, Institut de Civilization Indienne).

SCHARFE, HARTMUT 1977: Grammatical Literature. Wiesbaden (Otto Harrassowitz). A History of Indian Literature, Vol.V.

SCHMITHAUSEN, LAMBERT 1987: Ālayavijñāna. On the origin and the early development of a central concept of Yogācāra philosophy. Part I (Text) & II (Notes, Bibliography and Indices). Tokyo (The International Institute for Buddhist Studies). Studia Philologica Buddhica, Monograph Series, IVa & IVb.

SCHNEPEL, CORNELIA 2005: Odissi. Eine ostindische Tanzform im Kontext der Debatten um regionale Tradition und kulturelle Identität. Halle (Saale). Südasienwissenschaftliche Arbeitsblätter hrsg. v. Rahul Peter Das am Institut für Indologie und Südasienwissenschaften der Martin-Luther-Universität Halle-Wittenberg, Bd.6.

SEN, NILRATAN 1977: Caryāgītikośa. Simla (Indian institute of Advanced Study).

SHARMA, PREM LATA 1992: Bṛhaddeśī of Śrī Mataṅga Muni. Ed. by Prem Lata Sharma, assisted by Anil Bihari Beohar. Delhi (Motilal Banarsidas). Kalāmūlaśāstra Series 8. Indira Gandhi National Centre for the Arts New Delhi.

SHARMA, PRIYAVRAT 1996 → see Carakasaṃhitā among the *Original texts* in this bibliography.

SHIVARAM, GIRIDHAR MYSORE 2001: The biological and moral status of the human embryo in some Sanskrit texts: A survey. M.A. thesis. to the Department of Sanskrit and Indian Studies, Harvard Univ. Cambridge, Massachusetts.

SHRINGY R.K. & SHARMA, PREM LATA 1999: Saṅgītaratnākara of Śārṅgadeva. Sanskrit and English Translation with Comments and Notes. Chapter I. Vol.I. New Delhi (Munshiram Manoharlal Publishers).

— 1989: Saṅgītaratnākara of Śārṅgadeva. Sanskrit and English Translation with Comments and Notes. Chapter II–IV. Vol.II. New Delhi (Munshiram Manoharlal Publishers).
SINGH, JAIDEVA 2003: Śiva Sūtra. The yoga of supreme identity. Delhi (Motilal Banarsidas) 2003 (First ed. 1979).
SINHA, K. P. 1993: Thoughts on Tantra and Vaiṣṇavism. Calcutta (Punthi-Pustak).
SLAJE, WALTER 1995A: "*R̥tu-, R̥tv(i)ya-, Ārtava-* Weibliche "Fertilität" im Denken vedischer Inder" *Journal of the European Ayurvedic Society 4.1995*. Reinbeck, pp.109–148.
— 1995B: "*Aśubhasaṃjñā* und *pratipakṣabhāvanā* : Zur Tradition einer 'Vergegenwärtigung des Widerwärtigen' in der Soteriologie des Nyāya." *Zeitschrift der Deutschen Morgenländischen Gesellschaft Band 145 – Heft 1*. Stuttgart, pp.109–124.
— 2005: "A Note on the Genesis and Character of Śrīvara's So-Called *"Jaina-Rājataraṅgiṇī"*." *Journal of American Oriental Society 125.3 (2005)*, pp.379–388.
SPEIJER, J.S. 1993: Sanskrit Syntax. J.S. Speijer. With an introduction by H. Kern. Delhi (Motilal Banarsidas). Reprint 1993. (First edition, Leiden 1886).
STRANGWAYS, A.H. FOX 1914: The Music of Hindostan. New Delhi (Mittal Publications). Photographically reproduced 1995. (First edition 1914).
STRAUSS, OTTO 2004: Indische Philosophie. Hrsg. v. Andreas Pohlus. Aachen (Shaker Verlag). Geisteskultur Indiens. Klassiker der Indologie, Bd.5.
SUNESON, CARL 1991: "Remarks on Some Interrelated Terms in the Ancient Indian Embryology." *Wiener Zeitschrift für die Kunde Südasiens und Archiv für indische Philosophie 35.1991*. Wien, pp.109–121.
TĀRAḶEKARA 1975 → see Saṅgītaratnākara among the *Original texts* in this bibliography.
ṬHĀKUR, RABĪNDRANĀTH 1755: Rabīndra-racanābalī. Pañcabiṃśa khaṇḍa (Vol.15). Kalikātā (Biśbabhāratī) 1755. (The year is according to the Bengali calendar.)
TIVĀRĪ, PREMAVATĪ (Premvati Tivari) 1986, 1990: Āyurvedīy Prasūti-tantr evaṃ Strī-rog. Part I. Prasūti-tantr (Obstetrics). Part

II. Strī-rog (Gynecology). Varanasi/Delhi (Chaukhambha Orientalia) 1986, 1990. Jaikrishnadas Ayurveda Series 49.

TROTTIER, ANNE-HÉLÈNE 2000: Fakir. La quête d'un Bâul musulman. Paris (L'Harmattan).

VALLAURI, MARIO 1942: La «Śivagītā». Torino (R. Accademia delle Scienze). Memorie della R. Accademia delle Scienze di Torino. Serie II, Tomo 70, Parte II: Classe di Scienze morali, storiche e filologiche. N.9.

VARMA, K. M. 1957: Nāṭya, Nṛtta and Nṛtya, Their Meaning and Relation. Calcutta (Orient Longmans).

VATSYAYAN, KAPILA (general editor) 1988: Kalātattvakośa. A Lexicon of Fundamental Concepts of the Indian Arts. Indira Gandhi National Centre for the Arts (New Delhi) and Motilal Banarsidas (Delhi).

VAUDEVILLE, CHARLOTTE 1993: A Weaver Named Kabir. Selected Verses With a Detailed Biographical and Historical Introduction. Delhi (Oxford University Press). French Studies in South Asian Culture and Society VI.

WACKERNAGEL, JACOB 1957: Altindische Grammatik. Band II,1. Einleitung zur Wortlehre. Nominalkomposition. Neudruck der 2., unveränderten Auflage. Göttingen (Vandenhoeck & Ruprecht).

WAYMAN, ALEX 1982: "A study of the Vedāntic and Buddhist Theory of *nāma-rūpa.*" Indological and Buddhist Studies, Volume in Honour of Professor J.W. de Jong on his Sixtieth Birthday. Canberra, pp.617–642.

WEBER, ALBRECHT → see Śathapathabrāhmaṇa among the *Original texts* in this bibliography.

WEISSER, URSULA 1983: Zeugung, Vererbung und Pränatale Entwicklung in der Medizin des arabisch-islamischen Mittelalters. Erlangen (Hannelore Lüling).

WEZLER, ALBRECHT 1992: "Paralipomena zum Sarvasarvātmakatvavāda II: On the Sarvasarvātmakavāda and its Relation to the Vṛkṣāyurveda." *Studien zur Indologie und Iranistik 17.1992*. Reinbeck, pp.287–315.

WHITE, DAVID GORDON 1996: The Alchemical body. Siddha Traditions in Medieval India. Chicago/London (The University of Chicago Press).

WHITNEY, WILLIAM DWIGHT 1987 → see Atharva Veda (AV) among the *Original texts* in this bibliography.

WIDDESS, RICHARD 1995: The Ragas of Early Indian Music, Modes, Melodies, and Musical Notations from the Gupta Period to c.1250. Clarendon Press Oxford. Oxford Monographs on Music.

WINDISCH, ERNST 1908: Buddha's Geburt, und die Lehre von der Seelenwanderung. Leipzig (B.G. Teubner).

WOODROFFE, SIR JOHN (see also AVALON) 1990: Introduction to Tantra Śāstra. Madras (Ganesh & Company). Eighth edition.

YAMASHITA, TSUTOMU 1998: Indo dento igaku ni okeru kotai ron. Śārīrasthāna no kenkyu. (In Japanese. The title means: "The Human being in Indian traditional medicine. A study of the Śārīrasthāna.")

— 2001/2003: "On the Nature of the Medical Passages in the *Yājñavalkyasmṛti.*" *ZINBVN 36(2).2001/2003.* Kyoto, pp.87–129.

YANO, MICHIO 1988: Indo Igaku Gairon. Charaka sanhitā. Kagaku no Meicho: Dai II ki I (11). Tokyo (Asahi Shuppansha). (In Japanese. "Introduction to Indian Medicine").

ZIN, MONIKA 2004: "Die altindischen *vīṇās*." Orient-Archäologie Bd.15. Studien zur Musikarchäologie IV. Musikarchäologische Quellengruppen: Bodenurkunden, mündliche Überlieferung, Aufzeichnung. Vorträge des 3. Symposiums der Internationalen Studiengruppe Musikarchäologie im Kloster Michaelstein, 9.–16. Juni 2002. Hrsg., Ellen Hickmann/Ricardo Eichmann. Rahden/ Westf. (Verlag Marie Leidorf) 2004, pp.321–362.

ZYSK, KENNETH G. 1986: "The evolution of Anatomical Knowledge in Ancient India, with Special Reference to Cross-cultural Influences." *Journal of American Oriental Society 106.4 (1986).* New Haven, pp.687–705.

— 1990: "The Indian Ascetic Traditions and the Origins of Āyurvedic Medicine." *Journal of the European Āyurvedic Society 1.1990,* pp.119–124.

— 1991: Asceticism and Healing in Anient India, Medicine in the Buddhist Monastery. Oxford/New York (Oxford University Press).

— 1993: "The Science of Respiration and the Doctrine of the Bodily Winds in Ancient India." *Journal of the American Oriental Society, 113.2 (1993),* pp.198–213.

Index

WORLDS OF SOUTH AND INNER ASIA
WELTEN SÜD- UND ZENTRALASIENS
MONDES DE L'ASIE DU SUD ET DE L'ASIE CENTRALE

The aim of the series "Worlds of South and Inner Asia" of the Swiss Asia Society is to publish high-quality, representative work issuing from academic research on all aspects of South and Inner Asia. It comprises, and accepts, studies on historical and present-day South and Inner Asian cultures and societies covering the fields of history, literature, thought, politics and art as well as translations and interpretations of important primary sources. Furthermore the series intends to present studies that offer expert knowledge on current themes appealing not only to the academic public, but also to an audience generally interested in South and Inner Asia.

One important goal of the series is to establish a forum for academic work in the fields of the humanities and social sciences in Switzerland. However, the series is also committed to the rich variety of studies and writings in the international research community. The main publication languages for monographs and collections (by individual or several contributors) are German, French, and English.

The series is supervised and internally reviewed by an editorial board which is advised by leading scholars in the academic fields concerned.

Vol. 1 Andreas Bigger, Rita Krajnc, Annemarie Mertens, Markus Schüpbach & Heinz Werner Wessler (eds)
Release from Life – Release in Life. Indian Perspectives on Individual Liberation.
2010. VIII, 339 p. ISBN 978-3-0343-0331-6

Vol. 2 Maya Burger & Nicola Pozza (eds)
India in Translation through Hindi Literature. A Plurality of Voices.
2010. 304 p. ISBN 978-3-0343-0564-8

Vol. 3 Makoto Kitada
The Body of the Musician. An Annotated Translation and Study of the Piṇḍotpatti-prakaraṇa of Śārṅgadeva's Saṅgītaratnākara.
2012. 346 p. ISBN 978-3-0343-0319-4

Vol. 4 Marietta Kind
The Bon Landscape of Dolpo. Pilgrimages, Monasteries, Biographies and the Emergence of Bon.
Forthcoming. 491 p. ISBN 978-3-0343-0690-4

Vol. 5 François Voegeli, Vincent Eltschinger, Danielle Feller, Maria Piera Candotti, Bogdan Diaconescu & Malhar Kulkarni (eds)
Devadattīyam. Johannes Bronkhorst Felicitation Volume.
2012. XIV, 847 p. ISBN 978-3-0343-0682-9